PROGRESS IN WATER TECHNOLOGY

VOLUME 5

AF867549

INTERNATIONAL ASSOCIATION ON
WATER POLLUTION RESEARCH

President
Dr. G. J. Stander

Vice-Presidents
Professor B. B. Berger and Mr. B. E. Kuntze

Secretary-Treasurer Mr. P. E. Odendaal

Executive Editor Dr. S. H. Jenkins

PROGRESS IN WATER TECHNOLOGY

VOLUME 5

DESIGN—OPERATION INTERACTIONS AT LARGE TREATMENT PLANTS

(*Vienna Workshop Papers*)

Edited by

S. H. JENKINS

PERGAMON PRESS

OXFORD · NEW YORK · TORONTO · SYDNEY

PARIS · FRANKFURT

U.K.	Pergamon Press Ltd., Headington Hill Hall, Oxford OX3 0BW, England
U.S.A.	Pergamon Press Inc., Maxwell House, Fairview Park, Elmsford, New York 10523, U.S.A.
CANADA	Pergamon of Canada Ltd., 75 The East Mall, Toronto, Ontario, Canada
AUSTRALIA	Pergamon Press (Aust.) Pty. Ltd., 19a Boundary Street, Rushcutters Bay, N.S.W. 2011, Australia
FRANCE	Pergamon Press SARL, 24 rue des Ecoles, 75240 Paris, Cedex 05, France
FEDERAL REPUBLIC OF GERMANY	Pergamon Press GmbH, 6242 Kronberg-Taunus, Pferdstrasse 1, Federal Republic of Germany

Copyright © 1972 Pergamon Press Ltd.

All Rights Reserved. No part of this publication may be reproduced, stored in a retrieval system or transmitted in any form or by any means: electronic, electrostatic, magnetic tape, mechanical, photocopying, recording or otherwise, without permission in writing from the publishers

First published in *Water Research* Volume 6 Numbers 4–5, and supplied to subscribers of that journal as part of their subscription.

Reprinted 1978

Library of Congress Catalog Card No. 73-1162

Printed in Great Britain by Biddles Ltd., Guildford, Surrey

ISBN 0 08 018293 3 flexicover

DESIGN-OPERATION INTERACTIONS AT LARGE TREATMENT PLANTS

THE PROCEEDINGS OF A WORKSHOP ON 20–24 SEPTEMBER, 1971, ORGANIZED FOR THE INTERNATIONAL ASSOCIATION ON WATER POLLUTION RESEARCH BY A COMMITTEE UNDER THE CHAIRMANSHIP OF PROFESSOR W. v. d. EMDE AND HELD AT THE TECHNISCHE HOCHSCHULE, VIENNA, BY KIND PERMISSION OF THE RECTOR o. Prof. Dr. phil. ERICH BUKOVICS

ORGANIZING COMMITTEE

Chairman

Prof. Dr.-Ing. W. v. d. Emde, Vienna, Austria

Vice-Chairman

Prof. Dr. J. Andrews, Clemson, S.C., U.S.A.

Members

Prof. W. W. Eckenfelder, Nashville, Tenn., U.S.A.

Dipl.-Ing. H. Graefen, Chief, Wastewater Treatment Works, Emschergenossenschaft, F.R.G.

Prof. Dr.-Ing. R. Kayser, Braunschweig, F.R.G.

Mr. H. B. Tench, B.Sc., F.R.I.C., General Manager, Sewage Disposal Department, Sheffield, G.B.

ORGANIZING SECRETARIAT

Dipl.-Ing. H. Fleckseder
Mrs. E. Woller

TRANSLATOR AT ALL SESSIONS

Dr. R. Kayser

Water Research Pergamon Press 1972. Vol. 6, pp. 315–318. Printed in Great Britain

WASTEWATER TREATMENT WORKS PLANNING, ECONOMICS AND TECHNOLOGY—SOME NEW DIRECTIONS

RALPH C. PALANGE

Division of Municipal Waste Water Programs, Office of Water Programs; U.S. Environmental Protection Agency, Washington, D.C. 20460, U.S.A.

1. INTRODUCTION

THERE is a lack of effective communication between research and other workers which has resulted in an inability to effectively utilize the new technologies. This paper will deal with how we can utilize the results of research in order to more effectively control our water environment.

2. THE U.S. TREATMENT WORKS CONSTRUCTION PROGRAM

The first permanent U.S. Water Pollution Control Act (PL 84–660) was enacted in 1956. This legislation, along with a series of amendments, established and later significantly increased appropriations for construction of waste treatment works from $50 million in Fiscal Year 1957 to $800 million in Fiscal Year 1970 to $1.0 billion in Fiscal Year 1971.

The Federal Act authorizes grants to municipalities to assist in construction or improvement of sewage treatment works, including necessary interceptor and outfall sewers, but not collection sewers. The range of Federal participation in a given project varies between 30 and 55 per cent of the cost of construction, depending upon a number of factors. As of May 31, 1971, the program had funded 11,886 projects having a total construction cost of $10.2 billion, of which about $2.6 billion has been provided by the Federal Government.

3. ASSESSMENT OF NEEDS

A major problem is that of determining needs and related costs for waste treatment facilities. To obtain these we have established a system which is designed to provide reliable and creditable data on currently predictable needs.

In order to fulfill our needs we have structured a program comprising many elements —realistic criteria, practical plans, enforcement, and research. The President has proposed to the Congress (S.1013) a program designed to answer several key questions, including: (a) what is the cost of waste treatment facilities needed by municipalities through Fiscal Year 1974 to meet water quality standards? (b) how can the Federal grant monies be most expeditiously and effectively distributed to the States and municipalities? and (c) how can we prevent the recurrence of a backlog of needed construction?

Our best estimate of the total cost of construction of municipal treatment works to be committed by the end of Fiscal Year 1974, to meet water quality standards, is $12 billion (cost of clean water).

4. PLANNING

Basic to the accomplishment of our goals is effective planning. Effective water quality management planning must be a continuous process of systematic and coordinated development of an efficient and effective course of action. Planning should be accomplished for *each* of two distinct but related types of geographic areas: (1) urban concentrations and groups of small towns in which the great bulk of domestic, industrial and storm water wastes are generated and require treatment or other controls; and (2) total river basin hydrologic systems into which all liquid wastes are discharged or ultimately flow. Water quality management planning results in a course of action for establishing governmental and private policies and directions relating to prevention or control of water pollution. To achieve an effective program, such a course must be implemented through laws, ordinances and regulations. That course must be implementable, flexible and geared to action, not to looking attractive on a library shelf. Effective planning must consider all problems and waste sources—municipal, industrial, agricultural, storm runoff and combined sewers, oil and hazardous materials, mining and wells, and all other identifiable sources. In addition, planning must consider the relationships between water quality problems and other environmental problems. On July 2, 1970, the U.S. Government (U.S. FEDERAL REGISTER, 1970) published regulations which require that any municipal treatment facility receiving Federal grant-in-aid support must be included both in a basin-wide and in a metropolitan or regional plan for water quality management. States and local governments are assigned the basic responsibility for this required planning effort.

5. COST EFFECTIVENESS

Over the last decade, the United States has almost doubled its waste treatment capitalization and will double it again in the next 5 years. Towards realizing the maximum return for this investment it is imperative that funds be utilized in a cost-effective manner. This implies that those facilities which result in the greatest overall improvement in water quality and/or environmental improvement will be constructed before those which provide less environmental enhancement. Costs are considered to be overall, life-cycle costs including operation, maintenance and amortization as well as initial capital investment. The goal is to minimize *total public costs* without regard to the incremental cost to any level of government. The selected approach or plan must also minimize environmental damage. The U.S. ENVIRONMENTAL PROTECTION AGENCY (1971) is utilizing cost effectiveness methodology, by critically reviewing overall plans before specifications for individual treatment facilities are prepared. The States and municipalities have been prompted on several occasions to consolidate two or more systems in a basin into one large, improved treatment facility. In some cases, however, consideration of cost effectiveness has dictated separation rather than consolidation of systems. We have encouraged neighboring communities to treat their wastes jointly and to design treatment systems that minimize pumpage over hydrographic boundaries, and to design systems to jointly handle adjacent municipal and industrial loads.

6. NEW TECHNOLOGY

Over $2.6 billion in grants have been awarded by the Federal Government since 1956 for construction of waste treatment facilities and billions more will be spent before our goals of clean water are met. With this huge expenditure of public funds every effort must be made to assure that the best available technology is utilized. Research, both public and private, which has been undertaken in the past few years has produced new techniques which are now ready for widespread application. In the past, adoption of new technology by design engineers has occurred slowly. For us, it is urgent that new technology be adopted now in order to have an impact on design of new facilities. At this point, we cannot afford to invest in facilities which might soon become obsolete or cannot be readily modified to meet more stringent effluent quality criteria in the future.

In recognition of this goal, in September 1970, we published "*Federal Guidelines for Design, Operation and Maintenance of Waste Treatment Facilities*" and a month later, in October 1970, initiated a Technology Transfer Program. The guidelines reflect the necessity for deriving the greatest benefit from pollution control expenditures through ensuring that new treatment systems are designed and operated in the most cost-effective manner. Technical Bulletins are also being issued which amplify specific areas in the Guidelines, evaluate new advances in technology, and provide guidance for incorporation into new facilities.

The goal of the Technology Transfer Program is to bridge the gap between development and demonstration of new treatment methodology and its actual acceptance and full-scale use. The program includes dissemination of information through design seminars, technical publications, public information programs and displays, and process design manuals for new treatment processes.

Manuals are being developed on phosphorus removal, suspended solids removal, activated carbon adsorption, and upgrading capacity and efficiency of existing plants. Others are scheduled.

Upgrading of treatment facilities will be required at some point beyond 1976 to meet new water quality standards dictated by increased population, per capita waste contribution, and industrial growth. Use of best available technology at this time will facilitate upgrading in the future. Water re-use will eventually be necessary by industry for non-potable purposes.

We would be pleased to provide or exchange information on our Technology Transfer Program. Information may be obtained by writing to Technology Transfer, U.S. Environmental Protection Agency, Washington, D.C. 20460, U.S.A.

7. ENVIRONMENTAL IMPACT

In January 1970, the Congress of the United States enacted the National Environmental Policy Act (PL 91–190) which requires that all Federal agencies proposing action which may have a significant impact on the environment shall consult with other agencies having jurisdiction. The purpose is to build into the Federal decision-making process appropriate and careful consideration of the environmental aspects of proposed actions. The agency proposing such actions generally prepares a statement covering the environmental impact, and this is distributed to public and private agencies, citizens groups and others for review and comment.

The statements are not justifications for proposed funding or actions; rather they are detailed presentations of the environmental impact of Federal or Federally supported public works. They cover, as a minimum: (a) the probable impact of the project on the environment; (b) any probable adverse environmental effects which cannot be avoided; (c) alternatives considered, with evaluation for each proposal; (d) relationship between short-term uses of the environment and enhancement of long-term productivity; (e) any irreversible and irretrievable commitment of resources; and (f) public objections to the project, if any, and their resolution.

8. SUMMARY

We believe there are great and difficult tasks ahead of us, to understand and control the disturbing by-products of our economic system which though extraordinarily productive has sometimes misused basic resources. Today, we have the will and believe we are developing the governmental mechanisms in the field of planning, economics, and technology which will enable us to meet the challenges of our urban America, and have clean water in the 1970's.

The international community of science and the international community of nations must cooperate to solve mutual problems of conservation and management of our environmental resources. Much excellent work is now being done in the water field by international organizations. But the magnitude of the problem is so large there is wide opportunity for useful activity by all of us. We look forward to continuing interchange of ideas and mutual assistance in achieving our common aim of clean water.

REFERENCES

PUBLIC LAW 84–660, approved July 9, 1956, as amended. (Basic Act amended by Federal Water Pollution Control Act Amendments approved 1961, 1965, 1966 and 1970.)

PUBLIC LAW 91–190, approved January 1, 1970 (National Environmental Policy Act of 1969).

U.S. SENATE BILL 1013 (92nd Congress, 1st session) (1970).

U.S. FEDERAL REGISTER (1970) 18 C.F.R. 601.32-.36, July 2.

U.S. DEPARTMENT OF THE INTERIOR, FEDERAL WATER QUALITY ADMINISTRATION (1970) *Federal Guidelines for Design, Operation and Maintenance of Wastewater Treatment Facilities*. (The Federal Water Quality Administration is now the Water Programs Office in the U.S. Environmental Protection Agency.)

U.S. ENVIRONMENTAL PROTECTION AGENCY. WATER PROGRAMS OFFICE (1971) *Cost of Clean Water. Vol. I. Municipal Investment Needs*, 21 pp; *Vol. II. Cost Effectiveness and Clean Water*, 128 pp.

U.S. ENVIRONMENTAL PROTECTION AGENCY. WATER PROGRAMS OFFICE. DIVISION OF FACILITIES CONSTRUCTION AND OPERATION (1971) *New Waste Treatment Technology Is Available Now*. Charles L. Swanson. 11 pp. plus tables and charts (to be published).

U.S. ENVIRONMENTAL PROTECTION AGENCY, WATER PROGRAMS OFFICE (1971). *Guidelines to Water Quality Management Planning*.

U.S. FEDERAL REGISTER. COUNCIL ON ENVIRONMENTAL QUALITY (1971) *Statements on Proposed Federal Actions Affecting the Environment. Guidelines*. Vol. 36. No. 19, Part II, January 28. pp. 1398–1402.

Water Research Pergamon Press 1972. Vol. 6, pp. 319–322. Printed in Great Britain

DESIGN-OPERATION INTERACTIONS FOR WASTEWATER TREATMENT PLANTS

J. F. ANDREWS

Environmental Systems Engineering, Clemson University, Clemson, S.C., U.S.A.

INTRODUCTION

MANY of the wastewater treatment plants in the United States operate at overall efficiencies well below those for which they have been designed. Most treatment plants are in a primitive state with respect to process operation when compared with industrial processes and gross failures, such as the bulking of activated sludge and "sour" anaerobic digesters, are all too frequent. Many other operational problems have been reported; for example, MICHEL *et al.* (1969) in a study of 1500 performance audits, lists 499 *reported* instances of plant bypassing from such causes as stormwater and infiltration, shutdowns for routine maintenance, and mechanical failure. There are significant variations in treatment plant efficiency, not only from one plant to another, but also from day to day and hour to hour in the same plant. Daily variations from 60 to 95 per cent efficiency in BOD removal are not uncommon and THOMANN (1970), by a statistical analysis of the variation in effluent quality from eight treatment plants, has shown that these variations can have a significant effect on the water quality of the receiving streams.

The United States government has recognized the need for more adequate consideration of the interactions between design and operation for waste water treatment plants. In September of 1970, the COMPTROLLER GENERAL of the United States submitted a report to the Congress on the need for improved operation and maintenance of municipal waste treatment plants. This report established that operation and maintenance problems are widespread and have resulted in inefficient plant operations. It was the belief of the Comptroller General's office that these problems are primarily due to a lack of qualified operating personnel, inadequate controls over industrial wastes, and inadequate plant design or lack of adequate equipment. In September 1970, the UNITED STATES FEDERAL WATER QUALITY ADMINISTRATION (now a part of the Environmental Protection Agency) issued guidelines for the design, operation, and maintenance of wastewater treatment facilities. These guidelines place strong emphasis on the need for adequate consideration of operation and maintenance during plant design and adequate inspections by State agencies for the evaluation and improvement of plant performance. In May 1971, public hearings (*Clean Water Report*, 1971) on wastewater treatment and technology were conducted by a United States Senate panel on environmental science and technology. The need for increased attention to operations and maintenance will also be discussed by RADEMACHER (1971) at this workshop.

SPECIFIC DESIGN-OPERATION INTERACTIONS

Some of the specific factors which should be considered in obtaining a proper balance between design and operation are listed below.

(i) *Operating personnel*

The need for operation must be minimized through design when it is anticipated that plants will be operated by unskilled personnel or when labor problems, such as strikes, may be expected. One way in which this is frequently done for smaller plants is by using oversized units to damp out the effects of fluctuating inputs. However, oversizing, and therefore capital costs, should be reduced for those plants where qualified operating personnel will be provided. In larger plants where labor problems may be anticipated, more provision should be made for automation so that plant management personnel may operate the plants for limited periods of time.

(ii) *Process stability*

It is well recognized that some processes used for wastewater treatment are more stable than others and therefore require less attention to operation. Although this is of considerable importance, quantitative comparisons of the stability of different processes are not available. For example, most environmental engineers will accept the statement that the trickling filter is more stable than the activated sludge process but would be hard pressed to express this in a quantitative fashion. University research on dynamic models of processes could be of value in comparing the different processes with respect to stability. Improved stability can be obtained by improving operational techniques thus decreasing the need for oversizing of units.

(iii) *Availability of space*

Sizes of units can usually be decreased by increasing the operational effort involved. Decreases in sizes of plants through improved operation would be of importance for our larger cities where space is at a premium. The interaction of wastewater transport cost with plant size must also be considered. Smaller and well operated plants can frequently be located nearer to the major wastewater sources thus decreasing transport costs.

(iv) *Reliability*

It is well recognized that bypassing of raw wastewater and gross process failures frequently occur. The effects on the receiving waters of failure to obtain the desired degree of treatment either intermittently or continuously must therefore be considered. A higher degree of reliability will be required for a plant discharging into a stream with no excess assimilative capacity than for one discharging into a stream with considerable excess assimilative capacity. Additional reliability can be obtained through either design or improved operation.

(v) *Financial*

The balance between design and operation is influenced by the feasibility of investing in capital costs or operation and maintenance costs. Industry usually favors investments in operation and maintenance over capital costs because of the U.S. tax structure. Municipalities have normally favored investments in capital costs over operation and maintenance since bond issues have been easier to obtain than increases in operating revenue. It is the author's belief that an increase in the ratio of operating and maintenance costs to capital costs would result in better overall wastewater treatment for municipalities. The operation of treatment plants by public or private corporations with the authority to establish charges for wastewater treatment could be of assistance in this respect.

(vi) *Flexibility*

Another reason why industry usually favors investments in operation and maintenance over capital costs is the flexibility provided in decision making with respect to time. As technological advances in wastewater treatment occur, capital can then be used for plant modifications. This same advantage would also be open to public or private corporations operating municipal wastewater treatment plants. On a shorter time scale (seasonal), it is conceivable that treatment plants could be operated at variable efficiencies with respect to time in order to match the assimilative capacities of receiving bodies of water. Physical–chemical processes or biological processes followed by physical–chemical processes would be best adapted to operation at variable efficiency and good operation would be essential in order to obtain this flexibility in treatment efficiency.

LARGE TREATMENT PLANTS

This workshop is concerned primarily with large wastewater treatment plants. Although on a numerical basis there are many more small plants than large plants, it should be noted that the major portion of wastewater treated in developed countries is in the larger plants. In the United States, approximately 60 per cent (FEDERAL WATER POLLUTION CONTROL ADMINISTRATION, 1967) of all wastewater treated comes from plants with capacities exceeding 190,000 m^3 day^{-1} (50 mgd). It is expected that this trend toward large plants, or a number of small plants operated by a central agency, will continue since large plants cost less per unit of plant capacity. Improvement of plant performance by increased attention to plant operations is also more feasible for large plants since they can afford to hire professional engineers and scientists and to purchase and maintain modern control systems.

Unfortunately, the larger number of small plants and the more severe operational problems encountered with these plants has tended to overshadow the need for more attention to the interactions between design and operation for large plants. Many environmental engineering programs in the Universities have considered wastewater treatment operations to be a low level task which is not suited for education or research in Universities. Future programs should strive for a better balance between design and operation in order to prepare their students for careers in plant operations. The need of large plants for this type of engineer is obvious and the need of governmental agencies for operation engineers will be discussed by RADEMACHER (1971) at this workshop.

The special requirements of large plants also should be more adequately recognized by consulting engineers. Many of our current treatment plants have been designed to minimize the need for operation. Although this is an excellent policy for those plants which do not have good operation, it can penalize those where skilled operation engineers are in charge. For example, even in most of the larger activated sludge processes, there are only three possible controls which the operator may use; namely (a) return sludge flow rate, (b) sludge wasting rate, and (c) air flow rate. The intelligent operator could use much more flexibility than this and on occasion has modified the operation in order to obtain more control. GOULD (1953) for example, was able to improve operation of his plant by varying the amount and point of introduction of raw waste along the length of the aeration basin. This additional control feature has since been incorporated into many designs.

REFERENCES

MICHEL R. L., PELMOTER A. L. and PALANGE R. C. (1969) Operation and maintenance of municipal waste treatment plants. *J. Water Pollut. Control Fed.* **41,** 335.

THOMANN R. V. (1970) Variability of waste treatment plant performance. *J. sanit. Engng Div. Proc. Am. Soc. civ. Engrs* **96,** SA3, 819.

COMPTROLLER GENERAL'S OFFICE (1970) *Need for Improved Operation and Maintenance of Municipal Waste Treatment Plants*, Report B-166506, September.

FEDERAL WATER QUALITY ADMINISTRATION (1970) *Federal Guidelines—Design, Operation and Maintenance of Waste Water Treatment Facilities*, September.

Clean Water Report (1971) Vol. 8, No. 5, p. 41. Blair Station, Silver Spring, Md.

RADEMACHER, J. M. (1971) Pollution control regulation—a problem of filling gaps. *Workshop on Design—Operation Interactions for Large Wastewater Treatment Plants.* International Association for Water Pollution Research, Vienna, Austria, September.

FEDERAL WATER POLLUTION CONTROL ADMINISTRATION (1967) *Problems of Combined Sewer Faciiities and Overflows.* Water Pollution Control Research Series, WP-20-11.

GOULD R. H. (1953) Sewage aeration practice in New York City. *Proc. Am. Soc. civ. Engrs* **79,** 307–311.

Water Research Pergamon Press 1972. Vol. 6, pp. 323–325. Printed in Great Britain

DESIGN CONSIDERATIONS FOR LARGE TREATMENT PLANTS

W. VON DER EMDE

Institut für Wasserversorgung, Abwasserreinigung und Gewässerschutz, Technische Hochschule Wien, Karlsplatz 13, A-1040 Wien, Austria

DESIGN concepts are based on certain conditions which differ from country to country. I regard the following conditions decisive in Austria.

(i) *Increasing purification levels in the future*

Large cities are centers of water pollution. It is therefore reasonable to purify their waste waters extensively. In many small plants waste waters are better purified at present than in large plants.

Economic reasons also justify higher requirements for large treatment plants. The total initial cost (8 per cent on principal) and running costs for the same purification level (final BOD 25 mg l^{-1}) in Austria for cities with populations of

20,000 are about 4·0 \$/PE.a
200,000 are about 2·0 \$/PE.a
2,000,000 are about 1·5 \$/PE.a

(ii) *Increasing wage costs*

Wages increased considerably in many countries during the last decade (e.g. wages for plant operators in the Federal German Republic 1·8 times). Wages are expected to increase further. Wage costs affect future construction costs including cost of machinery, and particularly the operational costs.

(iii) *Increasing shortage of labour*

Qualified operators are already hard to come by in today's critical labour market. In the future it will presumably be even more difficult to compete with industry.

(iv) *Comparatively steady cost of electricity*

The cost of electric power remained comparatively steady during the last decade. The above conditions lead to the following conclusions.

(i) Increasing requirements can be met by:

(a) full utilization of existing technology, e.g. lightly loaded biological processes
(b) new technologies, e.g. replacement of biological purification by physico-chemical methods
(c) combination of existing and of new technologies, e.g. activated carbon treatment after biological purification.

The operation of existing works must be optimized by careful management. A valuable aid here is the continuous record of effluent quality. The digital presentation on a control panel of effluent TOC values appears particularly valuable. This has the advantage of agreeing quite well with the BOD_5 (e.g. TOC = 15 mg l^{-1} – BOD_5 = 10–15 mg l^{-1}). Such instruments should be considered at the planning stage. Their price in relation to the total plant costs for populations of over 100,000 is insignificant.

(ii) Increasing wage costs (construction and operational costs) can be reduced by:

(a) Few and preferably large units of tanks, containers and machines wherever possible; simple shapes should be adopted such as plane walls and level floors. Few connections but large sections are required for few units. In Austria 1 m^2 concrete floor (d = 25 cm) costs 15·0 \$/m^2 and 1 m^2 concrete wall (d = 30 cm and 3·0 m water depth) 35·0 \$/m^2. It is therefore more economical to adopt small areas of wall concrete and large areas of floor concrete.
For the large Vienna treatment plant (2·5 million pop. equ.), for instance, only four aeration tanks (of 10,000 m^3 each) are planned (l = 84 m, w = 48 m, d = 2·5 m). Unit costs at present prices are 22·0 \$/m^3 due to the relatively small wall areas. This principle applies to other tanks.

(b) Planning to reduce operator cost. Few, preferably large units, facilitate operation. The plant is easier to run and becomes more foolproof. Excess sludge has to be withdrawn at two points only if there are only two return sludge cycles. Fewer units also reduce cleaning work. Low operator costs are associated, e.g. with screw pumps for pumping raw waste water.

(iii) The increasing shortage of skilled labour may be met by:

(a) Replacement of skilled labour by computer, e.g. for;
Control of plant operations,
Recording of operational data,
Analytical quality control and continuously operating analytical instruments,
Control of machinery,
Programming operation and maintenance where computer maintenance service itself is to be provided by the supplier.

(b) When a computer is installed at a treatment plant the computer should not become a purpose of its own. The chance of breakdowns increases with the number of control and regulating units. The sources of breakdowns may be reduced by the provision of reasonably few regulating and control operations. Waste water quantities may for instance, be distributed to the various tanks with simple hydraulic devices (weirs, etc.), without locks or mechanical regulators, which are of quite adequate accuracy.
Slowly progressing reactions, e.g. lightly loaded activated sludge tanks and sludge digesters, are advantageous with respect to trouble-free operation and control. The purification process should, therefore, be designed to require little manipulational steps.
The minimal steps are: oxygen supply and excess sludge withdrawal for aeration tanks and the withdrawal of raw sludge from the first thickeners (further steps depend on sludge treatment).

(iv) The comparatively steady cost of electricity can be utilized by replacing wherever possible manual labour by electric energy: e.g. programme controlled air lift for raw sludge removal, simple sludge digester operation through intensive gas mixing, aerated crosschannels in front of the sedimentation tanks. Energy intensive processes, e.g. lightly loaded activated sludge plants, come to favour. The economy of power stations serving internal requirements only should be checked in each case.

Water Research Pergamon Press 1972. Vol. 6, pp. 327–330. Printed in Great Britain

DESIGN PERIOD

J. ZEPER

Dwars, Heederik and Verhey, Laan 1914 Nr. 35, Amersfoort, Netherlands

1. INTRODUCTION

IN THIS paper two aspects of the design period will be discussed. Firstly the period over which the various parts and items of a wastewater treatment plant should be written off and, secondly, the period between subsequent expansions of such plants.

Although nobody knows for certain what will happen to the rate of interest in the years to come, a return to the old days, when the rate of interest was 3–4 per cent, seems rather unlikely. The present-day design engineer is faced with a rate of interest of about 8 per cent. His way of thinking, however, is very often still based on the 3–4 per cent rate, even though he may not always be conscious of this.

It is the purpose of this paper to break through this unrecognized barrier by indicating some trends in the design periods, based on a high rate of interest.

2. DESIGN PERIOD AS DEPRECIATION PERIOD

Design engineers who received their training 10 or more years ago were always told that what they designed should be solid and safe and more or less for eternity. Depreciation periods of 40–50 yr for civil engineering works were normal practice. Indeed the civil engineer found it most annoying that mechanical and electrical equipment had to be written off in periods as short as 25 or even 20 yr. What was the reason for this real or pretended annoyance? Simply that at a rate of interest of 3 per cent, a reduction of the depreciation period from 40 to 20 yr meant an increase of the amortization from 4·3 to 6·6 per cent, a relative increase of over 50 per cent. The result was that the period of depreciation adopted was sometimes one of the major factors for the decision to proceed or not to proceed with a certain scheme. In the today's dynamic development of society, design periods of 40–50 yr are rather unrealistic. Nobody can see that far into the future. The forecasts for the year 2000 are rather confused; and it is even difficult to predict more than 10–15 yr ahead.

The manufacturers of mechanical and electrical equipment and, to an even greater degree, those of automatic control and computer control systems reckon with rather short periods of depreciation, mainly because their equipment will be "out of date" in a few years.

The design engineer of wastewater treatment plants will have to realize that this trend is not limited to mechanical, electrical and electronic equipment. It also applies to the civil engineering work *and* to the process selected for the plant.

If we are honest we must admit that when a treatment plant, constructed in the fifties, has to be expanded in answer to present day demands, it is nearly always very difficult to fit the existing plant into the new scheme. How often does the design engineer secretly wish that the existing plant were to disappear overnight so he could make a fresh start! But there it still is, solid and safe, and according to the books only having depreciated to 2/3 or 1/2 of its original value.

Strange though it may sound, at least a partial solution to this problem of the design engineer of the future may be found in the present—and possibly future—high rate of interest. This is because the financial attractiveness of long depreciation periods decreases as the rate of interest rises. A reduction of the depreciation period from 40 to 20 yr, with a rate of interest of 8 per cent, only means an increase of the amortization by about 20 per cent, whereas with a rate of interest of 3 per cent this increase was more than 50 per cent.

This makes all the difference, because:

(a) if the construction remains unchanged, a reduction of the depreciation period from 40 yr to 20 only results in a 20 per cent increase of the annual costs;

(b) if the annual costs are to remain constant, the doubling of the lifetime of a particular construction, or a part of it, from 20 to 40 yr may only require a 20 per cent higher investment whereas if the lifetime is halved from 40 to 20 yr the investment required may still be 85 per cent of the original cost.

A reduction of the depreciation period from 20 yr to 10—such as may be the case with mechanical, electrical and electronic equipment—still allows for a 70 per cent investment of the original costs if the annual costs are to remain constant.

All these examples point to a future situation where, partly due to uncertain forecasts of further developments and partly due to a shorter technical lifespan of equipment—and even of processes—a distinct preference has to be shown for relatively cheap designs, which can be written off in relatively short periods.

This conclusion may be upsetting to the solid, safe, for eternity design engineer, but let us not forget his colleague who will have the job in the eighties of expanding or replacing the plants that are being designed at present.

3. DESIGN PERIOD AS EXPANSION PERIOD

Against a background of expected changes in quantity and quality of the wastewater to be treated, the design engineer has to decide to what extent future growth should be taken into account in the design of today's plants. How much extra capacity should be installed, after how many years will the plant have to be expanded? Very often the answer to this question is to design the plant for the load expected in 15 yr time. One may wonder what the basis is for this calculated (?) guess.

Before an attempt is made to answer the above question, the considerations which follow have to be put in the context of the national—if not international—economy.

The treatment of wastewater concerns the entire population and therefore the optimalization of the period of expansion has to be considered as a part of the national economy and not as that of a private undertaking.

This means that the inflationary trends can be disregarded in the calculations as long as the inflation does not overtake the rate of progress of the national income.

In other words, if in the future the costs of construction and with them the costs of operation of the treatment plants increase and become twice as high as those at present, and at the same time the national income also doubles (or increases more due to progress), then such costs are the same in relation to the national income (or lower in case of increased progress).

An important factor in ever-increasing national prosperity may well be an optimum phasing of the construction of treatment plants.

The question of phasing can be put as follows: should the large unit be constructed at once or should a smaller unit be constructed now, to be followed at some time by the construction of an additional unit? To find the answer, the respective costs of investment—at present-day price level—have to be known:

$$I_1 = \text{cost of investment of large unit}$$
$$I_s = \text{cost of investment of small unit}$$
$$I_a = \text{cost of investment of additional unit.}$$

When $I_s + I_a = I_1$, it is obvious that construction in two phases is advantageous, because the interest on I_a can be saved over the period of the first phase.

Mostly, however, construction in two phases will prove to be more expensive than in one phase. Supposing

$$I_1 = 100 \text{ costs units } (U)$$
$$I_s = 70 \text{ costs units } (U)$$
$$I_a = 50 \text{ costs units } (U)$$

then the first impression is that construction in one phase has advantages to spreading over two phases. If, however, the rate of interest is 10 per cent and the additional unit is needed after 10 yr, the comparison of total investment costs after 10 yr is as follows:

$$\text{one phase: total } I = I_1\ (1{\cdot}10)^{10} = 260\ U$$
$$\text{two phases: total } I = I_s\ (1{\cdot}10)^{10} + I_a = 232\ U$$

from which it appears that in this case phasing also makes sense.

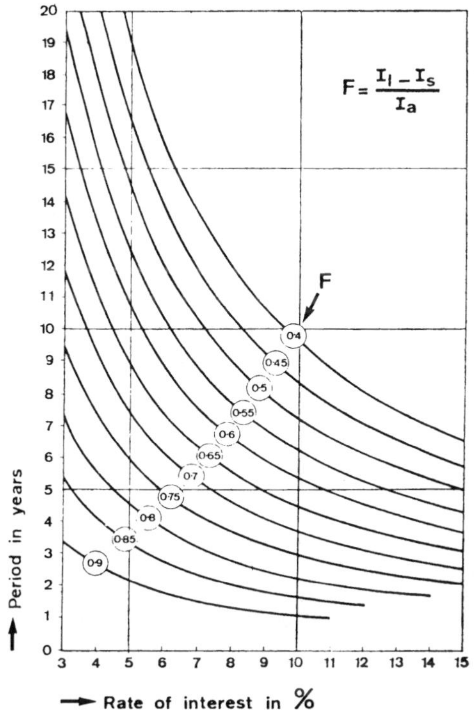

FIG. 1.

In order to find out whether phasing is advantageous use can be made of the graphs as shown in FIG. 1 (source: World Bank, Washington D.C.). Based on the known costs of investment, a factor

$$F = \frac{I_1 - I_s}{I_a} \text{ is determined.}$$

If in FIG. 1 the intersection given by the rate of interest and the period between the two phases lies above the line of F—as determined by the respective costs of investment—construction in two phases is to be preferred. If this point of intersection lies below the line of F, construction in one phase is the cheapest solution.
A check with the last example:

$$F = \frac{100\ U - 70\ U}{50\ U} = 0{\cdot}6.$$

The intersection 10 per cent/10 yr lies above the line $F = 0{\cdot}6$ construction in two phases.

Using the same example, and a rate of interest of 3–4 per cent, the turning point between construction in one or two phases lies at about 15 yr. So it seems that the calculated (?) guess originates from the old days when the rate of interest was low.

With the present rate of interest of 8 per cent, however, phasing should take place every 7 yr in order to make an optimum use of our economic resources. This means that, including a period of about 3 yr for design and construction, the design engineer needs a forecast for a period of about 10 yr, and every 7 yr thereafter a revised forecast for the 10 yr then to come.

So, just as with the depreciation period and the lifespan of processes and equipment, the high rate of interest helps to bring the horizon of the future nearer to the design engineer. It is to be hoped that this will improve the general visibility in this field, giving him a clearer view of today's financial realities.

Water Research Pergamon Press 1972. Vol. 6, pp. 331–333. Printed in Great Britain

ECONOMIC LIMIT FOR THE AMOUNT OF RESERVE CAPACITY ON CONSTRUCTION OF SEWAGE TREATMENT PLANTS FOR RAPIDLY GROWING MUNICIPALITIES

C. F. SEYFRIED

Institut für Siedlungswasserwirtschaft, Technische Universität Hannover, Am Welfengarten 1. D-3 Hannover

A SEWAGE treatment plant is usually planned for a period of 25–30 yr. Whether construction is carried out immediately for the planned final capacity or in sections, depends upon several factors, least of all economic or commercial considerations. The following points are made before discussing this question in detail.

(a) Determination of costs is based on the function of construction costs (SCHMIDT, 1964). This function can be plotted on a double logarithmic scale. It shows the cost function for activated-sludge ditches; the specific costs of the activated-sludge ditch, related to the population equivalent, decrease more rapidly than for other treatment systems.
(b) The average interest on capital is taken as 6·5 per cent.
(c) The maintenance cost of buildings *not yet* in operation is taken as 1·5 per cent.
(d) Depreciation is taken as 3·3 per cent.
(e) Building price increases are taken to rise linearly at 6·95 per cent from 1950 to 1970. Actually the increase of the index follows an *e*-function. For short periods up to 20 yr the assumption of linearity is of less importance than the uncertainty attached to forecasts of the increases.

The following cost classes can be compared:

(1) *For complete construction.* Interest, maintenance and depreciation costs for the parts not yet being in operation (reserve) for the period when these parts are not yet in use.

(2) *For construction in two sections.*
   (a) Additional costs for the first section; due to the smaller building volume, the specific costs are higher.
   (b) Additional costs for the second section, because the specific costs are higher.
   (c) Additional costs for the second section because building prices have risen.

Equalizing both cost classes the time of cost equality can be determined. If the reserve is required *before* this time, it is cheaper to complete immediately; if it is required later, the construction in two sections is more favourable.

According to SCHMIDT (1964) the general function of building costs for sewage treatment plants is:

$$y = A \;.\; x^z = y_\infty \qquad (1)$$

where y = costs, A = constant of costs of a treatment system for a certain year, x = population equivalent, z = exponent indicating the degression of the specific building costs when extending the plant.

For 1962 costs, Schmidt's functions are:

Different treatment systems	$y = 880\ x^{0 \cdot 73}$
activated-sludge ditches	$y = 16{,}220\ x^{0 \cdot 29}$
prefabricated sewage treatment plants (Schreiber).	$y = 3000\ x^{0 \cdot 56}$

The following signs are introduced for cost comparison:

a = reserve (%), exceeding a certain capacity
b = linear increase of the building price index (% yr^{-1})
n = time (yr).

The annual costs for the "non utilization" amount to: interest 6·5 per cent, maintenance 1·5 per cent, depreciation 3·3 per cent, total 11·3 per cent.

With simultaneous construction of the reserve a at the time $n = 0$ the costs amount to:

$$y_{0a} = A \cdot \left[x\left(1 + \frac{a}{100}\right)\right]^z. \tag{2}$$

After n yr and an increase of the building price index b the costs amount to

$$y_{na} = A \cdot \left[x\left(1 + \frac{a}{100}\right)\right]^z \cdot \left(1 + \frac{b . n}{100}\right). \tag{3}$$

With K_{na} = additional costs in n yr due to "non utilization" of the reserve a as part of the total plant (equation 2)

$$K_{na} = \frac{\frac{a}{100}}{1 + \frac{a}{100}} \cdot y_{0a} \cdot 0{\cdot}113 \cdot n \tag{4}$$

$$= \frac{a}{100 + a} \cdot A \cdot \left[x \cdot \left(1 + \frac{a}{100}\right)\right]^z \cdot 0{\cdot}113 \cdot n. \tag{5}$$

Building costs for extension of the plant by the reserve a after n yr approximately amount to:

$$y_{nc} = A \cdot \left(x \cdot \frac{a}{100}\right)^z \cdot \left(1 + \frac{b . n}{100}\right). \tag{6}$$

With K_{n0} = additional costs of construction in two sections after n yr

$$K_{n0} = y_{00} + y_{nc} - y_{0a} = (1) + (6) - (2) \tag{7}$$

$K_{na} = K_{n0}$ gives the two schemes at equal cost. Solving this equation according to n and simplifying this fraction by A and x^z, the time of cost equality is

$$n = \frac{1 + \left(\frac{a}{100}\right)^z - \left(1 + \frac{a}{100}\right)^z}{0{\cdot}113 \cdot \frac{a}{100 + a}\left(1 + \frac{a}{100}\right)^z - \left(\frac{a}{100}\right)^z \cdot \frac{b}{100}}. \tag{8}$$

Considering an average increase of the building cost index by 6·95 per cent/a and different treatment systems, one obtains a hyperbola.

Thus it is cheaper to build the sewage treatment plant of a municipality immediately for 20,000 inhabitants, if the 10,000-inhabitant limit will be exceeded within 15 yr.

The objection may be made that the increase in building costs cannot be fully considered, due to decreases in the purchasing power of money. It should be borne in mind, however, that the index increase was plotted only linearly.

On the other hand the treatment system also has an important influence on the decision whether to build in two sections or not. Curves of cost equality plotted against various exponents show that for activated-sludge ditches and lagoons provided with great reserves of capacity it is always cheaper to build immediately in one section.

CONCLUSION

In most cases complete construction is cheaper than constructions in two sections. For municipalities with an increase in population above average prefabricated sewage treatment plants or plants with extended aeration, such as activated sludge ditches and lagoons, are especially favourable for complete construction in which a large reserve capacity is required.

REFERENCE

SCHMIDT U. (1964) Über die Kosten der biologischen Abwasserreinigung (On the costs of biological sewage treatment). Heft 13 der Veröffentlichungen des Instituts für Siedlungswasserwirtschaft der Technischen Universität Hannover. The full text of this paper appears in a paper submitted to the Conference "*Sewage Purification and Pollution Control*". Haus der Technik, Essen, 1970.

Water Research Pergamon Press 1972. Vol. 6, pp. 335–337. Printed in Great Britain

EVALUATION OF DESIGN DATA

ROLF KAYSER

Institut für Stadtbauwesen, TU Braunschweig, Pockelstrasse 4, D-33 Braunschweig, F.R.G.

IN ORDER to determine the basic design data for the expansion of a treatment plant in Northern Germany a study was carried out during the period June–October 1963. The treatment plant consisted of primary tanks and trickling filters. The raw sludge was disposed in lagoons. The plant was loaded with the sewage of a city with about 45,000 inhabitants, the wastewater of a chemical plant producing biocides and a vegetable cannery. The town has a separate sewer system.

The typical flow distribution curve (FIG. 1) was obtained by determination of the average flow and the standard deviation for each hour of the day over the whole test period.

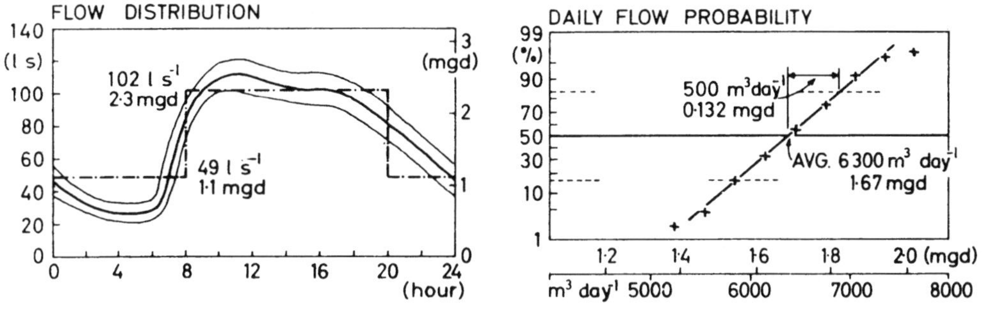

FIG. 1. Sewage flow analysis.

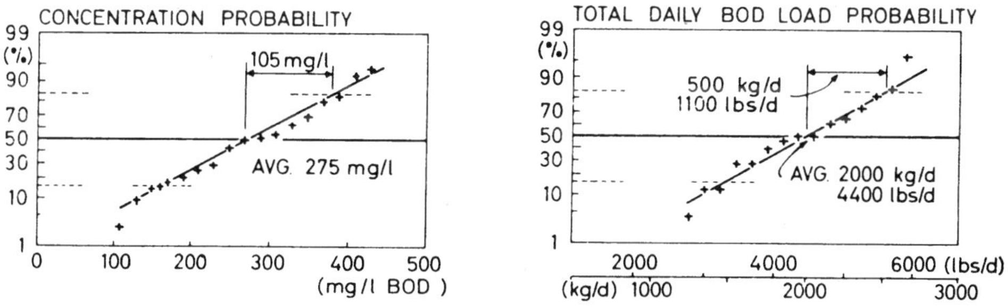

FIG. 2. BOD of primary effluent.

The total daily flows were plotted on probability paper (FIG. 1). The average was 6300 m^3 day^{-1} (1·67 mgd) and the standard deviation was 500 m^3 day^{-1} (0·132 mgd).

Composite samples were taken from the primary effluent. Probability plots of the BOD and the total BOD load are shown on FIG. 2. The average BOD was 275 mg l^{-1}, the standard deviation 105 mg l^{-1}. The total load averaged about 2000 kg BOD day^{-1} (4400 lb day^{-1}) and the standard deviation was 500 kg day^{-1} (1100 lb day^{-1}).

In order to take peak loadings into consideration we decided to use as basic design data average plus standard deviation. This means that only at 16 per cent of the days the load will be higher than the design value.

Future increase of sewage flow and total BOD loading was estimated to have a rate of 3 per cent yr^{-1}. During a 20-yr period flow and loading would increase by 80 per cent (Factor 1·8). The increase during the 8-yr period 1963–1971 would total 27 per cent (Factor 1·27).

The basic data are shown in the following table.

	1963	1971	1983
Total daily flow			
Average (m^3 day^{-1})	6300	8000	11,300
Average + stand. dev. (m^3 day^{-1})	6800	8600	12,000
Total BOD load			
Average (kg day^{-1})	2000	2500	3600
Average + stand. dev. (kg day^{-1})	2500	3200	4500

The main design data were a sewage flow of 12,000 m^3 day^{-1} (3·15 mgd) and a total BOD load of 4500 kg day^{-1} (10,000 lb day^{-1}). The plant finally was designed for a daily sewage flow of 15,000 m^3 day^{-1} (4·0 mgd), a peak flow of 260 l. s^{-1} (6·0 mgd) and a BOD load of primary effluent of 4500 kg day^{-1}. The master planning for the expan-

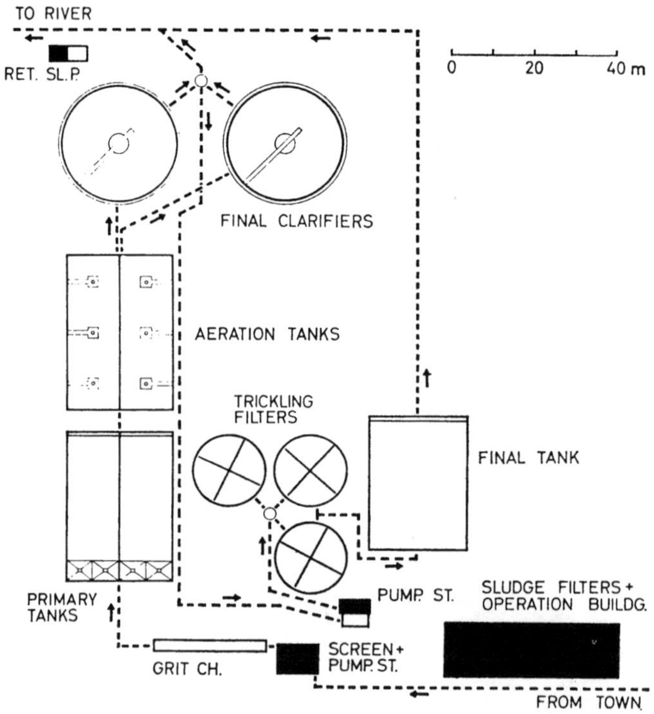

FIG. 3.

sion was completed in March 1969, the construction of the activated sludge plant was finished by the end of 1970. A flow diagram of the plant is shown below. The sewage passes a pumping station, screens, aerated grit chamber, primary tanks, aeration tanks and final tanks. Up to flow rates of 400 m^3 h^{-1} (2·5 mgd) the effluent of the final clarifiers is pumped to the existing trickling filters and than passes the existing final tank.

In summer 1971 measurements were carried out in order to determine the performance of the new plant. During two short periods from April 19 to May 4 and from June 16 to July 7 the following values were found:

	Apr. 19–May 4		June 16–July 7	
	Sew. flow (m^3 day^{-1})	BOD load* (kg day^{-1})	Sew. flow (m^3 day^{-1})	BOD load* (kg day^{-1})
Average	9400	4050	10,400	6700
Max.	10,400	5280	13,200	8700
Min.	8900	2750	8300	3500

* BOD load of primary effluent.

The higher loadings during the second period were due to the influence of the vegetable cannery, which operates from about June 15 to the end of September. The data indicate that the aeration tank was overloaded. The performance therefore, was only about 70–80 per cent BOD reduction.

Investigations at the chemical plant and the vegetable cannery indicate that the BOD since 1963 must have been increasing continuously with a higher rate than 3 per cent yr^{-1}. At the treatment plant the increasing loadings could not be observed, for no more analyses were made after 1963.

It is therefore recommended that analyses of sewage flow and concentration should be made on a continuous schedule. If the period between the evaluation of design data and plant construction is longer than 5 yr the design data may have to be modified.

Water Research Pergamon Press 1972. Vol. 6, pp. 339–340. Printed in Great Britain

DISCUSSION
STATISTICAL EXPRESSION OF EFFLUENT QUALITY STANDARDS

A. B. WHEATLAND

Water Pollution Research Laboratory, Elderway, Stevenage, SG1 1TH, Herts.

IN THE United Kingdom the Trent Research Programme had shown the need for the standards of quality to which the effluent from a sewage treatment plant would be required to conform to be expressed in statistical terms.

The Water Pollution Research Laboratory obtained data on suspended solids content and BOD of daily samples of effluent from 14 large sewage works in the River Trent area covering a period of about 1 yr. Calculations were then made for each parameter of (1) the overall annual mean for each works; (2) the concentration which was exceeded on (a) 5 per cent of the days (the 95 percentile), (b) 50 per cent of the days (the 50 percentile) and (c) 95 per cent of the days (the 5 percentile). When the data were plotted it was found that there was an essentially linear relationship between the annual mean and the corresponding percentiles. Regression equations relating these percentiles to the means are given in TABLE 1.

TABLE 1. COEFFICIENTS FOR EQUATIONS OF THE FORM $y = mx + c$ WHICH RELATE THE CONCENTRATION, y IN mg l^{-1}, CORRESPONDING TO A GIVEN PERCENTILE TO THE MEAN CONCENTRATION, x IN mg l^{-1}, FOR BOTH BOD AND SS

Parameter effluent quality	Percentile	Coefficients for equations		Correlation coefficients	No. of observations
		m	c		
	5	0·44	0·23	0·89	14
BOD	50	0·98	−0·2	0·98	14
	95	1·76	1·5	0·96	20
	5	0·64	−4·0	0·95	13
SS	50	1·04	−1·9	0·997	13
	95	1·5	5·2	0·96	19

It was apparent from the results obtained that in order to ensure that an effluent would have a biochemical oxygen demand < 20 mg l^{-1} for 95 per cent of the time it would be necessary to design a works to produce an effluent having an average BOD no greater than 11 mg l^{-1}. Also to ensure that the suspended solids content of the effluent was < 30 mg l^{-1} for 95 per cent of the time it would be necessary to design the sedimentation tanks and the rest of the plant so that the concentration of suspended solids would be on average no greater than 17 mg l^{-1}. The relation between the mean value and the 95 percentile for two effluent quality standards is indicated in TABLE 2.

TABLE 2. ASSUMED QUALITY OF EFFLUENTS AND THE CORRESPONDING MEAN CONCENTRATIONS FOR EACH PARAMETER

Parameter of effluent quality	Effluent quality (mg l^{-1})			
	ET2		ET4	
	Mean	95 percentile	Mean	95 percentile
BOD	11	20	5	10
Suspended solids	17	30	4	10
Ammoniacal nitrogen	20	30	1·0	3·5
Oxidized nitrogen	5	15	19	28

The works studied included both biological filtration and activated sludge plants. However, as similar relations had been observed at the Water Pollution Research Laboratory between the mean and 95 percentile values of effluents from pilot-scale biological filters, there was reason to believe that the variations in effluent quality from activated sludge and from biological filtration plants were not vastly different.

Water Research Pergamon Press 1972. Vol. 6, pp. 341–345. Printed in Great Britain

INFLUENCE OF THE PRESETTLING TANK SIZE ON DIMENSIONING AND COSTS OF A SEWAGE TREATMENT WITH ACTIVATED SLUDGE

T. SĘDZIKOWSKI

Politechnika Łódzka, Department of Civil Engineering, Ul. Zwirki 21, Łódz, Poland

INTRODUCTION

THE SYSTEM: presettling, aeration, secondary settlement should be capable of optimization and the minimum cost found according to:

$$K = K_w + K_k + K_e + K_{wt} + K_p \tag{1}$$

where

K = general expenditures in the competent period, K_w = capital cost of the presettling tank, K_k = capital cost of the aeration chamber, K_{wt} = capital cost of the secondary settling tank, K_p = capital cost of the recirculating station, K_e = cost of electric energy for aerating in the competent period.

From equation (1) one can eliminate the cost of the secondary settling tank K_{wt} and that of the recirculating station K_p which are constant.

The views of ELIASSEN and COBURN (1969) regarding preliminary settlements are extended.

2. THEORETICAL PRINCIPLES OF DIMENSIONING

BOD removal by presettlement and aeration (η) can be represented by

$$\eta = a\,(1 - e^{-kt}) \tag{1a}$$

or

$$\eta = \frac{at}{b + kt} \tag{1b}$$

where k = reaction rate factor, a = coefficient determining the maximum efficiency of the process, b = experimental factor, t = time, e = base of natural logarithms. For presettlement, for BOD, acg (IMHOFF, 1966) $a = 0{\cdot}35$ provided that for $t = 2$ h, $\eta = 0{\cdot}3$. For aeration chambers (ECKENFELDER, 1966) for $t = \infty$, $a = 1$. Taking the degree of sewage treatment as η_c for various times (t_w) for preliminary sedimentation and the corresponding efficiency η_w of the presettling tank, the required efficiency η_k of the aeration chamber can be calculated from the following equation (2):

$$\eta_k = \frac{\eta_c - \eta_w}{1 - \eta_w}. \tag{2}$$

For the calculated η_k the aeration time t_k and the corresponding volume capacity of the chamber can be calculated. The aggregate total volume of the presettling tank and aeration chamber (V) for a unit flow $Q = 1$ will be

$$V = Q\,t_w + Q\,t_k = t_w + t_k. \tag{3}$$

The known unit costs of cubitures being respectively k_w and k_k it can be stated that

$$K_w + K_k = t_w k_w + t_k k_k. \tag{4}$$

The electric energy demand is proportional to the concentration of clarified sewage and to the efficiency of the treatment. As an approximation one can write

$$N = \frac{OC}{E} = \frac{d\ C_s/1 - \eta_w/ \cdot \eta_k}{E} \tag{5}$$

where d = factor of oxygen demand for oxidizing the organic compounds as expressed by BOD, E = economy of the oxidizing equipment, kg O_2 kW^{-1}, N = power of oxidizing equipments, kW.

According to PASVEER (1957) d is ~ 1·5 and less according to ECKENFELDER (1966). E has been estimated to be about 2 kg O_2 kW^{-1}.

Assuming $d = 1{\cdot}5$ and $E = 2$ the power required for aeration will be:

$$N = 0{\cdot}75\ C_s\ (1 - \eta_w)\ \eta_k. \tag{6}$$

From N and the price of 1 kWh (k_e) one can estimate K_e for a given exploitation time.

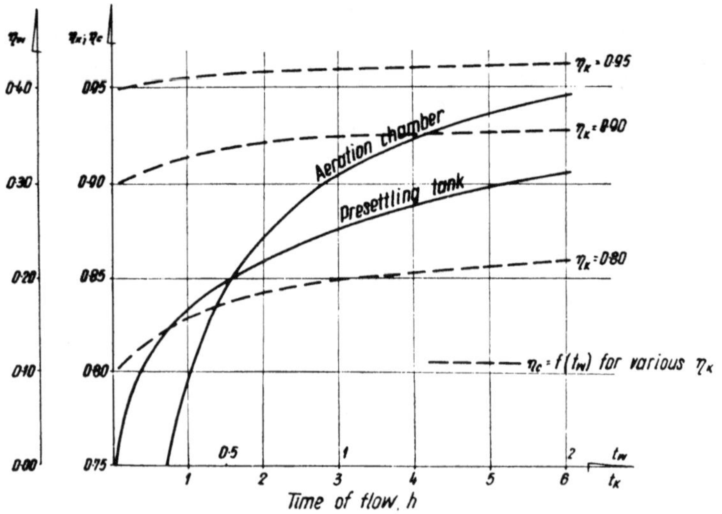

FIG. 1. Influence of the initial sedimentation and aeration time on the cleaning efficiency.

3. DIMENSIONING THE PRESETTLING TANKS AND THE AERATION CHAMBERS

Calculations of the effect of sedementation and aeration on the efficiency of treatment are shown in FIG. 1.

4. CAPITAL COST

For unit sewage flow intensity, e.g. 1 m^3 h^{-1} the times t_w and t_k of flow correspond to a cubic metre of settling tank and aeration tank capacity. Assuming equal unit capital cost for the capacity of the total cost of both was calculated.

The results of calculation for various values of η_c are given in FIG. 2. The capital cost of sewage treatment plants when $\eta_c \geqslant 0{\cdot}85$ is minimum for presettling tanks with the time of flow $t_w = 0{\cdot}5$ h. For $\eta_c < 0{\cdot}85$ the minimum cost will be obtained after eliminating the presettling tank. The cost was given as $K_{wzt} = (K_w + K_k)\, K_{w2}^{-1}$.

The dependence of capital cost on η_c for various times t_w of the preliminary sedimentation is given in FIG. 3, which shows that for $\eta_c < 0{\cdot}85$ presettling tanks result in an increase in sewage treatment costs.

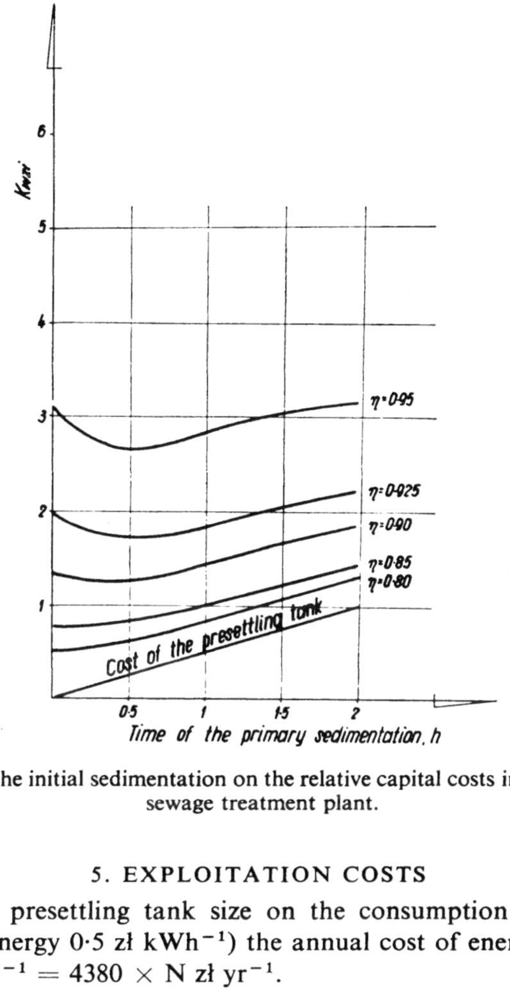

FIG. 2. Influence of the initial sedimentation on the relative capital costs in the flow section of a sewage treatment plant.

5. EXPLOITATION COSTS

The influence of presettling tank size on the consumption and cost of energy (with the price of energy 0·5 zł kWh^{-1}) the annual cost of energy is $K_e = N \times 365 \times 24 \times 0{\cdot}5$ zł kWh^{-1} $= 4380 \times$ N zł yr^{-1}.

6. SELECTING THE OPTIMUM VARIANT

In FIG. 4 the total aggregate relative expenditures on investments and energy (K_{wz}) of sewage treatment with activated sludge is shown. The relation was given as an example for $C_s = 200$ and 600 mg O_2 l^{-1} for $t_w = 0{\cdot}5$ and 2·0 and for different $\eta_c = 0{\cdot}8$, 0·9, 0·95. The straight lines for $t_w = 0{\cdot}5$ and 2·0 at the same degree η_c of treatment can intersect after a time, called "the time of balancing the costs" in the sewage treatment plants being compared.

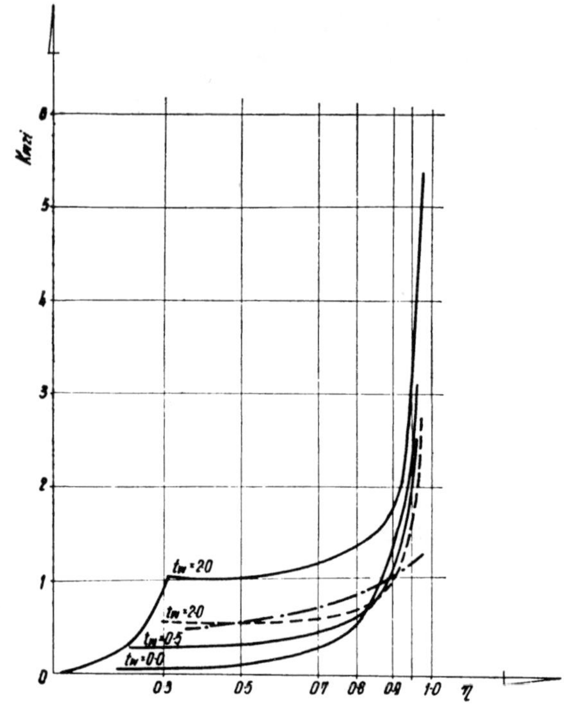

FIG. 3. Relative cost of presettling tanks and aeration chambers as a function of the treatment efficiency for BOD_5.

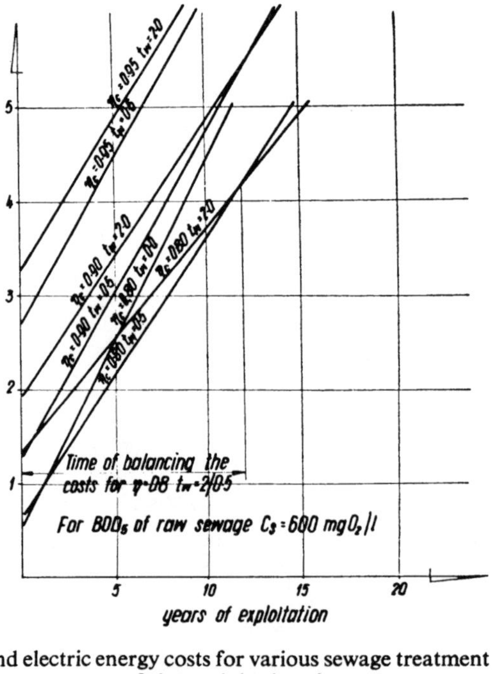

FIG. 4. Total capital and electric energy costs for various sewage treatment degrees as a function of the exploitation time.

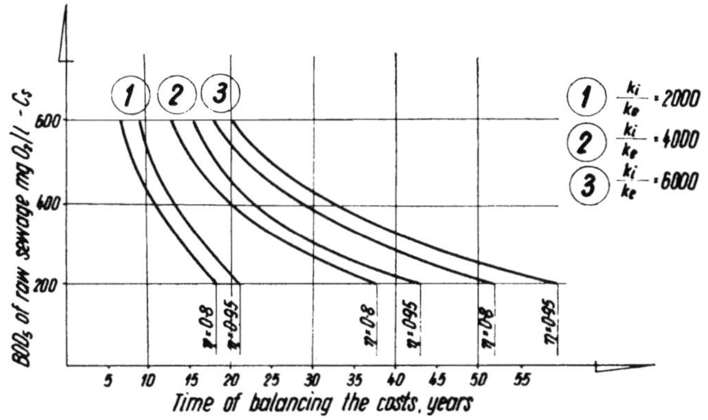

FIG. 5. Time of balancing the costs of a sewage treatment plant with presettling tanks for the 2 and 0·5 h flow time in relation to the sewage concentration and the ratio of the object cubature unit costs to that of 1 kWh k_i/k_e.

In FIG. 5 the times of balancing the costs are shown as a function of the raw sewage concentration for various η_c and for three variants of the ratio of 1 m^3 capacity net cost to that of 1 kWh. This shows that the time for balancing costs is dependent on k_i/k_e and on the raw sewage concentration.

7. CONCLUSIONS

(1) The time of preliminary sedimentation influences capital and running costs of sewage treatment.

(2) Capital costs are lowest when $t_w = \leqslant 0{\cdot}5$ h.

(3) Balancing the annual costs from capital and operating costs depends on the raw sewage concentration and on the ratio of the unit cost of capacity to the price of electricity and to a lesser extent on the degree of sewage treatment.

(4) Selecting the time of balancing in given circumstances is the basis of optimization.

(5) For sewage, with a cost balancing time of less than 15 yr the use of presettling tanks with $t_w > 0{\cdot}5$ h is not economical.

(6) Presettlement does not substantially increase the total degree of sewage treatment.

REFERENCES

SMITH R. (1969) *Proc. ASCE* **95,** SA 1, 117.

ELIASSEN R. and COBURN D. F. (1969) *Proc. ASCE* **95,** SA 2, 299.

IMHOFF K. (1966) *Taschenbuch der Stadtentwässerung*. Oldenbourg, Munchen.

ECKENFELDER W. W. (1966) *Industrial Water Pollution Control*. McGraw-Hill, New York.

PASVEER A. (1957) Über den Begriff OC/load. *Münchener Beiträge Bd.* **5,** 240.

ROMAN M. (1965) Ekonomiczna efektywność grupowych oczyszczalni ścieków dla miast i osiedli. Reports from the *Conf. of NOT* (*General Technical Organization*). Warszawa, January.

SĘDZIKOWSKI T. and BYLEDBAL J. (1969) Studia do założeń na budowę Grupowej Oczyszczalni Scieków Łódzkiego Okręgu Przemysłowego. Typescript in the Design Office of the Communal Engineering in Łódż.

SĘDZIKOWSKI T. (1966) Niektóre zagadnienia ekonomiczne zwiazane z etapowaniem stopnia oczyszczania ścieków na oczyszczalniach. *Gospodarka Wodna* **4,** 140.

Water Research Pergamon Press 1972. Vol. 6, pp. 347–350. Printed in Great Britain

BASIC PROBLEMS IN THE DESIGN OF LARGE TREATMENT PLANTS—ARAKAWA TREATMENT PLANT

SATORU TOHYAMA

Association of Ara-river Basin-wide Sewage Works, 48, 1-Chome, Nishihara, Shibuya-ku, Tokyo, Japan

INTRODUCTION

SAITAMA prefecture is located on the north of Tokyo metropolis and it has experienced a remarkable growth of population. Due to delays in the improvement of the sewerage system, domestic sewage and industrial wastewaters from this area are now discharged to the Ara river without treatment and pollute the middle stream of the river which is used as drinking water in the Tokyo and Saitama prefecture.

In the Ara-river basin, fifteen treatment plants are planned and some of them are being operated at present. To prevent the water of the Ara-river from being contaminated and to preserve the water resources, sewerage systems are necessary.

The recent reform of the law on sewerage has empowered each prefecture to carry out area-wide sewerage services. Saitama prefecture has decided to conduct such services along the four basins of Ara-river. First, the left side of the river was selected and this work was delegated to the Ara-river Basin Sewerage Works Association. This work has been undertaken since the year of 1966.

This paper describes the outline of Ara-river basin sewerage plans, particularly the design of treatment plant and presents material for discussion on the problems of large treatment plants.

ARA-RIVER BASIN SEWERAGE PLANS

The Ara-river basin sewerage works is intended to construct sewerage systems for 8 cities on the left side basin of the river.

The drainage area is about 200 km^2, and the population to be served is about two millions. The wastewater will be treated at one plant.

According to the municipal public sewage plan, the capacity of the conduit system shall be three times the maximum hourly flow for combined sewerage systems, and they must be able to receive the maximum hourly flow in case of separate sewerage systems, respectively.

The volume of the projected sewage flow was determined separately for domestic sewage and for industrial wastewater. The former is estimated on the future water demand.

Table 1 shows the projected wastewater flow.

ARAKAWA TREATMENT PLANT

Since the main objective of Ara-river basin sewage works is the prevention of pollution of the Ara-river water and preservation of the water resources, the sewage

TABLE 1. PROJECTED WASTEWATE FLOW

	Average dry weather flow	Maximum daily flow	Maximum hourly flow
Domestic sewage	360 l. day^{-1} $capita^{-1}$	500 l. day^{-1} $capita^{-1}$	620 l. day^{-1} $capita^{-1}$
Industrial wastewater	90 m^3 ha^{-1} of floor space day^{-1}	—	130 m^3 ha^{-1} of floor space day^{-1}

from this area has to be treated with great care, and well-treated stable effluents have to be discharged. To attain these objectives under all conditions, the newest techniques have been adopted in the Arakawa treatment plant.

Treatment is by the standard activated sludge process. The characteristics of Arakawa treatment plant are: (1) installation of pre-aeration tanks; (2) division of the plant into small units with equal loading to each tank; (3) combination of the unit tanks into a single structure in order to minimize hydraulic losses; (4) utilization of one-eighth of the facilities as a monitoring plant; (5) provision of storm water storage tanks; (6) systemization into central monitoring and controlling systems by use of a computer.

The Arakawa treatment plant, as shown in FIG. 1, consists of eight blocks of treatment facilities, and one block is further divided into 6 series of treatment units. Therefore, there are 48 series of treatment units in all. The reason for this sub-division is to minimize damage due to accidents. Stoppage of a unit has a considerable effect on other units and may cause overload, but with smaller divisions, it is possible to minimize the excess load to the units in operation. A problem arising from these smaller units is the difficulty of distributing the wastewater to each unit equally. It is impossible to control the flow entering the units by means of sluice gates at the inlets. In the Arakawa treatment plant, as each incoming pipe had electromagnetic flow meters and valves, the wastewater may be distributed equally.

In one-eighth of the facility, return sludge and air flow can be controlled independently. Since the units of this block can be operated independently it is intended to utilize this block as a monitoring plant and to establish the conditions for operating other blocks of the treatment plant. The monitoring plant has instruments such as a DO meter and pH meter at important points, and samples of the liquid treated may be sent to the laboratory continuously in order to ensure that optimum conditions are obtained.

The computer will collect and record processing data besides controlling processes.

Hydrographical features in this plant are freedom from short circuiting by dividing the facility into tanks with large length to width ratios, and combining the units from the pre-aeration tank to the final sedimentation tank into a single structure without connection conduits or pipes (FIGS. 2 and 3). This is intended to minimize hydraulic losses, to reduce the power for pumps, and to reduce construction costs by eliminating inlet and outlet sluice valves and connecting pipes.

The drainage area served by the Arakawa treatment plant is very large, so the sewage entering the plant is likely to be septic owing to the longer retention time. In addition, since supernatant liquids from sludge treatment plant will be returned to the sewer, it is intended to provide pre-aeration tanks to pretreat these wastewaters.

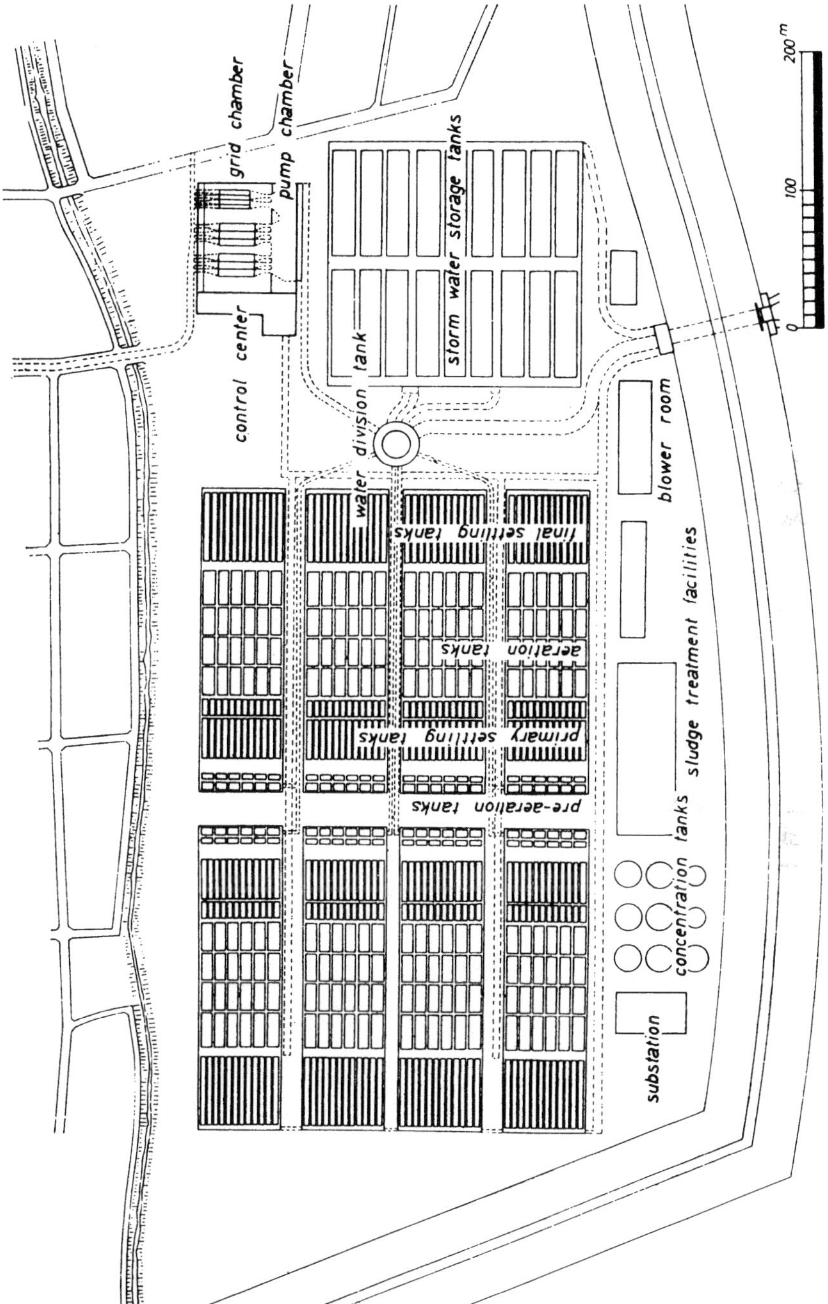

FIG. 1. Layout of Arakawa treatment plant.

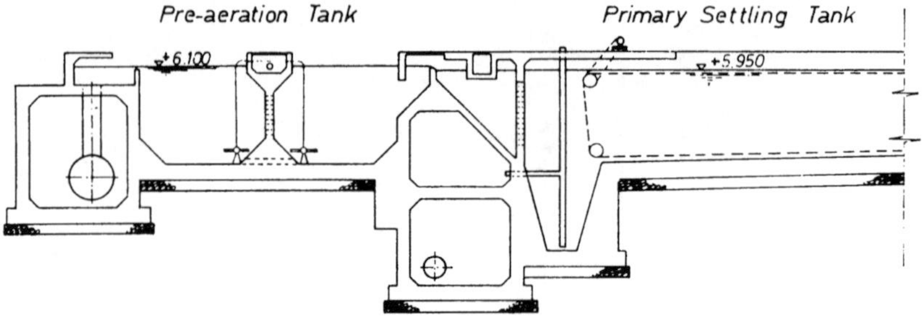

FIG. 2. Cross-section of tanks.

The Arakawa treatment plant receives 3 times the maximum hourly flow. The treatment plant is designed to treat the maximum hourly flow. In order to overcome the defects of the combined sewerage systems, flows in excess of maximum hourly flow are stored in tanks and returned to the treatment plant after the storm, i.e. after primary treatment.

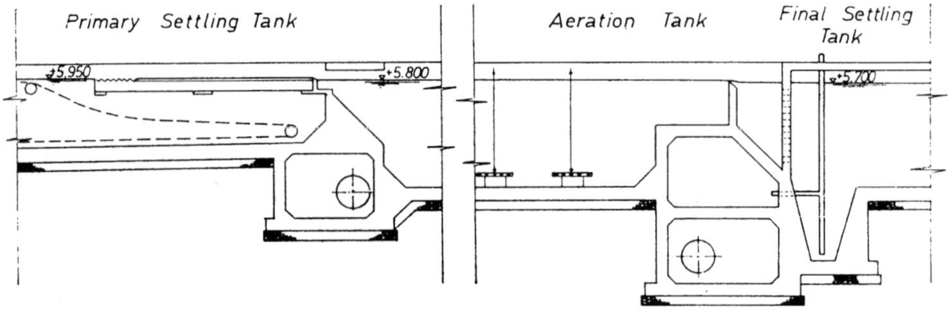

FIG. 3. Cross-section of tanks.

It is necessary to treat the maximum volume of wastewater since the flow of the river is small.

Although the newest techniques and some special features have been introduced, further studies are necessary to solve some of the technical problems.

Water Research Pergamon Press 1972. Vol. 6, pp. 351–356. Printed in Great Britain

DAN REGION SEWAGE RECLAMATION PROJECT

M. KREMER
Mekoroth Water Co. Ltd., POB 20 128, 9, Lincoln Street, Tel Aviv, Israel

1. INTRODUCTION

THE PURPOSE of the Dan Region Sewage Reclamation Project, Israel, is to solve the problem of sewage disposal and to reuse the reclaimed water of seven towns of the Greater Tel Aviv area. These have a population of 750,000. For this purpose they are organized in the Dan Region Association of Towns. Up until 1970 most of the untreated sewage from these towns was discharged into the Mediterranean Sea.

Studies have shown that reclamation of the sewage and its re-use would be cheaper than satisfactory discharge of sewage into the sea, with or even without treatment. Moreover, reclamation could provide water for unrestricted re-use and at a lower cost than most other water projects proposed for development in Israel.

The Dan Region Sewage Reclamation Project will operate as follows: Sewage will be collected by means of a coastal trunk sewer and delivered, with the assistance of two pumping stations, to stabilization ponds situated in the sand dunes to the south of the Dan Region. The sewage will then be treated in ponds with a recirculated ratio 1:1·5 (sewage:effluent).

The BOD load on the ponds will be 23 g m^{-2} day^{-1} (200 lb $acre^{-1}$ day^{-1}). The stabilization pond effluent will undergo lime treatment in an upflow solid contact clarifier for the removal of algae, suspended solids, phosphates, viruses and bacteria at pH = 11·0. The lime treated effluent will flow from the clarifier to detention ponds for ammonia stripping and pH balancing. The effluent will then be pumped to spreading basins in the sand dunes and recharged into the local aquifer. Recovery wells, drilled around the spreading basins at appropriate distances, will pump the reclaimed water from the aquifer and deliver it to the water supply system for general use.

The Project is being constructed in stages. The coastal trunk sewer and some of the stabilization ponds have already been completed and 10×10^6 m^3 yr^{-1} (million m^3 per annum) is being treated at present. The second stage of the project is under construction and in June 1972 the total annual quantity of sewage undergoing treatment will reach 30×10^6 m^3 yr^{-1}.

The plant is designed to meet the projected requirements of the 'nineties when, according to forecasts, the population of the Dan Region will number about one and a quarter million persons, and the raw sewage will amount to 110–120 $\times 10^6$ m^3 yr^{-1}.

2. EXISTING SYSTEM

In January 1970, the first stage of the project, consisting of a system of anaerobic facultative ponds was put into operation. During the running-in phase the residents of the locality raised objections to the plant because of the odours given off by the anaerobic ponds. A method had to be found to put an end to this nuisance.

It was decided to follow the example of the Cape Town (S. Africa) plant and install an additional system for the recirculation of effluent. This installation which has produced excellent results in Cape Town is proving equally effective at the Dan Region plant. The ponds, which previously were intended for anaerobic process, now operate

Raw sewage flow	Q	$= 27{,}000$ m^3 day^{-1}
Sewage concentration (B.O.D.)	C	$= 350$ mg l^{-1}
Ponds area	A	$= 400{,}000$ m^2
Organic load	L	$= 23$ g m^2 day^{-1}
Ponds volume	V	$= 670{,}000$ m^3
Recirculation ratio	R	$= 1:1{\cdot}5$
Detention time (sewage)	T_1	$= 25$ days
Detention time (sewage and effluent)	T_2	$= 10$ days

as aerobic facultative recirculated ponds, were put into operation in June 1970 and since then the plant has not caused objectionable odour or other nuisances. The ponds in operation at present are Ponds A, B, C, D and E (see FIG. 1) and are the first stage of the Dan Sewage Reclamation Project. A diagram of the ponds showing their operational system is given in FIG. 2.

3. FOLLOW-UP AND CONTROL

The effluent from the plant is planned to be recharged into the regional aquifer, after undergoing lime treatment. Special attention must therefore be paid to follow-up and control of the quality of the sewage and the effluent. Decisions on the future development of the plant will depend on how successfully it operates. The plant absorbs 27,000 m^3 day^{-1} ($7{\cdot}0 \times 10^6$ g day^{-1}) of sewage and this amount will be trebled within a year.

Quality control is carried out from two standpoints: (1) to check the efficiency of the system and (2) to detect potential environmental hazards or nuisances.

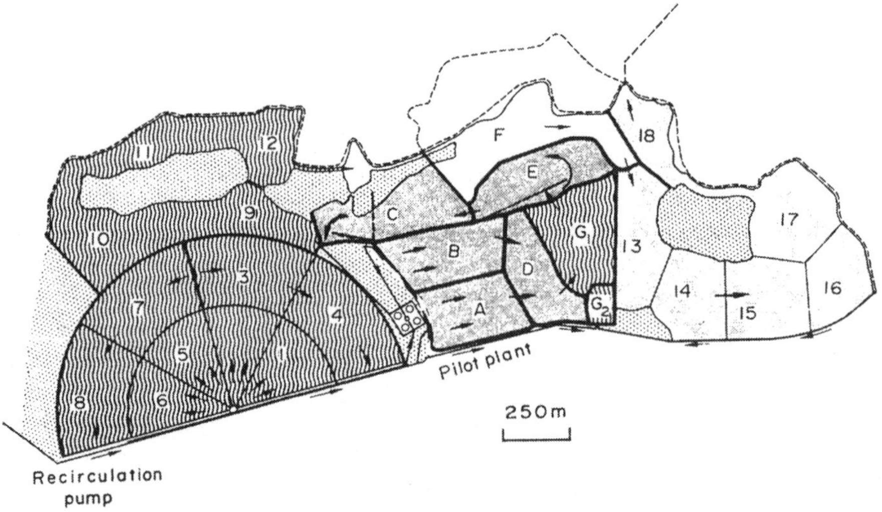

FIG. 1. Dan Region sewage project 30×10^6 m^3 yr^{-1} stage (1972).

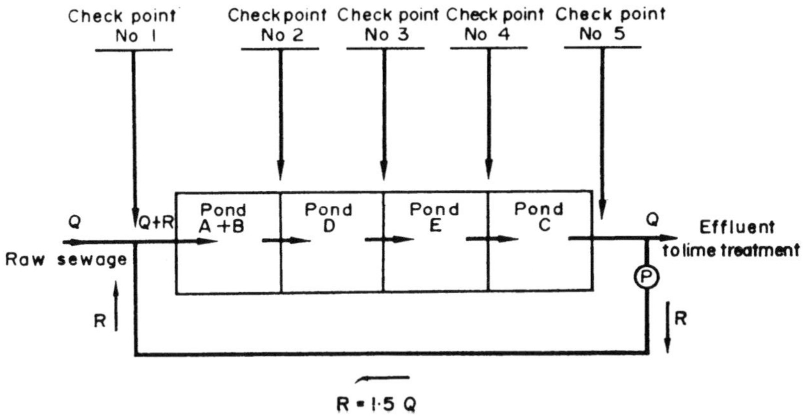

FIG. 2. Operational data.

3.1 *Efficiency of the system*

An important conclusion which can be drawn after one year's operation is that the main treatment of the sewage occurs in the first pond, as shown by the decrease in the concentration of organic matter and the removal of nutrient material by 80 per cent (see FIGS. 3 and 4). Moreover, in the first pond the main treatment takes place in the region of the raw sewage inlet; the load per unit area there is identical to that found in typical anaerobic ponds, while at the same time there is no evidence of the creation of objectionable odours or other nuisances. The direct cause of this is obviously the recirculation of the effluent which is done at a ratio of 1 : 1·5. The recirculated effluent carries with it to the first pond a high concentration of dissolved oxygen; the rate of saturation is 100–200 per cent during daylight hours and about 30–50 per cent during the night. The recirculated effluent also has high concentrations of micro-organisms and algae which are believed to prevent the creation of anaerobic conditions in the upper layer of the pond and even to increase the growth of algae and the supply of dissolved oxygen in the ponds.

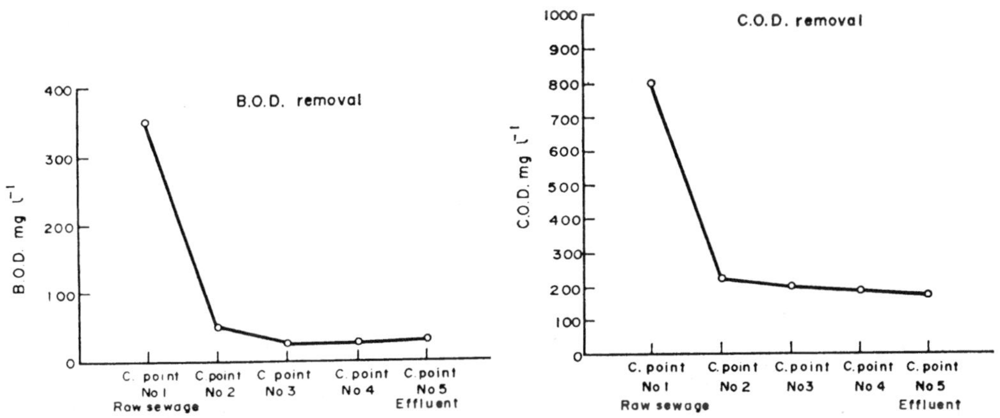

FIG. 3. BOD, COD removal.

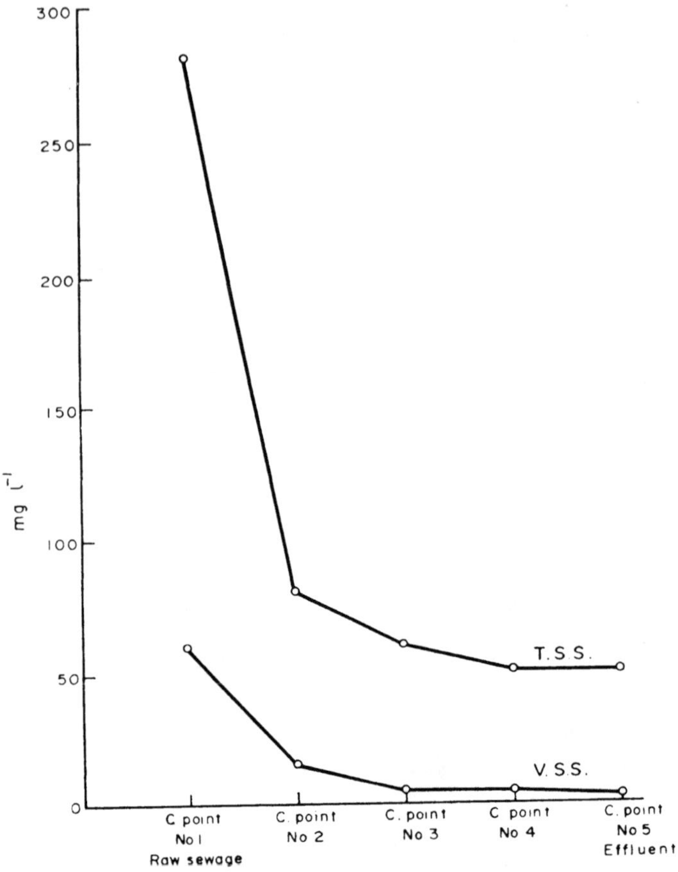

FIG. 4. Suspended solids removal.

Studies are now being made on the efficiency of the system, the optimal recirculation ratio, the influence of climatic factors and the effect of stirring the water in the ponds, with the aim of increasing the organic load to above 23 g m^{-2} day B.O.D., which is the normal load at present. Table 1 shows the average results of analysis of the sewage at various points in the system during January–April 1971.

3.2 *Location and prevention of environmental nuisances*

The principal potential nuisances for the local population, who live several kilometers from the plant, are odours and mosquitoes. In the operational regime described above permanent aerobic conditions are ensured throughout the pond system, as shown by the high dissolved oxygen concentration and the low concentration of sulphides, volatile acids and other malodorous substances. Very sensitive control instruments have been installed in the surrounding settlements. These detect H_2S in the air at a very early stage.

The problem of mosquito control is easily dealt with by antimalaria spraying from helicopters every 12 days from May to October.

TABLE 1. ANALYSIS OF RAW SEWAGE AND EFFLUENT AT VARIOUS POINTS IN SYSTEM (SEE FIG. 2), JAN.–APRIL 1971 (AVERAGES)

Analysis	Unit	Results				
		Check point 1 raw sewage	Check point 2	Check point 3	Check point 4	Check point 5 effluent
Temperature	°C	20	17	17	17	17
pH		8.3	7·8	8·0	8·2	8·3
B.O.D. (5 day, 20°C)	mg l^{-1}	350	50	25	25	30
C.O.D.	mg l^{-1}	800	220	195	180	170
Total suspended solids	mg l^{-1}	280	80	60	50	50
Volatile suspended solids	mg l^{-1}	60	15	5	5	2
Alkalinity	mg l^{-1} as $CaCO_3$	470	270	270	265	230
Sulphate	mg l^{-1} as SO_4	45	38	35	35	35
Phosphate	mg l^{-1} as P	11	5	5	5	5
Nitrogen (Kjeldahal)	mg l^{-1} as N	78	30	28	28	29
Nitrogen (ammonia)	mg l^{-1} as N	55	17	17	17	16
Sulphide	mg l^{-1} as H_2S	0·1	0·2	0·1	0·1	0
Volatile acids	mg l^{-1} as Acetic Acid	130	70	75	65	20
Dissolved oxygen (morning)	mg l^{-1}	0	2	5	6	6
Dissolved oxygen (evening)	mg l^{-1}	0	5	8	15	20
Chlorides	mg l^{-1} as Cl	130	110	115	120	120
Synthetic detergents	mg l^{-1} as A.B.S.	20	8	8	7	7
Fats	mg l^{-1}	180	10	8	7	4
Algae—total count	units in mm^3	0	2000	2500	3000	3700

4. SUMMARY AND CONCLUSIONS

In plants for the reclamation of wastes which treat raw sewages in areas similar to that of Israel (sub-tropical), good results can be expected from a system operating on the basis of faculative recirculated aerobic ponds. The method is also cheaper than with plants using mechanical equipment where both the investment and running costs are high. An outstanding advantage of this system is the fact that the long detention time of the sewage permits accurate control and a quality check before the effluent reaches the stage of tertiary treatment and recharge to the groundwater.

It is intended to operate this system with a BOD load of not more than 23 g m^{-2} day^{-1}. Where the effect of stirring the liquid in the ponds may be observed, this is recommended in order to prevent stratification in the ponds. This can encourage the growth of algae with an accompanying increase in the organic load to the ponds of 20–30 per cent.

Water Research Pergamon Press 1972. Vol. 6, pp. 357–364. Printed in Great Britain

AN "OPTIMIZED MODULE" APPROACH TO WASTE WATER RECLAMATION DESIGN

ROBERT A. RYDER* and MICHAEL T. MCGOVERN†

*Kennedy Engineers, 657, Howard Street, San Francisco, California 94105, U.S.A. and
†Consulting Engineer, 46-139 Punalei Place, Kaneohe, Hawaii 96744, U.S.A.

1. GENERAL

PLANNED urbanization of an area in the Ka'u District of Hawaii could result in a population of more than 100,000 persons within the next 15 yr. This area is characterized by infrequent rainfall, rather costly water supply development, and a distinct need to minimize pollution and eutrophication of the coastal waters.

The reclamation facility concept which will be discussed is a modular "building block" design, which provides high quality, secondary-plus treatment at very low flows [500 m^3 day^{-1} (0.125 mgd)], which can be increased in capacity within a single module up to 4 times [2000 m^3 day^{-1} (0.5 mgd)] with equivalent efficiency, and whereby identical modules can be added as demand requires. The treatment process is designed to provide optimum effluent quality for reuse as irrigation water for golf courses, parks, pastures, orchards and urban greenbelt.

2. CRITERIA AND CONSIDERATIONS

In addition to basic treatment requirements of solid separation, waste organics oxidation, and disinfection, the criteria of process and mechanical equipment selection included; (1) minimum capital cost, (2) minimum operational and maintenance cost, (3) maximum process flexibility, (4) minimum structure and area requirements, (5) readily obtained and easily-maintained equipment, (6) minimum odor from warm, sometimes septic wastes, (7) maximum operational stability under shock loads, (8) maximum expansion capability under unpredictable growth rates, (9) optimum power demand to system demand relationships, and (10) maximum treatment effectiveness with component or power failure. In concept, the modular approach immediately satisfies two of the primary objectives: maximum standardization of process and mechanical equipment, and unlimited capability for expansion.

Table 1 lists the process objectives, together with the anticipated raw wastewater quality.

Process efficiency objectives include removal of color, 80 per cent; grease, 90 per cent; turbidity, settleable solids and detergents, 95 per cent; suspended and volatile solids and BOD, 96 per cent; bacteria and virus, 100 per cent; and nitrogen, phosphorous, dissolved solids, and refractory organics, 0 per cent.

The intent of the current work was to optimize the matrix created by these 10 objectives and their relative degrees of dependent improvement.

3. MODULAR CONCEPT

The concept of a number of relatively small wastewater reclamation facilities rather than provision of centralized collection and treatment is particularly attractive when

TABLE 1. WASTEWATER QUALITY AND TREATMENT OBJECTIVES

Water quality characteristic	Units	Anticipated wastewater quality	Reclaimed wastewater type A*	Reclaimed wastewater type B*
pH	—	7.0–8.0	7.0–8.0	7.0–8.0
Color	PCU	75–150	20	75
Turbidity	JTU	50–200	10	50
Settleable solids	$Ml\ l^{-1}$	10	<0.5	<1.0
Suspended solids	$mg\ l^{-1}$	240	<10	<45
Volatile solids	$mg\ l^{-1}$	200	<10	<40
BOD (5 day)	$mg\ l^{-1}$	240	<10	<20
Grease	$mg\ l^{-1}$	50	<5	<15
M.B.A.S.	$mg\ l^{-1}$	10	<0.5	<2
D.O.	$mg\ l^{-1}$	0–2.0	>2.0	>2.0
T.D.S.	$mg\ l^{-1}$	540	540	550
Calcium	$mg\ l^{-1}$	30	30	30
Magnesium	$mg\ l^{-1}$	45	45	45
Sodium	$mg\ l^{-1}$	140	140	140
Potassium	$mg\ l^{-1}$	15	15	15
Iron	$mg\ l^{-1}$	2	<0.03	<0.03
Manganese	$mg\ l^{-1}$	0.1	<0.02	<0.02
Carbonate	$mg\ l^{-1}$	0	0	0
Bicarbonate	$mg\ l^{-1}$	190	190	190
Chloride	$mg\ l^{-1}$	160	165	175
Sulfate	$mg\ l^{-1}$	50	50	50
Silicate	$mg\ l^{-1}$	50	50	50
Nitrate	$mg\ l^{-1}$	2	10	10
Nitrogen (Total)	$mg\ l^{-1}$	30	10	10
Phosphate (PO_4)	$mg\ l^{-1}$	30	25	25
Fluoride	$mg\ l^{-1}$	0.75	0.75	0.75
Boron	$mg\ l^{-1}$	0.25	0.25	0.25
Arsenic	$mg\ l^{-1}$	<0.05	<0.05	<0.05
Hardness ($CaCO_3$)	$mg\ l^{-1}$	270	270	270
Alkalinity ($CaCO_3$)	$mg\ l^{-1}$	155	155	155
Coliform (MPN)	MPN $100\ ml^{-1}$	200×10^6	<2.2	<23
Sodium (%)	% cations	53.5	53.5	53.5
Stability index (Ryzner at 70°F)		8.0	8.0	8.0
Irrigation water class (California)		Class 1	Class 1	Class 1

* Type A. Reclaimed wastewater for irrigation of public access areas. Type B. Reclaimed wastewater for irrigation of non-public access areas and ground spreading.

there is a broad regional need for irrigation with the reclaimed wastewater, and concurrent development and growth of contiguous areas is not likely to occur.

The result of the application of these over-all objectives and the process criteria has been the creation of the "building block" module utilizing aerobic biological treatment. FIGURE 1 is a schematic flow diagram of the module process, and FIG. 2 is a plan and cross-section of one reclamation module. At full design capacity, the major components of the module include comminutor, twin aeration basins, a peripheral feed sedimentation tank with tube settlers, a post-chlorination system, a polyelectrolyte coagulation system, an air-washed multi-media filter, and aerobic sludge digestion.

It was determined that the largest treatment module that could be constructed considering initial cost, projected growth, service area, and nearby land requiring irrigation would provide the most economic overall selection. On this basis, a module

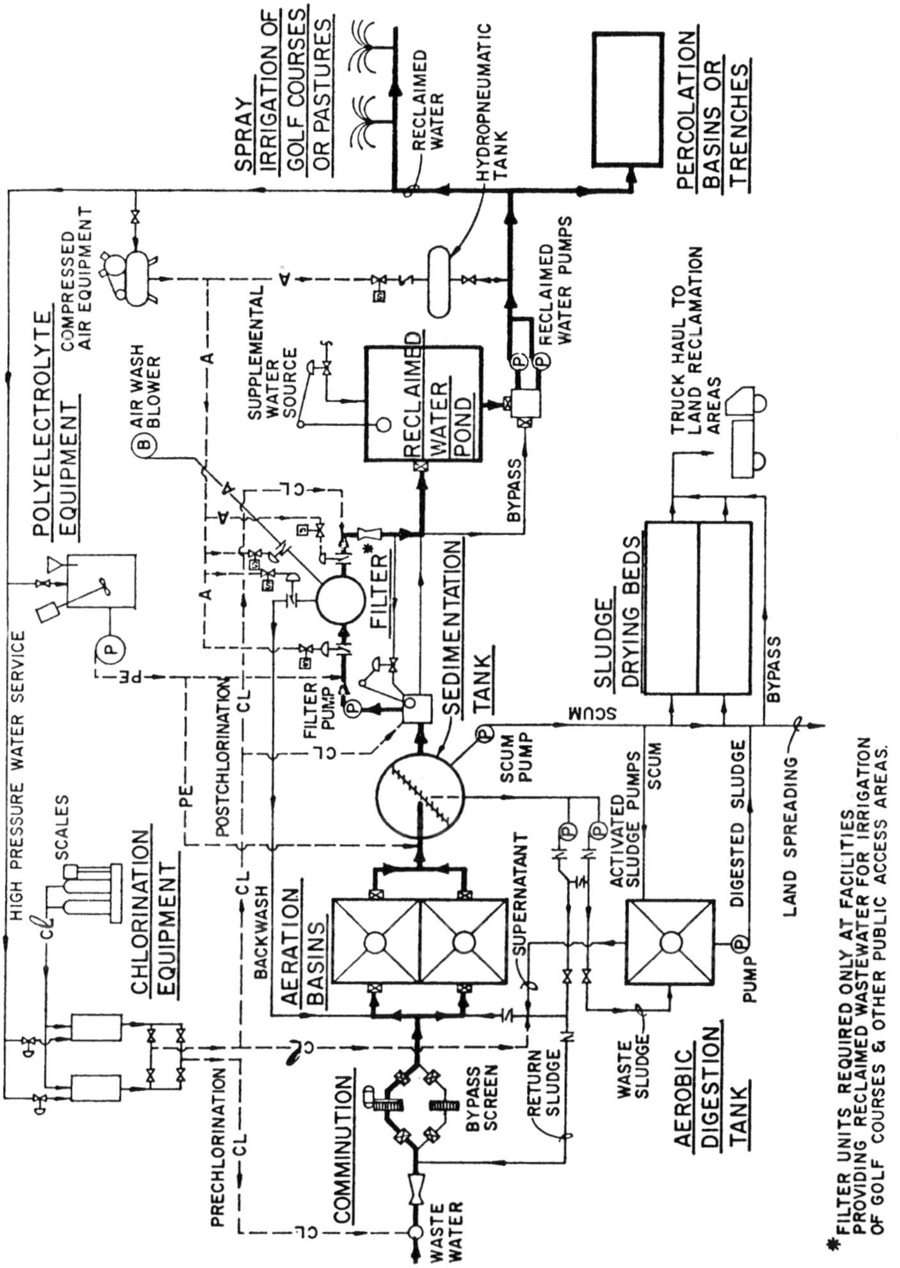

FIG. 1. Wastewater reclamation schematic flow.

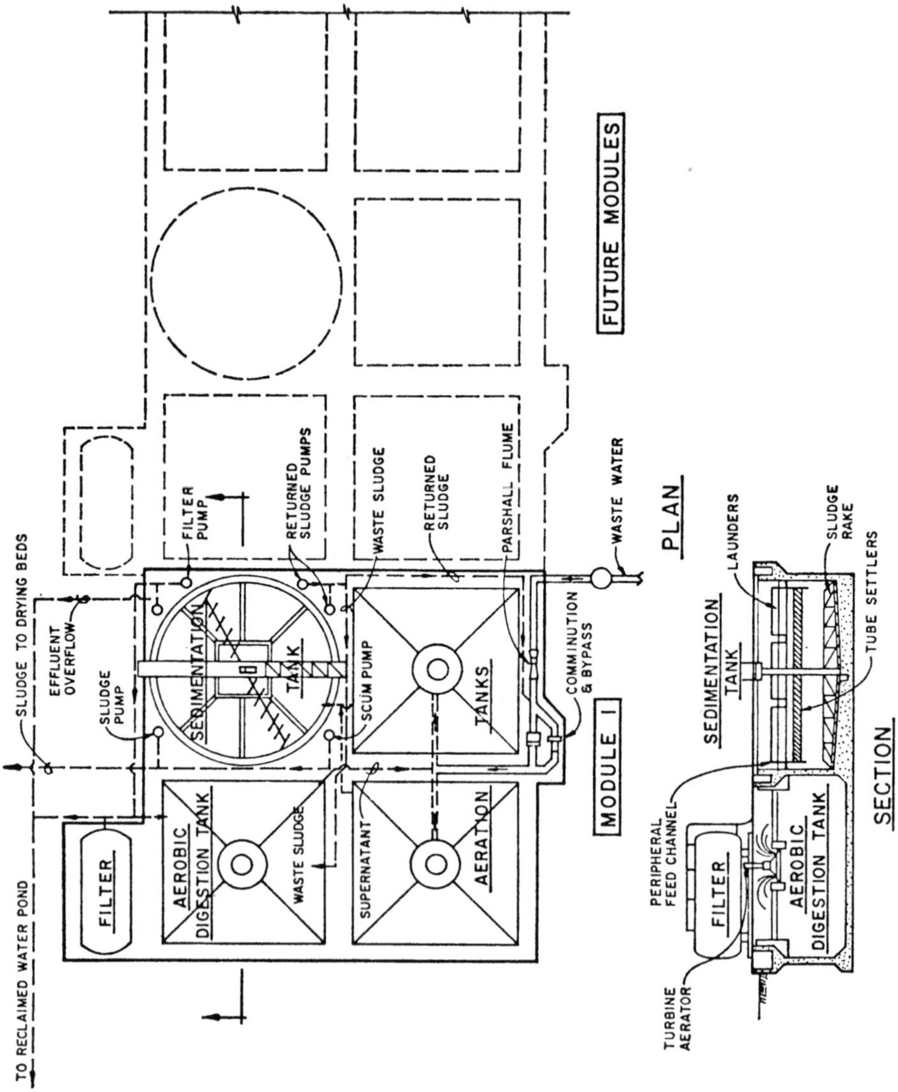

FIG. 2. Wastewater reclamation process module.

for treating an average daily wastewater volume of [2000 m^3 day^{-1} (0.5 mgd)] was selected. A group of these modules would ultimately comprise a complete wastewater reclamation facility, which could have a capacity of as large as [10,000 m^3 day^{-1} (2.5 mgd)]. These facilities would be constructed approximately 8 km (5 miles) apart by the building block concept as needed by the growth in population. The overall result is a regional system of wastewater reclamation plants, all exactly alike, and functioning like single units although dispersed to serve the entire region.

In order to provide for growth flexibility and initial savings in cost, a concept design was developed which would allow partial construction of a module. This is proposed for the first facility to be constructed late in 1971. At low initial flow rates [500 m^3 day^{-1} (0.125 mgd)], only one aeration tank and the sedimentation tank will be constructed to provide extended aeration activated sludge treatment with excess sludge being wasted directly to drying beds. The combination of high aerator capacity, a detention time of about 17 h in aeration, and the lower overflow rate in the sedimentation tank, are sufficient to provide a high degree of wastewater treatment. At medium flow rates [1000 m^3 day^{-1} (0.25 mgd)], with the addition of the second aeration basin, the aerobic sludge stabilization tank and polyelectrolyte coagulation, the module can operate as a plug flow or contact stabilization process. As flows approach the maximum design flow [2000 m^3 day^{-1} (0.5 mgd)] with addition of the filter and the tube settlers, the process can be converted to a completely mixed activated sludge system. With this extreme flexibility, major economies and efficiencies are effected in power requirements and operation costs. Also, capital cost is minimized during the early stages of resort development. The first construction phase is estimated to cost \$360,000, exclusive of irrigation facilities. This is equivalent to approximately \$270,000 on the west coast of California and in West Germany. Construction of the second phase would require an additional capital outlay of \$120,000, and the final phase would require another \$85,000 in capital. The total estimated cost, therefore, for the 2000 m^3 day^{-1} (0.5 mgd) module would be \$565,000 in Hawaii or about \$424,000 in California. It is obvious that, at the initial flow rates, the unit cost of waste reclamation by this process is relatively high. It should be recognized, however, that much of the facility for the full capacity module is constructed or provided at the first stage and, in fact, such items as the control building (\$60,000) will be used, regardless of the number of modules ultimately connected.

One important point is that there is no requirement, other than convenience that the modules be grouped at one location, with the exception of multiple use of the control building. The module could be placed anywhere in the sewerage system where collection is convenient and area for irrigation exists. Thus, it is conceivable that a large metropolitan area could have many such modules, or groups of modules, on somewhat of a "neighborhood" basis. This has the great advantage of reducing the size and length of trunk sewers, as well as reducing the magnitude of facilities to return reclaimed water to the land.

The processes and mechanical equipment selected for the module generally meet the design objectives listed above. When a clear-cut choice between capital cost and operating cost was presented, the lower operating cost alternative was selected. Wherever possible, standardization of equipment was accomplished. For example, the same type and size of turbine aerator was selected for use in the aeration sludge stabilization tank as was used in the aeration basins. Thus, it is economically justifiable to keep one full replacement unit in stock which can be used wherever one unit might fail, and down time is therefore minimized.

4. PROCESS ALTERNATIVES

Of substantial importance to the success of the modular concept are the details of process and equipment selection. It is perhaps most interesting to look first at the reasons for selection of an aerobic biological oxidation process for removal of organic constituents.

The alternative processes evaluated fall into two categories: biological treatment and physical–chemical treatment. In addition to the aerobic biological oxidaion process, five alternative biological treatment systems were studied; (a) plastic or redwood slat media trickling filters, (b) rotating biological discs, (c) oxidation pond with algae stripping, (d) aerobic lagoon with clarification, and (e) pure oxygen-activated sludge. Four physical–chemical treatment systems were evaluated and include chemical flocculation–clarification systems with; (a) activated carbon adsorption, (b) ultra filtration, (c) ion exchange, and (d) ozone oxidation.

None of these alternatives are as economical, based on construction and operation cost, as the selected activated sludge process, followed by effluent polishing, coagulation and filtration. Nor do the alternatives provide the same assurance of meeting the requirement of 96 per cent removal of biochemical oxygen demand and suspended solids. The normally assumed advantages of treatment by oxidation pond or aerobic lagoon are largely negated in Hawaii because of space limitations, a steep terrain, high land costs and the highly permeable soils.

Currently, physical–chemical processes are considered practical only for single units of 20,000 m^3 day^{-1} (5 mgd) capacity or larger because of complexities involved with smaller parallel units. Furthermore, the relatively high costs of electric power and chemicals, as well as gas for supplemental fuel for carbon regeneration or lime calcining substantially increase costs of such processes in Hawaii over conditions on the U.S. mainland or other more industrialized areas.

5. AEROBIC BIOLOGICAL TREATMENT

Criteria for design of the activated sludge system to provide the objective of 96 per cent BOD reduction were based on an organic BOD loading in the range of 0.14–0.16 kg day^{-1} kg^{-1} (0.30–0.35 lb lb^{-1} day^{-1} $MLVSS^{-1}$) of mixed liquor volatile suspended solids, a mixed liquor suspended solids sludge concentration in the range of 2500 mg l^{-1}, and a return sludge rate of 50 per cent of throughput. The result of computations based upon these criteria was an aeration tank with a retention time of 8 h and an oxygen supply capability equivalent to 1.3 kg of oxygen per kilogram of BOD removed. For the module design, it was considered desirable to provide two aeration tanks for either series or parallel operation to allow operational flexibility to utilize continuous mixing, conventional plug flow, or sludge reaeration as available operational flow patterns.

A comparative cost analysis of diffused air and mechanical turbine aeration was made to determine their respective economic advantages. The basis of comparison included an aeration allowance for aerobic digestion, an allowance of 1.64 kg of oxygen per kilowatt-hour (2.7 lb O_2/HPh) transfer rate for the mechanical aeration, and a 5 per cent oxygen transfer rate for diffused air aeration. The results of this comparison show that for a 2000 m^3 day^{-1} (0.5 mgd) module, equivalent annual cost for entrainment aeration is \$8500 as opposed to an equivalent annual cost for diffused aeration of \$16,500, a dramatic comparison.

A further comparison of fixed-bridge mounted and floating turbine aerators was made with the result that low-speed, large pumping rate turbine aerators are preferable for this installation to direct-drive high-speed units. These floating units have an initial cost of approximately half that of fixed units. They also provide long-life operation and low wear on bearings and gears. The floating aerators will permit variable level operation of the aeration and sedimentation tanks and thereby reduce peak flows which could result in excessive losses of solids overflow from the sedimentation tank. A further benefit is that the aeration level can be set to always maintain the maximum oxygen transfer efficiency, and the turbine blades not exposed or submerged to such extent to cause surge or motor overload difficulties. The aerators can be operated intermittently to optimize energy requirements for aeration in response to either pre-set cycle timers or dissolved oxygen control.

The floating aerators can be attached by cables to the corners of the aeration and digestion tanks in a manner to permit several feet of vertical travel with variation in water level.

As part of the modular design concept, it is quite desirable to make the side dimensions for each of the aeration tanks, the aerobic digestion tank, and the sedimentation tank equal. With selection of a circular sedimentation tank the four corner areas, resulting from insertion into a square module, provide for pump sumps to handle return sludge, scum, aerobic-digested sludge and settled water. The result is a square, compact module, requiring a minimum of land area and having a low profile for esthetic reasons. The exterior dimensions of one module is 20.5 m^2 (67 ft^2), and the nominal water depth is 3.7 m (12 ft). Additional modules may most easily be added in a row, although other arrangements are possible.

6. SOLIDS SEPARATION

A peripheral feed sedimentation tank was selected because it is considered that this will be most effective with the light sludges produced by turbine aeration, the capability to operate over a several-foot variation of water surface, and be easily adaptable for installation and support of tube settlers below the inner effluent collection launder.

A pressure filter system was selected for the modular filtration concept because of higher permissible head losses before backwashing and that the filter component can be pre-assembled by the supplier, thus minimizing construction costs as well as operation and maintenance requirements. A horizontal multi-cell filter unit was selected primarily on the basis of considerations of initial economy. A multi-cell filter is continuously on-line, even during backwashing, and backwash flow rate is reduced because of sequential cell washing. An 0.9 m (36-in.) bed of coal, sand and garnet providing a uniformly graded media was selected.

Cost and functional comparisons of other elements and components of the wastewater reclamation module were made in a manner similar to the foregoing as part of the "optimization" approach.

7. ENVIRONMENTAL ADAPTATION

Finally, the module lends itself to architectural treatment because of its uniform shape and single-unit appearance from the exterior. The module may be set deeply into the ground so that only enough height is exposed to prevent accidents by persons falling into the tank. The Hawaii modules are being designed with a concrete parapet wall around the perimeter in place of metal handrails. This parapet, and all exposed structural concrete, is planned to receive lava rock facing on the exterior in order that it will appear to the casual passerby to be merely a native rock wall similar to old corrals and fences in this area.

This concept of providing modular wastewater reclamation units to meet a wide variety of objectives appears particularly applicable to widely-scattered, low-density developments in arid and semiarid regions. The authors foresee that this concept also has applicability in certain areas of high population and density. Population centers in areas of multiple drainage basins may be one such example. A more specific example might be the urban areas of Southern California, Australia and the Mediterranean where there are many areas suitable for irrigation with reclaimed water, yet collection of sewage into a single main treatment facility with deep ocean disposal is extremely expensive.

Water Research Pergamon Press 1972. Vol. 6, pp. 365–367. Printed in Great Britain

POLLUTION CONTROL REGULATION—A PROBLEM OF FILLING GAPS

J. M. RADEMACHER

Region V11, EPA, 911 Walnut Street, Kansas City, Missouri 64106, U.S.A.

IN THE United States from July 1, 1956, to date almost 12,000 waste treatment projects costing in excess of $9 billion have been aided through Federal construction grants of more than $2 billion. Many miles of streams have been improved by this massive effort, *but* the return on this huge investment still has to be rated as only marginal. Because of serious deficiencies related to plant and system design and in operation and maintenance of these facilities, the clean water goal is still to be reached. Plants which can not maintain design capability or which are maintained and operated to produce a quality effluent only 50–60 per cent of the time will prevent that goal from being reached unless programs to overcome these deficiencies are developed and implemented.

H. Seidel estimated that almost 60 per cent of the plants in the United States have significant design and maintenance and operation deficiencies. The problems ahead will be considered by reference to FIG. 1.

DESIGN GAP

During the past year, we have seen real efforts in bridging the gap between research and field use of new ideas through the medium of a series of well prepared and presented "Technology Transfer" seminars. EPA's construction grant program is presently developing strong standards for use by the Agency and State pollution control programs. The regulatory agencies can provide standards and disseminate technical information but it remains up to the consulting engineer to educate himself in the waste treatment plant maintenance and operation mysteries.

OPERATION AND MAINTENANCE GAP

The operation and maintenance gap is made up of two key elements—formal training and expert assistance. We have made a start in providing operator training.

The decision makers—the mayors, city managers, city councils, even the public works or utility directors—have been conditioned to consider the physical construction of a waste collection and treatment system as an end unto itself. Until the decision makers know what their responsibilities are the whole training program for operators will only have a marginal return in clean water.

This is brought home by looking at the pay scales for operators. The minimum hourly wage in the United States is $1.60. By comparison, a management consulting firm in California a few years ago determined that a skilled plant operator for an activated sludge type treatment system was equivalent in training to a highly skilled automotive mechanic making $10,000–$12,000 yr^{-1} (approximately $5.00 h^{-1}).

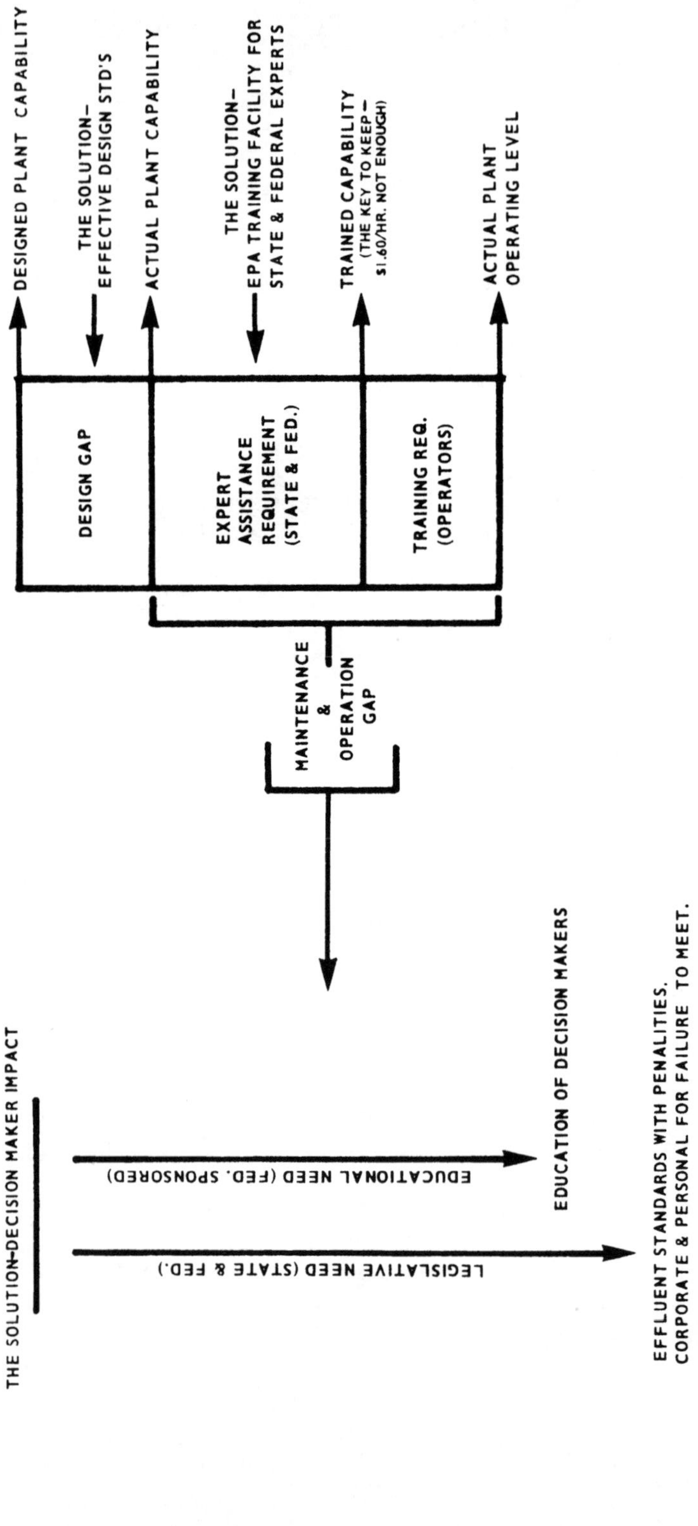

FIG. 1. Pollution control regulation—a problem of filling gaps.

MOTIVATION GAP

Up to this point the roles of consultants, the regulatory agencies, operators and the trouble shooters have been discussed. Yet if we stop here we will have missed the most important facet of this clean water effort—the motivation of the decision makers. To prevent our waste treatment systems from being perpetuated as "monuments to inefficiency" this key "motivation" gap must be closed. One tool must be education and we are instituting trial efforts in education of the decision makers. The University of Missouri has a grant from EPA to sponsor, develop and institute a series of information seminars across the State of Missouri geared to motivate the invited elected municipal officials and their key staff on the need for their informed support in plant operations. A companion effort by EPA working with the State pollution control agency in Iowa, is scheduled for a similar type of group this Fall. Both efforts will be evaluated and hopefully a workable effective program can be developed for use on a national scale.

However effective the educational effort is, the fact still remains that political decision makers will only respond on a continuing basis of their corporate and personal identity may be in jeopardy. We need effluent standards with punitive legal recourse to supplement the stream standards and to make "believers" out of the decision makers. Some States have this tool and, depending upon the effort expended, have met with varying degrees of success, but a national effort is required. This need is important since the massive construction effort to build waste treatment plants within the past decade and in the years to come could result in those same plants becoming serious sources of pollution.

SUMMARY

Unenlightened design, poor operation and maintenance condoned and abetted by the decision makers is a threat to our water environment. In addition to research and construction we also need a concurrent program to close the design and the "decision" gaps and to provide genuine expert help to plant operators in order to achieve really clean water in our time.

Water Research Pergamon Press 1972. Vol. 6, pp. 369–370. Printed in Great Britain

DISTILLATION OF M AND O PROBLEMS. EPA—REGION VII

J. M. Rademacher

Region VII, EPA, 911 Walnut Street, Kansas City, Missouri 64106, U.S.A.

1. Piping and equipment installations frequently do not allow adequate flexibility, to allow, for example, activated sludge to be washed and returned simultaneously. Usually return pumps are not variable speed and the flow not metered which results in guessing at the return flow volumes.

2. Plants located in cold climates have not been given adequate sludge disposal capability. Vacuum filtration or large aerated holding tanks should be considered as alternative methods of disposal.

3. Soil analysis by consulting engineers has not been adequate to limit erosion problems on lagoon berms, to eliminate leaks in lagoon bottoms and soil settlement under structures at mechanical plants.

4. Recurring problems are those of no skimmers on secondary clarifiers, inadequate flow metering on influents and inplant flows. To these problems we should add combination grit and bar screen structures. If the bar screens are not raked frequently, flows back up the grit channel with the result that organic matter as well as grit settle out in the channels.

5. Inadequate review or projection of hydraulic and organic loads by consulting engineers haunts many cities. Consulting engineers design on faulty parameters resulting in eventually overloaded facilities.

6. Plants being operated in cold climates need tanks covered or provided with wind breaks. Mechanical plant in North Dakota had surface mechanical aerators. They had a tendency to freeze since they were exposed to wind. Wind-breaks constructed out of plywood solved that problem. An oxidation ditch in northern Nebraska was built so the effluent weir was exposed to the wind. To prevent freezing the City built a box over the weir and installed heat lamps.

7. Bypasses have been built at elevations such that bypassing occurs before plant capacity is reached. Many of these bypasses can either be eliminated or could have been designed to take full advantage of treatment plant capability before bypassing occurs.

8. Some design engineers work only as designers and do not spend enough time around plants.

9. Inadequate fuel supply to run standby power equipment.

10. In general engineers have not taken full advantage of current technology.

11. Telescoping valves are used to control return sludge flows and water surface elevations in lagoons. The valves in the lagoon applications are sometimes rusted such that the valve is inoperative. Valves used to control volumes of return sludge flows at activated sludge plants may clog when flow is reduced below a certain point.

12. Alarms are not provided on equipment to indicate power or mechanical failure, which may result in odors, poor treatment and frequently raw bypassing.

13. Bar screens and comminutors have been installed in deep pits with inadequate

lighting and/or ventilation. The operator is asked to go down into this environment and he is frequently required to carry the screening back up a vertical ladder.

14. State inspection is generally inadequate. Defects that have been observed include lack of weed control and poor operation of treatment plants.

15. State personnel sometimes go out to inspect facilities without equipment to make even simple tests or to verify the results of operator testing.

16. Running water is not laid on at some mechanical plants, so that plants cannot be washed down and laboratory testing is impossible.

17. Utah has a policy of requiring chlorination after secondary treatment facilities. The experience of our Denver office has been that State inspectors do not require the facilities be opeiated and in the author's opinion this is poor economy.

18. Service charges by manufacturer's representatives are expensive. Many of the problems encountered could be referred to State personnel for routine assistance but these personnel are so shorthanded or inexperienced that the problems do not get solved.

19. Plant operators are sometimes given wrong information or contradictory advice.

20. It is found that after a few years essential equipment is no longer available to replace plant originally intended to last many years.

21. The status of the operator is downgraded. He frequently receives inadequate salary and an inadequate budget for repairs and maintenance.

22. Town politics create a situation where industrial waste ordinance or agreements are either not established or if they are they are not enforced.

23. Good design should consider that all mechanical equipment will fail eventually and will require disassembly but this is often not allowed for.

24. Infiltration during wet weather creates hydraulic overloads at the treatment plant or causes bypassing to occur. Infiltration in newly constructed lines is inexcusable.

25. Poor record keeping is the downfall of many operators. The keeping of mechanical and maintenance records aids the operator in justifying new equipment or modification of existing equipment and explaining frequent equipment outages.

26. Many plants do not operate at full potential because the importance of proper laboratory testing is ignored.

Water Research Pergamon Press 1972. Vol. 6, pp. 371–373. Printed in Great Britain

DIVISION OF WASTEWATER FLOW

WOLFGANG STALZER and WILHELM VON DER EMDE
Institut fur Wasserversorgung, Abwasserreinigung, und Gewässerschutz, Technische Hochschule Wien, Karsplatz 13, A-1040 Wien, Austria

AT LARGE treatment plants there are, among others, the following possibilities:
(1) simple division by separating channels
(2) division by flowmeters and gates controlled by the flowmeters
(3) hydraulicly by using weirs (e.g. circular distribution weirs) before the tanks
(4) hydraulicly by utilizing the head losses that exist at the inlet or outlet to or from the primary or secondary clarification tanks.

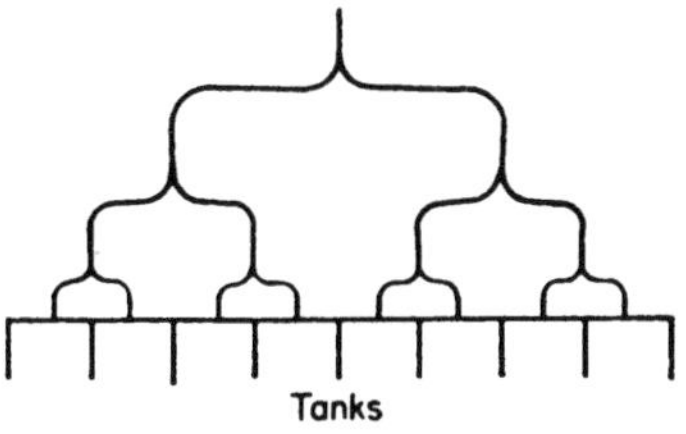

FIG. A.

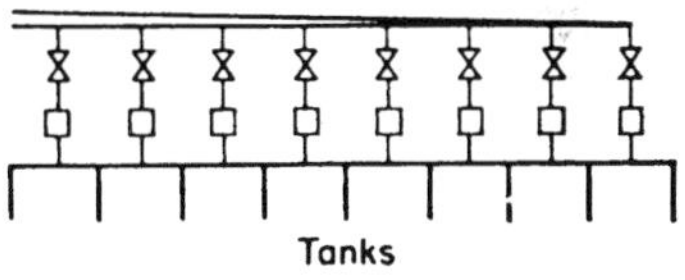

FIG. B.

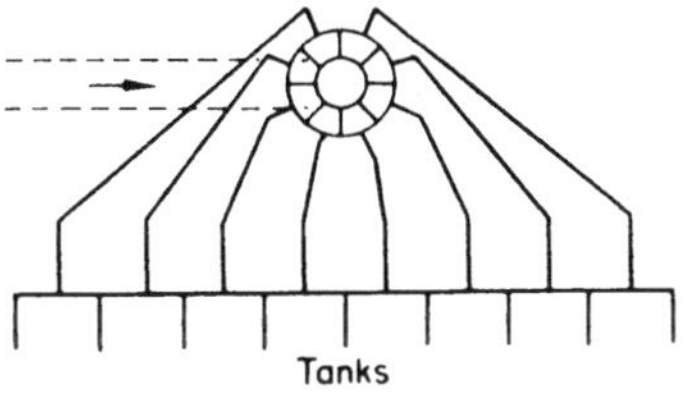

FIG. C.

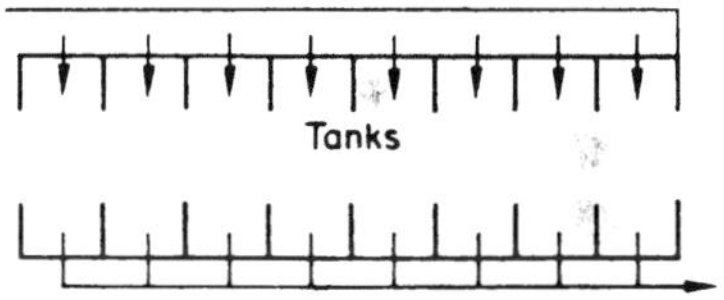

FIG. D.

(1) Simple division by a symmetrical separation of the channels has the advantage of a hydraulicly clear situation—whereas three-fold and multiple partition offers no clear situations—nevertheless long inlet conduits are required for appropriate radii of curvature and long exit channels are required to ensure uniform flow in the cross section before the separation. (Much space and expensive conduits are necessary.)

(2) A division of the wastewater by flowmeters often cannot be avoided in the case of numerous tanks; this solution is expensive and intensive maintenance will be necessary. Where flow measurement involves significant head losses and where large fluctuations of the flow exist, large head losses have to be considered.

(3) Division by weirs before the tanks (e.g. circular distribution weirs) has the advantage of a hydraulicly clear solution. Additional head losses and long lengths of the conduits have to be accepted. When applying circular distribution weirs, sufficient radii and uniform flow towards the weir has to exist.

(4) Division by utilizing the head losses at the inlet and the outlet does not cause additional head loss. A prerequisite with rectangular tanks is a wide distribution

channel which is situated in front of the battery of tanks and where low flow velocities exist. By this means small friction losses and a small regain of the velocity head, respectively, will lead to a small difference between the water level at the beginning and the end of the distribution channel. The advantage of such a distribution channel is that large particles will settle out or float to the surface. The disadvantage is that it is necessary to collect the separated particles or to aerate the channel.

Furthermore, an exact and equal level of all outlet weirs has to be established since this forms the basis for the hydraulic distribution (a check by filling the tanks is absolutely necessary). This method of hydraulic distribution only works if free flow exists over the outlet weirs.

The change in water level caused by different flows can be applied for flow measurement.

(a) *Utilizing the inlet head losses*

The velocity distribution in a sedimentation tank has to be uniform. By accelerating the wastewater and its contents (large velocity head), successive dispersion of the water contents (head loss) and equalization of flow across the section of the tank in the inlet construction. The objective of attaining uniform flow can be achieved. The head loss in the inlet construction will be much larger than the difference in the water level between the beginning and end of the distribution channel, and this is why the influence of the difference in the water level will be negligible with respect to the flow distribution e.g. "Stuttgarter Inlet" according to WEIDNER (1967).

Velocity in the distribution channel:	$v_1 = 0{\cdot}1 \text{ m s}^{-1}$
Velocity in the acceleration part of the inlet construction	$v_2 = 1{\cdot}0 \text{ m s}^{-1}$
Head loss according to	$h_v = \zeta V_2^2/2g$
"Stuttgart Inlet"	$\zeta = 1{\cdot}6$
Head loss:	$h_v = 8$ cm.

Velocity head and friction losses in the distribution channel around 1 cm. Such inlet structures are used with advantage in final clarifiers.

(b) *Utilizing the outlet head losses*

When applying a relatively large weir load a relatively large difference in the water level of the tanks will be established, too, if there is a fluctuation of flow. Where an uneven distribution of the flow to the tanks occurred, this would be caused by a relatively large difference in the water level at the beginning and end of the distribution channel. This difference will not exist when the distribution channel is properly designed.

Example

Velocity head and friction losses in the distribution channel:	0·22 m
Weir load	144 l. $\text{m}^{-1}\text{s}^{-1}$
Necessary head for free flow	0·18 m

When head for free flow 0·19 m, flow 156 l. m^{-1} s^{-1}

When head for free flow 0·17 m, flow 132 l. m^{-1} s^{-1}

Maximal deviation in the flow distribution:

$$\pm 12{\cdot}0 \text{ l. m}^{-1} \text{ s}^{-1} = \pm 8 \text{ per cent.}$$

Such outlet structures are used with advantage in primary clarifiers, since small entrance openings can clog. The clogging is avoided by this solution.

REFERENCES

SCHRÖDER-PRESS (1966) *Hydromechanik im Wasserbau.* Wilhelm Ernst u. Sohn, Berlin.

WEIDNER S. (1967). *Zufluß, Durchfluß und Absetzwirkung zweckmäßig gestalteter Rechteckbecken Stuttgarter Berichte zur Siedlungswasserwirtschaft*, Heft 30. R. Oldenbourg, München.

Water Research Pergamon Press 1972. Vol. 6, pp. 375–376. Printed in Great Britain

TREATMENT PLANTS WITH SMALL QUANTITIES OF SEWAGE DURING THE FIRST PHASE OF OPERATION

G. FRIESECKE

F. H. Kocks KG, Consulting Engineers, Duesseldorf, Fed. Rep. of Germany

1. INTRODUCTION

IT IS sometimes necessary to design large treatment plants which at first have to treat no more than 25–30 per cent of the design quantities. In such cases, a certain procedure of treatment can be determined according to criteria not to be discussed here, besides the specific design data for the different stages of treatment and also the capacity of the different phases of enlargement which result out of the expected increase of sewage. The different phases of enlargement mostly are designed in such a way that they can be extended by adding similar units—inside the same treatment stage—to that ones which are already in operation. However, this principle of simple "adding" mostly is not applicable for structures like gutters partly also not for grit chambers because of technical and economic reasons.

For towns with an extraordinary development of population and accordingly of sewage, the above mentioned principles are not always applicable, mainly if measures against pollution have to be taken as early and as intensively as possible. This applies especially to communities, the development of which is mainly influenced by fast growing industries or if mineral resources like ore, mineral oil, natural gas, etc. are disclosed. The simple "adding" of analogous units in treatment plants often can be uneconomical, for instance, if an initial quantity of sewage of 5–10 per cent of the final quantity has to be considered. In such extraordinary cases special criteria should be considered.

2. DESIGN CRITERIA

2.1 The cross-sectional design of the gutters and the design of the grit chambers have to consider especially that in the case of very small quantities of flow still acceptable hydraulic conditions can be achieved. For the gutters, trapezoid or triangular sections etc. are well-known means. For the grit chambers, the respective design criteria are basically usual because the optimum time of flow has to be met in any case.

2.2 The distribution gutters of rectangular aeration and settling basins have to be designed—if necessary by means of installing provisional dividing and guide walls—in this way that there are no "dead corners" and also no time of flow which leads to sedimentation. Also minimum filling heights have to be secured.

2.3 It should be possible to run all units of treatment individually without any trouble—or at least without additional essential means—or in common with an optional number of the other analogous units.

2.4 It should be possible to operate each stage of treatment individually or combined with any other stages. This applies also to the stages of sludge treatment. The realization of this requirement presumes that each stage has its own by-passes or its own by-pass-system. Also suitable lateral connections of the system can be of good use.

2.5 In certain cases, possibilities have to be found to reduce the volumes to be treated by provisionally installing of walls in at least one of the (rectangular) aeration basins. The design of the concerned basin has to take care that these walls can be installed and removed with a minimum of cost and time.

2.6 If it is necessary, the design must provide for possibilities based on which the treatment plant can be operated with a different mode of treatment in the beginning than in the case of larger quantities of sewage later on. A practicable solution is for instance the preliminary use of the activated sludge process with a stabilization of the sludge which will be replaced by low-rate—in the end by high-rate activated sludge processing.

This criteria mentioned under paras 2.1–2.5 have to be considered as an essential presupposition for such kind of alteration.

3. SUMMARY

Some larger treatment plants may have to start operation with about 25–30 per cent of the sewage quantity provided for in the final phase. In areas with extraordinary increase of population this percentage can be as low as 5–10 per cent. The design of treatment plants has to observe special criteria to achieve economically and technically sound solutions.

These criteria relate to the implementation of the step-wise start of operation of the treatment plant even with small quantities of sewage. These criteria include a special design of channels and the different structures as well as of their interrelations and interconnections. Eventually they include also to provide for a treatment procedure which differs from the designed "final procedure".

Water Research Pergamon Press 1972. Vol. 6, pp. 377–380. Printed in Great Britain

DESIGN OF INDUSTRIAL WASTE TREATMENT PLANTS

A. M. M. BUYSSE, R. A. BEARDSLEY and S. J. THOMSON
Fluor Nederland N.V., P.O. Box 634, Haarlem, The Netherlands

1. INTRODUCTION

THE USUAL contaminants to be removed from waste water are phenols, oil, nitrogen, phosphorus, suspended solids, heavy metals, coliform and toxicity.

2. DESIGN CRITERIA

A waste treatment plant is generally an expense item with no method of recovering the expense. This simplifies the optimization of the plant because the optimization is strictly an evaluation of the operating cost against the capital investment. The plant design must be one that meets the product specification on a 24-h day basis. Provisions must be made to store feed or effluent during treating plant upsets so that they can be rerun once the plant is again operating satisfactorily. The design also has to be such that future changes in effluent standards can be handled without creating an expensive alteration of the waste treatment system.

The following is a review of some important design criteria.

2.1 *Segregation*

Segregation of the various sewers to provide the various feed streams through the waste treatment plant will often avoid passing all the streams through the entire plant. Considerable economy can be realized by proper segregation of the waste water streams from the plant.

2.2 *Elimination of waste water*

An important step in the design of a plant waste treatment study is an in-depth study to find out which waste streams may be re-used, or eliminated by proper selection of inplant designs.

2.3 *Oil removal*

The most popular methods of free oil removal are based on three principles. (a) Gravity separation by allowing the oil to float on the top of the water (API and CPI separators); (b) gravity separation by assisting the oil to flow to the top more completely and rapidly by use of flotation air to increase the difference in gravity; (c) gravity separation by coagulation of the oil with a heavy additive such as lime, forcing the oil to settle to the bottom of a settling device where it is removed by underflow and bottom rakes.

2.4 *BOD–COD removal*

BOD and COD removal are effected either by biological treatment, which removes 80–90 per cent of the BOD or this treatment followed by carbon adsorption in order to remove 90–100 per cent.

BOD removal processes usually require prior oil removal down to less than 50 ppm of free oil.

There are other physical–chemical methods of BOD and COD removal, but to date these have proved inadequate or too expensive to receive serious consideration.

3. SPECIFIC PROBLEMS

3.1 *Nutrient removal (nitrogen and phosphate)*

The bulk of the ammonia/nitrogen should be removed from process waste water streams before they enter the waste water effluent by such methods as ion-exchange and steam stripping.

Ammonia can be converted to nitrate in a Biological Treating unit, preferably by a second step with additional aeration. The use of a single-stage process (such as extended aeration) has been the usual practice for this purpose in industrial waste treating systems.

After nitrification, the nitrates can be converted to nitrogen by anaerobic biological treatment.

Phosphorus in small amounts is absorbed in the sludge of the biological treating unit and is usually not sufficient in quantity to give a problem in refinery operations. Phosphate in boiler blowdown streams is usually the only contaminant source. Most oil refinery biological oxidation systems may require the addition of phosphate as a nutrient. Positive removal of excess phosphate, if required, can be accomplished within the biological oxidation system by the addition of alum. Phosphate can also be removed in a separate reactor by precipitation with alum or lime.

3.2 *Heavy metals*

Chromate is usually the only heavy metal that gives a problem in refinery effluents. Recovery by ion exchange for re-use is one method which shows a chance of economic payout. The use of zinc in cooling water treatment is increasing. The effluent standards for zinc are less stringent than chromate. Zinc can be precipitated to oxide or sulfide and will be absorbed in various sludges formed in the waste treating system.

3.3 *Dissolved solids removal*

The construction of large impounding storm water reservoirs for use as the plant water supply or the elimination of brines from the waste effluent by mechanical and natural evaporation are among the choices available for large inland plants to solve the disposal of concentrated brines. Eventually the accumulation of solids must be disposed of at a location which will not contaminate water supply sources.

3.4 *Auxiliary waste processes*

Auxiliary waste processes to the main objective of meeting effluent quality standards can also create problems that take considerable planning including the disposal of solid wastes. The simplest solution, if available and economical, is the out-haul for disposal and land fill. Land fill sites, where any type of solid waste can be disposed, are becoming scarce. This is especially true of land fill sites for inorganic solids including lime sludge, as leaching can contaminate ground water supplies and even surface supplies if the land fill site is near a water course.

Rehabilitation of strip mine sites with organic sludges including that from biological oxidation processes is becoming more attractive.

With large volumes, it may be more economical to dewater at the plant site including incineration.

3.5 *Petroleum coke wastes*

Most effluent quality standards state that there shall be no settleable solids which also means floatable solids. Coke fines have the peculiarity of sometimes floating and sometimes clumping and settling.

Except for fine strainers or precoat filters, other systems such as clarifiers or ordinary pressure filters that require less attention are not 100 per cent reliable. Storm drainage from a coke storage area can result in coke fines entering the receiving stream if precautions are not taken to retain the run off so coke fines are retained.

3.6 *Monitoring instruments*

As a general statement, every known contaminant that is listed in the effluent standards for the area must be monitored.

Most instrument manufacturers supply automatic monitoring instrumentation. One instrument may record several constituents with a probe package for monitoring conductivity, pH, dissolved oxygen, specific ions such as fluoride, etc. Even without automatic monitoring, automatic sampling devices should be installed at critical points. These sampling devices can come equipped with refrigeration for collection of samples for biological tests.

4. OPERATING VARIABLES

The hydraulic loading of the system should be sized for normal maximum flow plus normal rainfall. A surge basin should provide storage capacity for at least maximum 2 h storm. The size of such a surge basin would depend on the rainfall pattern in the vicinity of the plant. If the rainy season is extended and heavy, it may be necessary to separate highly contaminated waste water streams from normal oily system. The latter would include normal drips and drains as well as spillage in oil refinery systems.

Major primary and secondary treating facilities including pumps should be in duplicate. Not necessarily 100 per cent spared based on design flow but at normal flows, the system should be capable of producing the required quality effluent at all times. In multiple units such as surface aerating equipment or diffused air systems, there should be an installed spare ready to operate if another unit is out of service. The flow to duplicate units in parallel operation should be evenly divided.

Generally, temperature limits have been set for all discharges into receiving waters. There is considerable controversy in regards to temperature effects on fish life. Thermal pollution standards have placed a limit in some locations on the use of once-through cooling water for large surface condensers.

The use of once-through cooling water or the so-called clean water sewer effluent for dilution of more concentrated wastes is no longer approved in many inland areas.

Acknowledgement—We wish to thank the Fluor Corporation of Los Angeles, and Fluor Nederland N.V., of Haarlem, for their permission to present this paper.

Note by Editor. In its present form this paper has been considerably abbreviated. The full paper, illustrated by five diagrams showing treatment units and flow schemes, may be obtained from the authors

Water Research Pergamon Press 1972. Vol. 6, pp. 381–385. Printed in Great Britain

LARGE PUMPING STATIONS

EBERHARD SICKERT
Baubehörde Hamburg, Stadtentwässerung, Neuer Wall 72, D-2 Hamburg 36, FRG

DURING the last few years a phenomenal development in the technology of the single shovel pump has taken place which makes it possible to install pumping stations of capacities as much as 1·5 $m^3 s^{-1}$ that would at once be fully automatic in operation, making screens or prechambers redundant. Pumps of smaller capacities which work on the same principle have been built over the past two decades and have proved their worth in reliability and low maintenance. Out of 93 sewage pumping stations that work without any manual attention 59 have been equipped with single shovel pumps. The entire inspection, repairs, and maintenance of the widely strewn pumping stations in Hamburg are carried out by a staff of 78.

For higher capacities, i.e. for discharges greater than 1·0–1·5 $m^3 s^{-1}$ mixed flow pumps are used. These are centrifugal pumps with a half-axial rotor equipped with curved impellers. They have a better efficiency than the single shovel pumps. They are suitable for parallel operation and have capacities much higher than the maximum available among the single shovel pumps in the market. But they require screens before the inlet to prevent clogging and damage in pumping sewage.

Screw pumps are particularly suitable where sewage has to be raised to a higher level, such as a high level sewer or to treatment plants. This is true also in cases where sizable amounts are to be pumped and are more advantageous than mixed flow pumps. The operational advantages of the screw pumps are as follows.

Continuous operation. This means without extra control the pumping operation balances itself to the fluctuation of the inflow up to full capacity. The efficiency is affected only to a limited extent by widely ranging quantities of inflow.

Neither screens nor grit chambers are necessary before the pump and without damage all foreign materials, even sand and stones, are carried.

Less deterioration through a smaller number of revolutions. Less susceptibility to damage. Damages if any is small and can be located in time. Less maintenance.

Sewage is pumped to two large treatment plants Köhlbrandhöft and Stellinger Moor in Hamburg. Whereas the Hafenstrasse pumping station which serves the treatment plant Köhlbrandhöft is situated 1·7 km away, separated by the river Elbe in between, the pumps for Stellinger Moor are located within the plant area itself. The Hafenstrasse pumping station is equipped with 4 mixed flow pumps while in Stellinger Moor 2 screw pumps are in operation. As auxiliary equipment at Hafenstrasse there are 4 screens with 60 mm bar width, 4 comminutors with 15 mm bar width, and 2 grit chambers. This equipment is located before the inlet of the pumps. In Stellinger Moor 2 screens with 25 mm bar width and 2 rectangular grit chambers are situated after the screw pumps. Due to the proved susceptibility and dislocation arising therefrom, the pumping station at Hafenstrasse requires constant supervision. Maintenance of the pumping station at the Stellinger Moor treatment plant is done by its staff in addition to their other duties.

TABLE 1. DATA OF THE PUMPING STATIONS HAFENSTRASSE AND STELLINGER MOOR

	Hafenstrasse	Stellinger Moor
Installed	1958	1965
Present average inflow	370,000 m^3 day^{-1}	40,000 m^3 day^{-1}
Maximum capacity, lift	7 m^3 s^{-1}, 16 m	1·6 m^3 s^{-1}, 7·65 m
Maximum capacity	600,000 m^3 day^{-1}	138,000 m^3 day^{-1}
Equipment	mixed flow pumps 1 No. a 3·0 m^3 s^{-1} 2 Nos. a 2·5 m^3 s^{-1} 1 No. a 1·5 m^3 s^{-1}	screw pumps 2 Nos. a 0·8 m^3 s^{-1}
Staff	33	3
Cost of construction (market price)	5,100,000 DM	1,160,000 DM
Machinery part	1,440,000 DM	550,000 DM
Operational costs	1,350,000 DM	110,000 DM
part of the staff	580,000 DM/a	52,000 DM/a
part of power	520,000 DM/a	46,000 DM/a
part of repairs, materials etc.	250,000 DM/a	12,000 DM/a
out of which for pumps alone	130,000 DM/a	6000 DM/a

TABLE 1 shows the main data of both large pumping stations. In both cases cost details have been calculated in relation to pumping, including screens and grit chambers. It is evident that the maintenance and labour costs in the case of the Hafenstrasse pumping station is considerably higher than in the case of Stellinger Moor. It is due to the susceptibility of the mixed flow pumps and their connected equipment for repairs, and also to the expensive maintenance of the screens (comminutors). For the pumping equipment alone the annual cost for repairs and materials comes to 9 per cent of the present market price of the entire machinery at the pumping station, whereas the cost for repairs and materials of the whole pumping station comes to 17·3 per cent. The corresponding figures for Stellinger Moor are only 1·1 and 2·2 per cent.

The economy and the advantages referred to earlier for the screw pumps were stated not only for Stellinger Moor but also for other treatment plants in Hamburg where screw pumps are used, especially for pumping return sludge. The most important of the advantages is the low expenditure for maintenance, repair, and personnel attention.

Due to the encouraging results obtained with screw pumps it was interesting to examine how big the dimensions of a screw pumping station would become if we were to handle 10 m^3 s^{-1} sewage, which would be the case at the Köhlbrandhöft treatment plant. The problem here is to raise sewage from a low level outfall sewer (—15·52 m NN) to the level of the treatment plant area (+5·70 m NN). So far as our knowledge goes screw pumps have neither been designed nor built for a lift of nearly 21 m and for a discharge of 10 m^3 s^{-1} as in this case.

Two possible arrangements were examined to prove the economy of different pumps and the results obtained are: shown in TABLE 2. The pumping stations examined are:

(1) A circular construction (simple static construction and foundation) with mixed flow pumps and screens in front.

(2) A two-stage, rectangular pumping station with screw pumps with length 18·5 m and diameter 3·5 m each followed by screens.

TABLE 2. PUMPING STATION KÖHLBRANDHÖFT—COST COMPARISON

	Mixed flow pumps	Screw pumps
Equipment	4 pumps at 2·5 $m^3 s^{-1}$	3 pumps at 3·3 $m^3 s^{-1}$
Screens	in front of the pumps	at the back of the pumps
Necessary staff	15	6
Costs of construction		
Civil	9·8 mill. DM	9·6 mill. DM
Mechanical	2·0 mill. DM	2·5 mill. DM
Electrical	0·5 mill. DM	0·5 mill. DM
Total	12·3 mill. DM	12·6 mill. DM
Annual expenditure		
Personnel	275,000 DM/a	108,000 DM/a
Power	900,000 DM/a	950,000 DM/a
Repairs, material etc.	375,000 DM/a	240,000 DM/a
Capital annuity		
Depreciation	361,000 DM/a	338,000 DM/a
Interest 6/2 = 3 per cent	369,000 DM/a	378,000 DM/a
Total	2,280,000 DM/a	2,014,000 DM/a

With the background of our present stage of experience with regard to the two existing large pumping stations, the construction costs and especially the operation costs were calculated. The annual costs for repairs and materials came up to 15 per cent of the installation costs in the case of pumping station with mixed flow pumps.

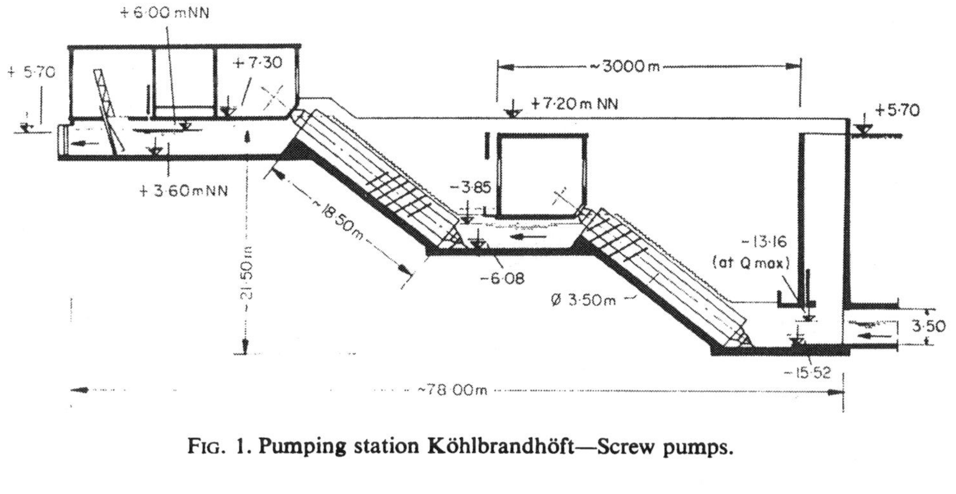

FIG. 1. Pumping station Köhlbrandhöft—Screw pumps.

That is a little more than in the case of the Hafenstrasse pumping station. In the case of screw pumps the figure was 8 per cent which is deliberately on the high side, even though it came to only 2·2 per cent in Stellinger Moor. The number of years generally taken into account for purposes of amortization in the case of machinery was 15 yr whereas for screw pumps it was 25, civil works 50 and electrical installations 20 yr.

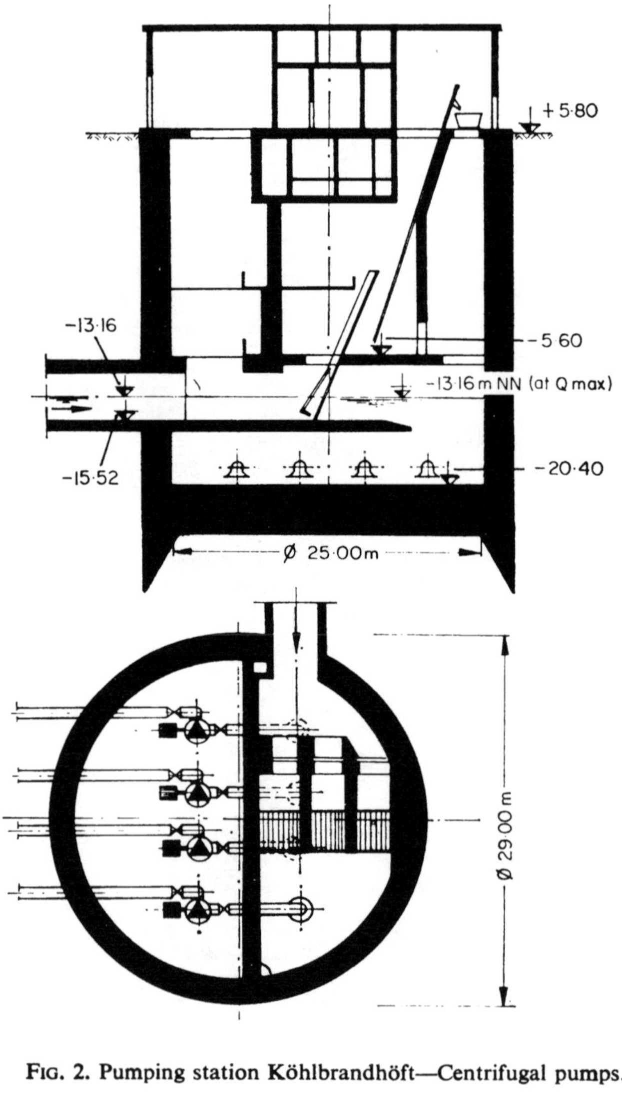

FIG. 2. Pumping station Köhlbrandhöft—Centrifugal pumps.

The low operation cost, not least the low personnel requirement have led to the decision to choose screw pumps for new construction. The expectation that the total costs for a works with screw pumps is not going to be much greater than that with mixed flow pumps helped us to reach the decision in its favour quickly.

Water Research Pergamon Press 1972. Vol. 6, pp. 387–389. Printed in Great Britain

PUMPING STATIONS

H. WEINRICH

Klein, Schanzlin und Becker AG, Johann-Klein-Strasse 9, D-6710 Frankenthal/Pfalz, FRG

1. LIFTING SYSTEMS

THE IDEAL case for favourable settling conditions during the process of pre-settling is given with normal gravity sewage flow towards the pre-settling tank. However, lifting will always be necessary. The pumping station then has to be arranged either behind the pre-settling tank or behind the post-settling tank. The decision depends on ground levels and the costs of construction involved.

Ideal conditions as mentioned above can only be realized in a few cases. Thus the question arises as to how the ideal case can best be reached if the sewage has to be lifted up to the level of the sewage treatment plant. As suspended matter in the sewage begins to coagulate in the main sewer, steps must be taken to preserve this coagulation as far as possible. It has been proved that the settling velocity in the pre-settling tank decreases if the coagulation is reversed. This in turn leads to a requirement for larger pre-settling tanks. This point must therefore be taken into account in the pump design. There are two ways of handling the untreated sewage carefully:

(1) The use of screw conveyors.
(2) The use of slow-speed pumps.

Screw conveyors are not only slow-running, but also have a low water flow velocity. They are therefore well-suited to this particular application.

The same effect is achieved with slow-speed centrifugal pumps. The impeller shape can be either a non-clogging impeller (FIG. 1) or a propeller, according to capacity.

Furthermore, it has to be considered whether the pumping station is to be arranged either directly at the sewage treatment plant or—depending on the length of the main sewer—some kilometers in front of the sewage treatment plant.

If it is at the sewage treatment plant the pumping station is included in the operation, control and maintenance of the complete plant. A disadvantage may be in the very deep position of a main sewer and the nearness of the pre-settling tank. Recoagulation after the process of pumping may then be impossible.

If the pumping station is arranged some distance away from the sewage treatment plant, recoagulation can occur if the coagulation had been reduced by pumping. Furthermore the main sewer to the sewage treatment plant can be arranged at a lower depth—i.e. at reduced costs.

The two pump designs mentioned above—screw conveyors and non-clogging pumps —do not require any pre-screening. The latter therefore may be arranged at reduced costs between a pumping station and grit trap settling tank.

With non-clogging pumps the pumping station is economically designed with the common suction pit for all the pumps of the station subdivided into a pre-chamber and a main suction pit (FIG. 2). One or two pumps may be controlled by variable speed in order to guarantee a continuous feed to the sewage treatment plant.

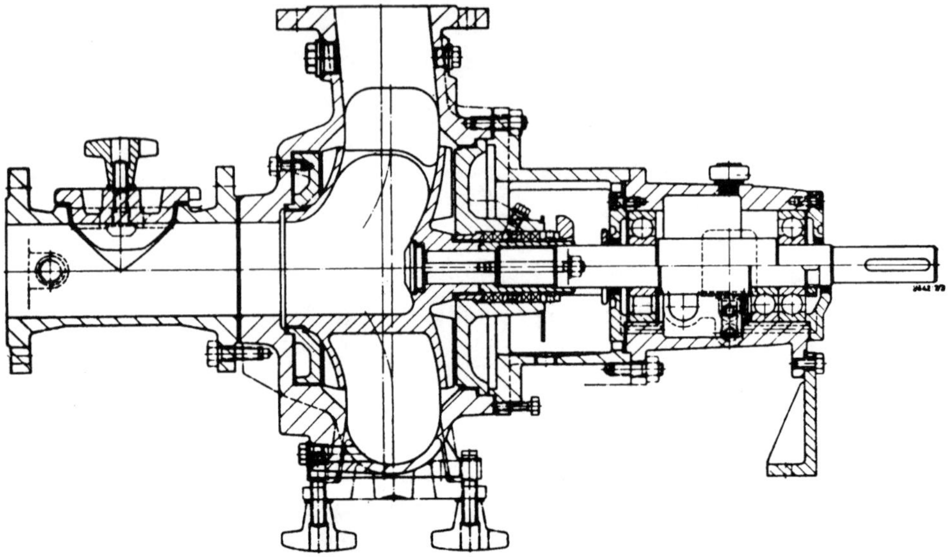

FIG. 1. Single-vane impeller pump. This pump is equipped with a single-vane impeller having a free passage of the same size as that of suction and discharge branch. Since the flow through the impeller is not split up by two or more vanes, long-fibrous parts in the sewage are unlikely to twist or bunch. The single-vane impeller pump has the advantage of being able to still operate at partial flow. This means the pump will continue to operate even when the suction line is not completely filled with sewage. This feature is important because the pump adapts itself automatically to varying suction conditions. The pump sump is emptied completely and no sludge deposits are formed. For the purpose of purification in a sewage treatment plant it is quite important to have a steady flow of sewage. This can be realized by means of a single-vane impeller pump because of the continuous adapting of the flow according to suction conditions, even when several pumps are working in parallel. In the latter case parallel pumps will be started up when the suction branch of same is only partially filled.

2. SLUDGE CONVEYING

When conveying sludge, it is important that the gas trapped in the sludge should be pumped through without causing operating trouble. Also, long-fibred or stringy material in the sludge should be prevented from twisting or bunching. These requirements are met by open-impeller pumps.

It is sometimes necessary to pump sludge over a distance of several kilometers. On account of the high pipe friction the total head required in such cases exceeds the available head of single-stage pumps. Therefore two pumps are arranged in series. If one of these pumps is driven by a speed-controlled motor, surge effects (water hammering) on start-up and shutdown will be avoided.

3. CONVEYING ACTIVATED SLUDGE

In activated sludge installations, activated sludge has to be returned to the aeration tank. In the aeration tank a splitting-up of the flocs is desirable. However, in most cases the activated sludge not required for the process is delivered to the pre-settling tank with the same pump. In this case the flocs should be large and therefore handling of this activated sludge is required.

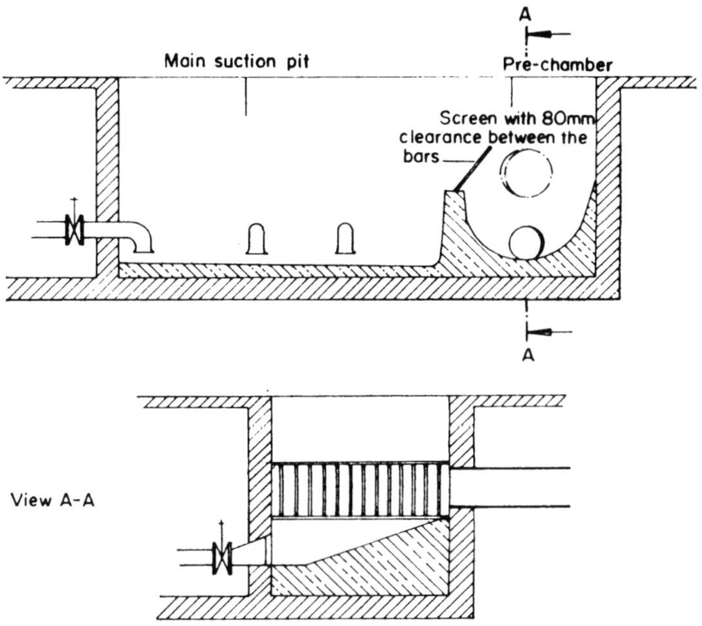

FIG. 2. Pumping station with suction pit subdivided into a pre-chamber and a main suction pit. In this pumping station the sewage from the main sewer flows into the pre-chamber, and from there direct into a single-vane impeller pump which—as already known—does not require a screen. In case of rising flow the sewage passes over a weir into the main suction pit. A maintenance-free screen is arranged on the overflow weir in order to prevent solids from entering the main suction pit. As the water in the main suction pit has been screened it can be handled by pumps with conventional impellers having a higher efficiency. Thus a pumping station designed as described above combines trouble-free operation without maintenance at best economy.

Here the same considerations for the pump selection apply as in the case of lifting the sewage to the treatment plant. Either slow-speed centrifugal pumps or screw conveyors are required.

Water Research Pergamon Press 1972. Vol. 6, p. 391. Printed in Great Britain

ON THE CONSTRUCTION OF LARGE SCREENS

GERNOT. W. ANNEN
Ruhrverband, Kronprinzenstr. 37, D-43 Essen, FRG

IN EUROPE the question whether screens shall be erected in a special building or not is often discussed. A building raises problems of ventilation, installation and explosion protection. Even in the case of a simple type of construction the costs are substantial. The Ruhrverband therefore prefers free-standing screens if there are not climatic or aesthetic reasons against this solution.

The screen at the sewage treatment plant in Duisburg–Kasslerfeld has a width between the bars of 2 cm and the screenings are thrown on to a concrete slab. A truck with its own hydraulic grab lifts the screenings and transports them to a nearby pit where they are dumped and covered with sand. This method is not perfect but it is easy, safe and cheap. The concrete slab can be cleaned by sprinkling. Difficulties due to frost have not yet arisen. The plant is sheltered from wind and is continuously loaded with about 8000 $m^3 h^{-1}$ wastewater.

The conditions at Witten sewage treatment plant are less favourable. During night the two screens installed on this plant get an inflow of about 800 $m^3 h^{-1}$ only and have long intervals between operation. They transport the screenings into elevated boxes. An hydraulic elevating mechanism inclines the boxes so that the screenings fall directly on a truck and they can be immediately removed. This construction has proved satisfactory. At first operating difficulties arose during wintertime due to frost, as the screen is extremely high and therefore exposed to the wind. The screen was therefore coated and equipped with electrical heating. The inclinable boxes were protected. These measures have proved to be successful and cheaper than a special building.

Water Research Pergamon Press 1972. Vol. 6, pp. 393–394. Printed in Great Britain

EFFICIENCY OF A GRIT CHAMBER

GERNOT. W. ANNEN
Ruhrverband, Kronprinzenstr. 37, D-43 Essen, FRG

THE SEWAGE treatment plant of the Ruhrverband in Duisburg–Kasslerfeld has two unaerated grit chambers for two sewers at different level. The larger grit chamber has a length of 40·50 m and treats the wastewater of about 270,000 inhabitants. During storm flows the overflow rate may rise to 39 m h^{-1} and the longitudinal velocity is 0·33 m s^{-1}. The corresponding values for normal inflow are 17 m h^{-1} and 0·24 m s^{-1}. During dry weather organic material settles and is washed out in a stationary silo. The silo overflow contains sand of small grain size and is therefore returned to the influent of the grit chamber.

The sand retained weighs 1800 kg m^{-3} and contains 85 per cent solid, 92·3 per cent of which is mineral and 63·4 per cent HCl-insoluble. The amount varies but corresponds to the curve of the monthly precipitation as demonstrated in FIG. 1. The yield of sand averages 4·5 l./inhabitant/annum or 0·02 l. m^{-3} wastewater. These amounts are low so that the efficiency of the grit chambers was questioned.

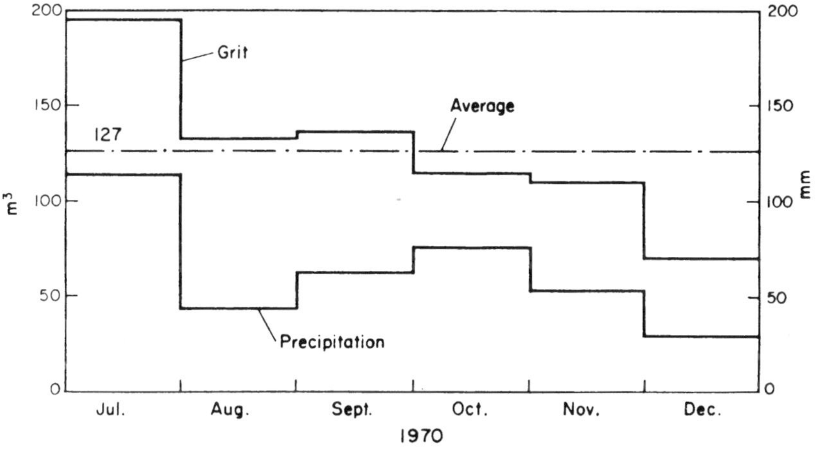

FIG. 1. Grit removal and precipitation in Duisburg–Kasslerfeld.

We investigated this question and first determined the grain size distribution (FIG. 2, curve a). This method however does not help one to determine total efficiency of grit removal. In particular the quantity of the sand contained in the sludge and its grain size were required.

For this purpose we investigated the content of the digester with a volume of 7200 m^3 and a digestion time of 24 days. The analysis of the recycled sludge showed a dry solids content of 70 kg m^{-3}; 28·4 per cent of this was HCl-insoluble, i.e. sand. Assuming that in the digester no sand accumulation occurs, the sand found in the digester corresponds to a 24-day load. Converting this to a month shows a throughput of 182,500 kg inert matter.

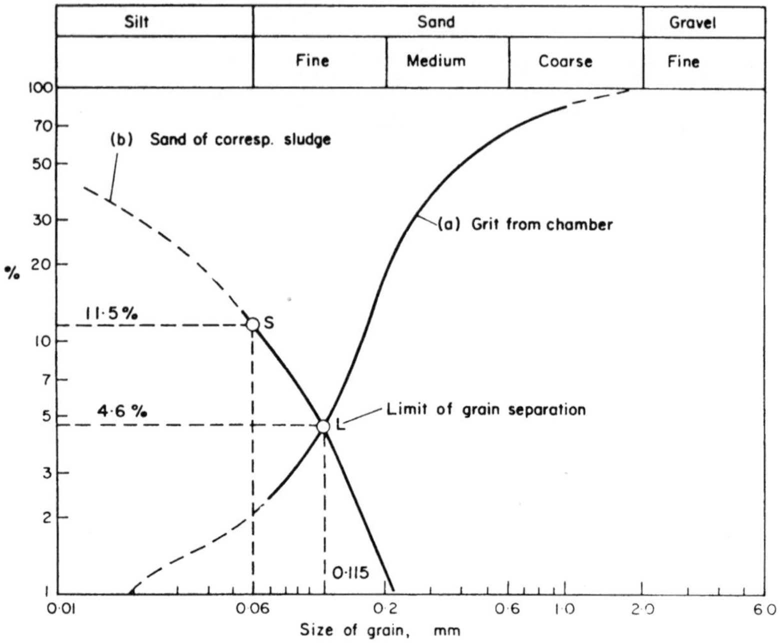

FIG. 2. Efficiency of a grit chamber in Duisburg–Kasslerfeld.

This material consists mainly of small grain fractions as demonstrated by curve b in FIG. 2. The curve is the graphic representation of an integration starting at large grain sizes. All material with a size of more than 0·06 mm is called sand. By the location of point *S* it is concluded that 11·5 per cent of the inert matter in the digester is sand. This amounts to about 21,000 kg month^{-1}.

The corresponding calculation for the weight of material retained in the grit chamber permitted the overall efficiency of the grit chamber to be determined. The results were:

removal in the grit chamber

$127 \times 1800 \times 0{\cdot}85 \times 0{\cdot}634 =$	123,000 kg month^{-1}	(85·4 per cent),
sand concentration in the sludge . .	21,000 kg month^{-1}	(14·6 per cent),
total	144,000 kg month^{-1}	(100·0 per cent).

Thus the overall efficiency of the grit chamber was 85·4 per cent. This result is satisfactory especially as the grain fractions below 0·1 mm do not disturb sludge treatment.

Intersection *L* in FIG. 2, resulting from the graphic representation, is interesting. Its coordinates mark the separation effect of the grit chamber. If as in this case the abscissa is smaller than 0·16 mm and the ordinate smaller than about 7–8 per cent, the grit chamber works well.

In our opinion the method described can be generally applied to determine the efficiency of a grit chamber under operating conditions.

Water Research Pergamon Press 1972. Vol. 6, pp. 395–396. Printed in Great Britain

TO WHAT EXTENT ARE PRIMARY TANKS REQUIRED?

W. VON DER EMDE

Institut für Wasserversorgung, Abwasserreinigung und Gewässerschutz, Technische Hochschule Wien, Karlsplatz 13, A-1040 Wien, Austria

1. THE FIRST activated sludge works of Ardern and Lockett operated without primary sedimentation. Only enlarged grit chambers (detritus tanks) were used ($t \sim 30$ min). Works were soon provided, apart from a few exceptions, with primary tanks to overcome operational difficulties, particularly with fine bubble aeration and also to open the way to sludge thickening.

2. An increasing number of smaller activated sludge plants have been operating successfully over the last 15 yr without primary tanks and with simultaneous aerobic sludge stabilization (oxidation ditches).

3. The Vienna-Blumental treatment works (150,000 population equivalent at present) was built to relieve a 15-km long waste water channel to the Danube (which will be connected later to the main Vienna–Simmering works). Primary tanks were omitted and the excess sludge will be led into the waste water channel because the sludge will be treated centrally at the planned main treatment works.

4. Two years plant experience at the Vienna–Blumental works indicate:

(i) The degree of purification corresponds, for the same aeration time and organic load, to works with primary tanks.

(ii) The activated sludge has good settling properties due to the settleable solids in the raw waste water. A sludge content up to 8 kg m^{-3} (SV640 ml—SVI 80) at $q_F = 0{\cdot}5$ m h^{-1} or 1·0 m h^{-1} (slight floc loss) with a return sludge ratio of 100–200 per cent can be readily maintained in the existing final sedimentation tanks.

(iii) The plant is easy to operate and possesses, due to its high sludge content potential a large buffer capacity against peak loads.

(iv) Operational difficulties appeared in the form of excessive floating sludge formation up to 3 cm thick in the final sedimentation tank (ice formation risk in winter). This was remedied by cleaning belts on the water surface in conjunction with an air pressure lift (attached to bridge on circular tank).

(v) Fine sand deposit in the aeration tank (up to 30 cm high mainly under large rotors) requires occasional (6–12 monthly) removal.

(vi) Accumulations of textile shreds at the final tank inlets require occasional (6–12 monthly) removal; a second fine screen was installed later.

(vii) The energy consumption in kWh kg^{-1} BOD_5 removal corresponds to works with primary tanks (therefore higher for BOD_5 of settleable solids).

5. The higher energy consumption arising from the partial oxidation of settleable solids is offset by the savings in the construction costs. The BOD_5 of the settleable solids 20 g PE^{-1} day^{-1} (at Blumental only 10–15 g PE^{-1} day^{-1} = 10 Wh PE^{-1} day^{-1} or 4 kWh PE^{-1} annum^{-1} at 150,000 PE and 2·0 cent kWh^{-1} 12,000 \$ annum^{-1}).

Construction cost of primary tank	$V_{PT} = 3{,}600$ m^3 ($t_{min} = 1$ h)
Construction cost:	144,000 \$ (40 \$ m^{-3})
Interest charge 8 per cent	11,600 \$ annum^{-1}.

6. Conclusions—deletion of primary tanks appears to be advantageous:

(i) For works without sludge treatment (e.g. central sludge treatment at mainworks).
(ii) For works where difficulties with light sludge are to be expected.
(iii) For septic waste waters, where odours could develop in the primary tanks.
(iv) For typical long period aeration works with aerobic sludge digestion (large oxidation trenches) particularly suited for warm climates.

7. For works with sludge treatment:

Advantages:
(i) Simple operation.
(ii) Sludge will be equalized in aeration tanks (with respect to consistency and quantity).
(iii) Storage for sludge in aeration tanks without odour nuisance development.
(iv) Sludge withdrawal at one point only.
(v) Further processing of sludge while fresh, possible.

Disadvantages:
(i) Higher energy consumption for aeration.
(ii) Lower gas yield on sludge digestion, thereby lower energy production in works owned electricity generators.
(iii) Operational difficulties (fine sand deposits, textile shred removal, floating sludge).
(iv) Reduced capacity for thickening of the excess sludge from plants without primary tanks (e.g. laboratory tests show only 4–5 per cent solid content) in contrast to mixtures of primary tank sludge and excess sludge. This disadvantage would be eliminated if the excess sludge could be economically (without chemicals) thickened (e.g. jet separators).

8. Where primary tanks are needed (probably the rule for large works), particularly in combined systems (storm water purification), the size of tanks should be designed so as to avoid operational difficulties in the successive units of the works. The higher energy consumption due to the oxidation of more suspended matter is certainly no longer important today.

Water Research Pergamon Press 1972. Vol. 6, pp. 397–399. Printed in Great Britain

BASIC PRINCIPLES OF AERATION TANK DESIGN

W. VON DER EMDE

Institut für Wasserversorgung, Abwasserreinigung und Gewässerschutz, Technische Hochschule Wien, Karlsplatz 13, A-1040 Wien, Austria

(1) THE "Activated Sludge Process" committee of the ATV presented in the *Text- and Handbook of Waste Water Technology*, amongst others, the following design data:

BOD-effluent (mg l^{-1})	B_R (kg m^{-3} day^{-1})	B_{TS} (kg kg^{-1} day^{-1})	t_{min} (h)	$OC_R : B_R$ (kg$O$ BOD^{-1})	$\dot{U}S_R : B_R$ (kgTS kg^{-1}BOD)
40	3·6	1·0	0·75	1·2	1·0
25	1·8	0·5	1·5	1·5	1·0
Nitrification	0·7	0·2	3·0	2·5	0·8

B_R = specific BOD load (kg m^{-3}day^{-1})
B_{TS} = sludge load (kg kg^{-1}day^{-1})
t_{min} = min aeration time (without return sludge, h)
OC_R = oxygenation capacity (kg m^{-3}day^{-1})
$\dot{U}S_R$ = excess sludge production (kg m^{-3}day^{-1})
Nitrification = complete Nitrification.

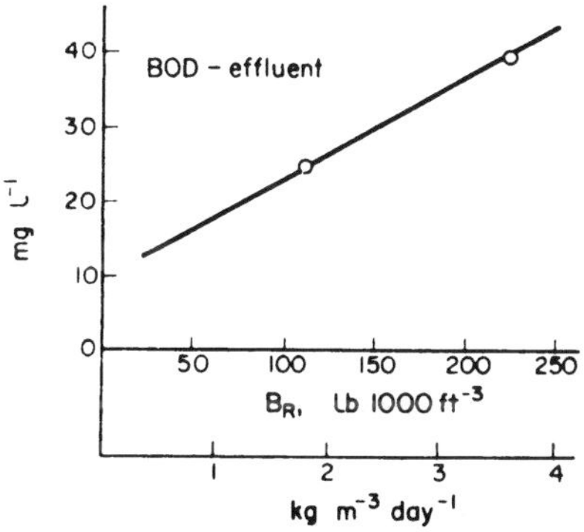

FIG. 1.

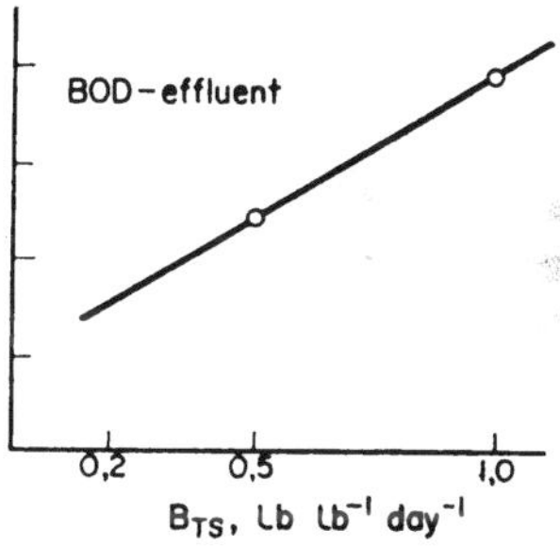

FIG. 2.

(2) The design may be based on the specific BOD load (B_R) or on the sludge load (B_{TS}). The design according to the sludge load should be checked to avoid exceeding the specific BOD load (the assumed dry weight of the sludge should not exceed 3·5 kg m^{-3}). The same procedure should be adopted in the design according to the specific BOD load. The minimal aeration time should not be reduced for diluted waste waters.

(3) Eighty-five per cent of the daily average of final BOD_5 values of domestic waste waters (BOD of the waste water after sedimentation about 200 mg l^{-1}) are assumed to fall below the specified final BOD value. (Assumptions: adequate dissolved oxygen concentration, design of aeration according to OC_R: B_R-values,

Comparison of costs

Example of a city with a population equivalent (PE) of 2·5 millions inhabitants.
Waste water quantity (dry weather) 650,000 m^3/day^{-1}
Max. Q (dry weather) 32,500 m^3/h^{-1} 900 m^3/s^{-1}
Waste water for one PE 260 l. PE^{-1} day^{-1}
BOD_5-load (settled) 125 t day^{-1}
BOD_5-load for one PE (settled) 50 g PE^{-1} day^{-1}
BOD_5-waste water (settled) 192 mg l^{-1}

BOD_5-effluent	40	25	Nitrification	(mg l^{-1})
Aeration tank				
Volume	14	28	72	l. PE^{-1}
B_R	3·6	1·8	0·7	kg BOD_5 m^{-3} d^{-1}
min t ($D.W.$)	1·1	2·2	5·5	h
Aeration	1·4	1·7	2·9	W PE^{-1}
Energy cons.	7·3	9·6	15·2	kWH PE^{-1} a^{-1}
Final tank				
Volume	26	39	52	l. PE^{-1}
min t ($d.W.$)	2·0	3·0	4·0	h
Q_{24} (dW) : RW	1:1·6	1:2	1:2·4	
max. q_F	1·2	0·8	0·6	m h^{-1}
Digester				
Volume	20	24	27	l. PE^{-1}
min t	15	15	15	day
Solids load	4·7	4·0	3·4	kg m^{-3} day^{-1}
Solids content, raw sludge	7	6	5	%
Q-digested sludge	0·86	0·92	1·0	l. PE^{-1} day^{-1}
Solids content, digested sludge	7·5	7	6	%

Construction costs

BOD_5-effluent	40	25	Nitrification	mg l^{-1}
Pumps, screens, grit chambers, primary tanks	0·83	0·83	0·83	\$ PE^{-1}
Aeration tanks	0·51	0·83	1·55	\$ PE^{-1}
Final tank	1·08	1·62	2·12	\$ PE^{-1}
Sludge digestion, etc.	2·70	2·86	3·03	\$ PE^{-1}
Works building, etc.	1·72	1·97	2·20	\$ PE^{-1}
Total	6·84	8·11	9·73	\$ PE^{-1}
Interest 8%	0·55	0·65	0·78	\$ PE^{-1}

Operational costs

BOD_5-effluent	40	25	Nitrification	mg l^{-1}
Labour	0·20	0·21	0·22	\$ PE^{-1} a^{-1}
Electricity	0·26	0·31	0·42	\$ PE^{-1} a^{-1}
Sludge handling	0·31	0·34	0·37	\$ PE^{-1} a^{-1}
Maintenance	0·06	0·07	0·09	\$ PE^{-1} a^{-1}
Total	0·83	0·93	1·10	\$ PE^{-1} a^{-1}

sufficiently large final sedimentation tanks, distributed waste water inflow or complete mixing and a return sludge ratio exceeding 50 per cent). Slightly higher daily peak values are accepted.

(4) The standard design value of 25 mg l^{-1} for the final BOD_5, with $B_R = 1{\cdot}8$ kg m^{-3} day $^{-1}$ and $B_{TS} = 0{\cdot}5$ kg kg^{-1} day^{-1} is regarded as adequate for the majority of receiving waters (in accordance with German standard requirements).

It appears to be advisable, however, to design with a view to higher water quality requirements in the future.

(5) What are the higher construction and operational costs for greater purification? The following total costs were found from the construction costs of the main Vienna treatment works:

BOD_5 (mg l^{-1})	B_R (kg m^{-3} day^{-1})	Unit costs $ PE^{-1} a^{-1}	ct m^{-3}	Relation
40	3·6	1·4	1·5	0·88
25	1·8	1·6	1·7	1·00
Nitrification	0·7	1·9	2·0	1·20

(6) The very extensive purification to nitrification (final BOD_5 = 10–15 mg l^{-1}) requires 35 per cent higher costs against a partial treatment to a final BOD_5 = 40 mg l^{-1}. The difference in cost between final BOD_5 = 25 per cent and nitrification amounts to about 20 per cent. The difference in costs is further reduced if the extension of the main collector usually required for newly built treatment works is also considered.

Water Research Pergamon Press 1972. Vol. 6, pp. 401–406. Printed in Great Britain.

CARROUSEL, A NEW TYPE OF AERATION-SYSTEM WITH LOW ORGANIC LOAD

A. C. J. KOOT

Professor of Sanitary Engineering, Technological University, Delft, Netherlands

and

J. ZEPER

Dwars, Heederik Verhey, Laan 1914 Nr. 35, Amersfoort, Netherlands

1. INTRODUCTION

OVER the last 10 yr a large number of oxidation ditches have been put into operation with great success. Its design parameters are:

(a) organic load 0·225 kg BOD m^{-3} day^{-1}.
(b) M.L.S.S.—concentration 3·5–4·5 kg m^{-3};
(c) sludge loading 0·05 kg kg^{-1} day^{-1};
(d) O.C./load 2 (operation) to 2·5 (maximum).

Practice has shown that production of excess sludge amounts to 0·75 kg kg^{-1} BOD, that is, 0·15 kg m^{-3} day^{-1}.

The sludge age in the system is M.L.S.S./sludge production =

$$\frac{4{\cdot}5 \text{ kg m}^{-3}}{0{\cdot}15 \text{ kg m}^{-3} \text{ day}^{-1}} = 30 \text{ days.}$$

With 30 days retention of sludge under aerobic conditions, the excess sludge is highly mineralized and no further treatment, apart from dewatering, is needed.

The advantages of the oxidation ditch process are:

(a) shock loads are absorbed by the large buffering capacity of the system, without disturbing the process;
(b) production of excess sludge is low and sludge treatment is reduced to dewatering only;
(c) operation of the system is simple and requires a limited staff;
(d) B.O.D. reduction is 95–98 per cent.
C.O.D. reduction is 90–95 per cent.
NH_4^+ reduction is significant.

In several countries additional considerations are:

(e) annual costs per organic load removed are lower than those of other known processes;
(f) the plants harmonize very well with their environment.

2. THE ECONOMY OF THE OXIDATION DITCH PROCESS

DURING the last decade, the tendency in the Netherlands has been towards a gradual increase in the size of plants, using the oxidation ditch process, mainly because of economic developments. This can be explained with reference to FIG. 1, which gives the relation between plant size and capital investment *per capita*, in the Netherlands, for traditional plants as well as for oxidation ditches. From FIG. 1 two trends can be derived:

(a) for a plant which is 10 times larger, the capital investment *per capita* is 1·8 times lower;
(b) for plants of the same size, the capital investment *per capita* for an oxidation ditch is 75 per cent of that for a traditional plant.

Ten years ago interest, depreciation and maintenance amounted to 6½ per cent annually, today this is 10 per cent. Moreover the ratio between the indices for construction costs and energy costs which was a : 1 10 yr ago, has now become 2ab : b. That is to say construction costs have increased twice as fast as those of energy.

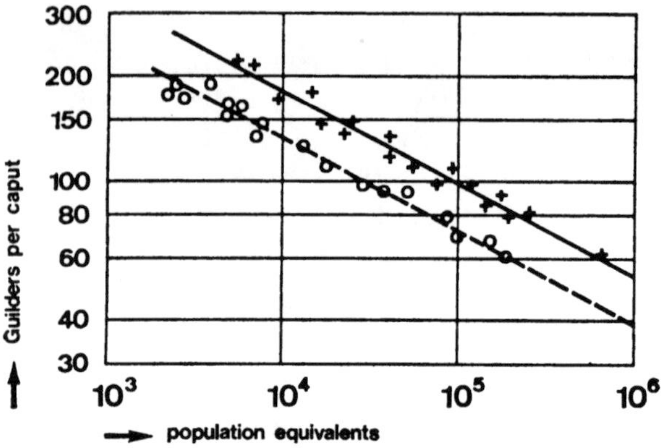

FIG. 1. Capital cost of sewage treatment plants (price-level 1971).

Taking the present difference between the capital investment *per capita* for a traditional plant and an oxidation ditch—both of the same size—to be I, then the break-even point between the two systems lies where 10 per cent of I equals the present cost E of the extra energy required for the oxidation ditch process, or $I = 10E$. Ten years ago the break-even point lay at 6·5 per cent of $(a/2ab) \times I$ equalling E/b or $I = 30E$. This means that today the break-even point lies at a value of I which is three times smaller than that of 10 yr ago. Given the sloping trend of the graphs in FIG. 1, the break-even point has moved to a size of plant which is 50 times as large, that is from 6000 population equivalents in the past towards 300,000 population equivalents at present. If construction costs keep rising at the present rate in relation to the costs of energy, the break-even point will move towards even larger plants.

However, to fulfil the trend as given in FIG. 1—that large plants are cheaper to build—it is obvious that all elements of the plant should be increased in size. For the oxidation ditch process this means deeper and wider aeration-channels and larger units for mechanical aeration.

A 300,000 population equivalents plant constructed to a conventional oxidation ditch design would need 100 cage rotors (ϕ 700 mm) of 6 m length each and, with a depth of 1·5 m, would cover an area of 5 ha. Apart from the considerable surface area required, the total costs of cage rotors becomes an important factor, requiring an investment of about N fl. 6 *per capita*. Alternative aeration apparatus should comply with the following requirements:

(a) low initial costs per kg oxygen transfer;
(b) high rate of aeration efficiency;
(c) adequate rate of flow, preferably in a deep circuit.

Surface aerators with a vertical axis fulfil the first condition. They were designed, however, for aeration tanks with organic loads of at least 0·7–1 kg m^{-3} day^{-1}. When these aerators are fitted in large tanks, the aeration efficiency is drastically reduced.

It appears necessary, therefore, to restrict the aeration zone of the oxidation ditch process to a part of the liquid content of the aeration channels. This still leaves the question of how adequate turbulence is obtained in the remaining part of the aeration channels.

3. CARROUSEL

The answer to this question is an aeration circuit of the type carrousel. This is a name given to a purely hydraulic solution to the problem. It proves to be a simple solution, which can be outlined as follows:

(a) take a square tank dimensioned in accordance with type and size of the surface aerator;

(b) move one side of the tank and extend the tank until the liquid capacity required is obtained;

(c) build a dividing wall lengthwise in the middle of this extended tank—which is now rectangular—leaving two sections open at either end of the tank;

(d) install a surface aerator at what would have been the centre of the original square tank.

As soon as the aerator is started up, oxygen is transferred into a limited aeration zone of the tank in the normal way, but also in addition a turbulent flow moves through the aeration channels in which the remaining part of the tank has been divided (FIG. 2).

For an explanation of this phenomenon, and for more information on the model

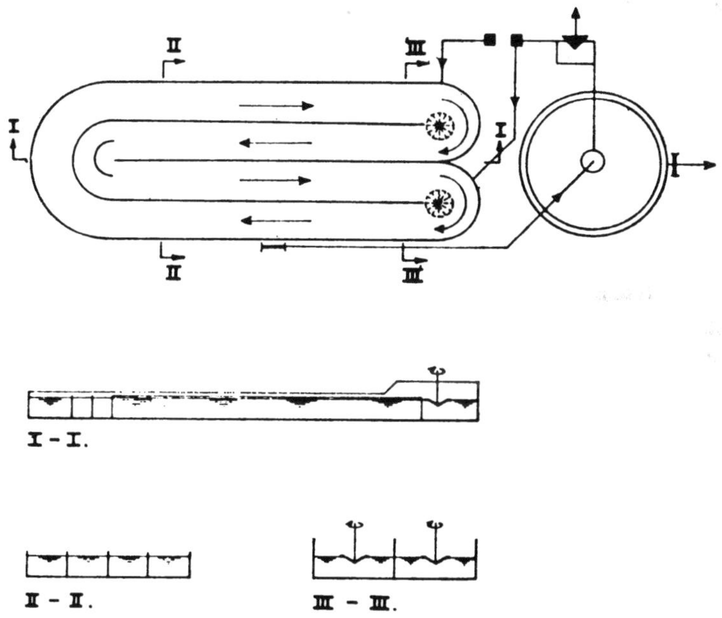

FIG. 2. Basic form of Carrousel.

tests which were carried out and on the results in the prototype carrousel at Oosterwolde (population equivalents 14,000) which was put into operation in September 1968 (see Ref. 1). A second carrousel was put into operation at the "Prinses' dairy works at Ursum (N. Holland), treating 8000 population equivalents.

The aeration circuit is 2·5 m deep and 5 m in width.

In the Netherlands there are at present 32 carrousels in the design stage for a total of about 2,600,000 population equivalents, varying in size between 6,000 and 400,000 population equivalents, with 8 plants for 100,000 population equivalents or over. Apart from these designs eight carrousels are already under construction, mostly with aerationcircuits of 4 m deep and 8 m wide, using the largest surface aerators available with an aeration capacity of 140 kg O_2 h^{-1}. This is sufficient for the aeration of 25,000 population equivalents with a O.C./load of 2·5.

The layout of the plant at Kerkrade (population equivalents 150,000) is shown in FIG. 3.

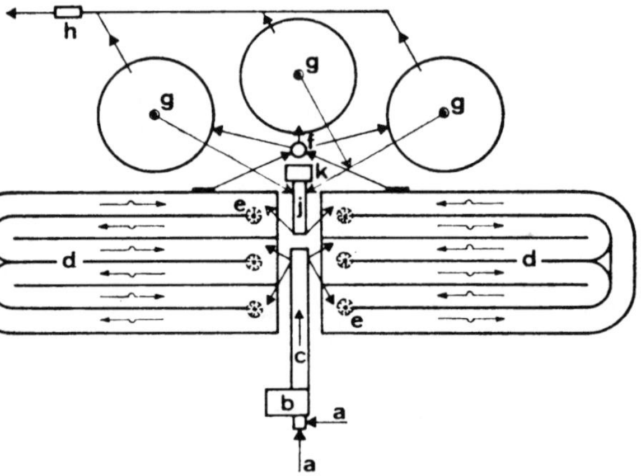

FIG. 3. Carrousel at Kerkrade for 150,000 population equivalents.

4. ENERGY REQUIREMENTS FOR AERATION AND FLOW IN THE CARROUSEL

A well designed surface aerator, working in an aeration zone which is reasonably adapted to the particular aerator, has an aeration efficiency of about 1·8 kg O_2 h^{-1} per net shaft horse power or about 2·45 kg O_2 kWh^{-1} (net); deviations from these figures are mostly within + and − 5 per cent.

The gross energy demand—depending on the efficiency of the reduction gear and the electric motor—is only of interest in regard to the total cost of energy consumption, but not for the performance of the aerator as such, nor for theoretical considerations of energy requirements within the aeration tank itself.

With an organic load of 0·255 kg BOD m^{-3} day^{-1} and a maximum O.C. load of 2·5 the maximum oxygen demand of the carrousel amounts to:

$$\frac{0{\cdot}225 \text{ kg BOD m}^{-3}\text{ day}^{-1}}{24 \text{ h day}^{-1}} \times 2{\cdot}5 \frac{\text{kg O}_2}{\text{kg BOD}} = 23{\cdot}5 \times 10^{-3} \text{ kg O}_2 \text{ m}^{-3} \text{ h}^{-1}.$$

This brings the maximum energy at:

$$\frac{23{\cdot}5 \times 10^{-3} \text{ kg O}_2 \text{ m}^{-3} \text{ h}^{-1}}{2{\cdot}45 \text{ kg O}_2 \text{ kW h}^{-1}} = 9{\cdot}5 \text{ W m}^{-3} \text{ (net at the shaft).}$$

Under normal operation conditions, with an O.C./load of 2 and a BOD loading between 2/3 × and 1 × its design load, the normal energy (net on the shaft) lies between 5 W m^{-3} and 7·5 W m^{-3}, related to the total volume of the carrousel. In the aeration zone, which occupies about 1/6 of this total volume, the net energy input is between 30 W m^{-3} and 45 W m^{-3}, with a maximum of about 60 W m^{-3}.

According to various investigators this higher energy input p m^{-3} of volume is advantageous for good aeration efficiency.

The energy needed to obtain sufficient flow in the aeration channels, can be calculated by multiplying the flow Q with the loss of head ΔH over the entire circuit. Q = velocity $\times$ cross-section $= v \times F\ m^3\ s^{-1}$ (in Newtons $= 10^4 \,.\, v \,.\, F\ N\ s^{-1}$).

With de Chézy's formula $v = C\sqrt{RI}$, $\Delta H = v^2L/C^2R$ in which

L = the length of the circuit

R = the hydraulic radius of the cross-section.

The energy required for the flow amounts to

$$Q \times \Delta H = \frac{10^4\ v.F. \times v^2.L}{C^2R}\ \text{Nm s}^{-1} = 10^4 \cdot \frac{v^3.F.L.}{C^2R}\ \text{W.}$$

As the volume of the carrousel equals $F \times L$ the energy m^{-3} of volume, needed for a velocity of flow $= v\ m\ s^{-1}$ is $10^4 \times v^3/C^2R$ W.

In FIG. 4 this energy is plotted against the velocity of flow for various values of R. It appears that in shallow oxidation ditches when the velocity increases to 0·4–0·5 m s^{-1} the energy needed for the flow may amount to about 10 per cent of the energy

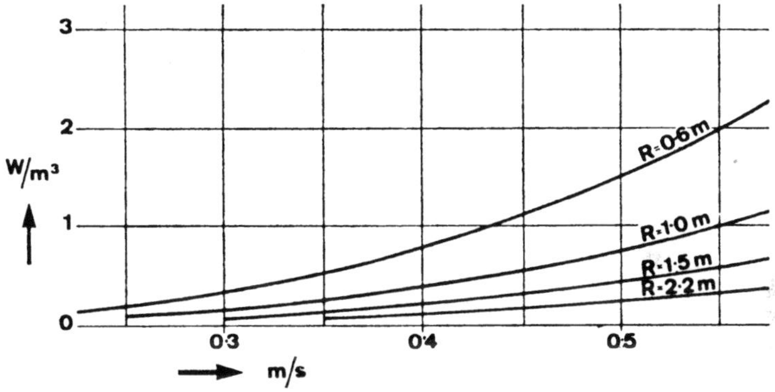

FIG. 4. Energy for flow in Carrousel.

needed for aeration. On the other hand the energy required for adequate flow in the deep aeration channels of the carrousel only amounts to about 1 per cent of the energy input for aeration purposes, and is less than 0·1 W m^{-3}.

5. AREA NEEDED FOR OXIDATION DITCHES AND FOR THE CARROUSEL

Based on detailed designs for oxidation ditches for respectively 500; 4000 and 7000 population equivalents and for carrousels for 22,000 and 120,000 population equivalents all using the same basic design criteria, the area needed for the plant and the buildings has been carefully determined allowing for roads and paths and a green zone of 10 m wide around the plant.

With regard to sludge treatment, plants up to 10,000 population equivalents have sludge drying beds; between 10,000 and 25,000 population equivalents a sludge-thickener is added. For plants between 25,000 and 75,000 population equivalents sludge is mechanically dewatered with a 3-month storage facility. For plants of 75,000 population equivalents and over, a thermal drying plant is added to the mechanical dewatering plant.

Drawings have also been made for a conventional activated sludge treatment plant with some kind of sludge treatment facility. A comparison of all these drawings has shown the gross areas needed for the various plants.

FIGURE 5 gives the gross area per population equivalent as a function of the size and type of plant. From this figure it appears that a carrousel for 100,000 population

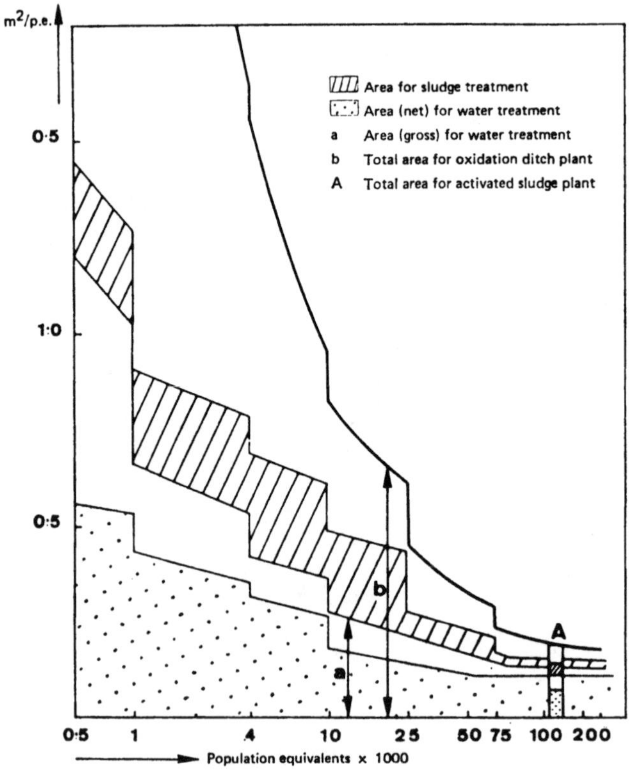

FIG. 5. Area needed for sewage treatment plants.

equivalents and over needs about the same gross area as a conventional activated sludge treatment plant with the same capacity, i.e. about 0·2 m^2 population equivalent or 5 population equivalents m^{-2} for both types.

REFERENCE

ZEPER J. and DE MAN A. (1970) New developments in the design of activated sludge tanks with low B.O.D. loadings. I.A.W.P.R. Conference. July 1970, San Francisco.

Water Research Pergamon Press 1972. Vol. 6, pp. 407–412. Printed in Great Britain

BASIC CRITERIA AND DESIGN ASPECTS FOR DEEP AERATION TANKS

THEODORE L. MOORE

Gibbs & Hill, Inc. 393 7th Ave., New York N.Y. 10001, U.S.A.

GENERAL

AN ACTIVATED sludge treatment plant for New York City has been designed with construction scheduled to start within a few months. The original site for this plant was determined by City engineers (ENVIRONMENTAL PROTECTION ADMINISTRATION). Various studies for possible sewer interceptors and extremely limited space on the Island of Manhattan showed the only available site to be on City property, partially on land with the greater area on a platform over the Hudson River supported by caissons.

In 1961 consulting engineers (ASSOCIATE CONSULTING ENGINEERS) were engaged by the City to start studies and conceptual design of this plant. Progress was slow, primarily due to interests of various public groups with a concern of property values and general environment. In the beginning a primary treatment plant met the standards for discharge. Subsequently with water quality standards changes this plant went through the design phase of "High-Rate" treatment. The final design is for standard aeration time.

PROCESS REQUIREMENTS

(1) AVERAGE flow and raw sewage characteristics recommended for design are:
Design average flow 84×10^4 m^3 day^{-1} (220×10^6 gal day^{-1})
BOD, mg l^{-1} 137
Suspended solids, mg l^{-1} 163.
These values are based on study of water use and population in the area tributary and on characteristics of raw sewage at Wards Island plant.

(2) Required degree of treatment, 90 per cent BOD removal, can be achieved by step aeration.

Recommended process design criteria, based on experience at New York City plants are:
Preliminary settling—7.7 m^3 day^{-1} 0.093 m^{-2} (2000 gal day^{-1} ft^{-2})
Aeration — 22.65 kg BOD day^{-1} 28.3 m^{-3}
3 h detention
Final settling —3.1 m^3 day^{-1} 0.093 m^{-2} (800 gal day^{-1} ft^{-2}).
The BOD load to aeration is 86,523 kg day^{-1} (191,000 lbs day^{-1}) including thickener overflow. For aeration tanks sized at 3 h detention for Q + 25% recycle, the resulting load is 17.7 kg (39 lbs) BOD 28.3 m^{-3} (1000 ft^3) of aeration.

Design performance for North River, based on Wards Island operating experience is as follows:

	BOD	SS
Raw waste, mg l^{-1}	137	163
Primary treatment		
% Removal	30%	40%
Effluent mg l^{-1}	96	98
Biological treatment ("step")		
% Removal	90%	87%
Effluent, mg l^{-1}	10	13
Plant removal	93%	92%

(3) By a combination of deep aeration tanks and double decking of final sedimentation tanks over chlorine contact tanks and equipment, as well as double decking preliminary tanks over piping and equipment, it is feasible to accommodate the required treatment units within the site limits. The recommended plant layout includes 8.5 m (28 ft) liquid depth aeration tanks.

(4) Although there appear to be no prototype sewage treatment aeration tanks of 8.5 m (28 ft) liquid depth, it is rationalized that such tanks present no problem for designing to achieve circulating transverse velocities required for solids suspension.

DEEP AERATION

Since there are very few, if any, sewage treatment aeration tanks of this depth, there was some concern as to whether deep aeration would create process or operating problems. Therefore, the question of deep tanks was considered for possible adverse affects on each of the important aeration system parameters, as follows: (a) oxygen transfer; (b) mixing; (c) higher pressure environment for biological solids; and (d) availability of suitable equipment to accomplish aeration and mixing.

(a) *Oxygen transfer*

For conventional submerged diffuser type system, it is well accepted that oxygen transfer efficiency increases with depth of diffuser. Assuming 6 per cent oxygen transfer efficiency for a 4.6 m (15 ft) deep tank, with diffusers at 4.1 m (13.5 ft) deep, efficiency in a 8.5 m (28 ft) deep tank with diffusers at 7.9 m (26 ft) would be:

$$E_2 = E_1 (D_2/D_1)\, 0.7$$
$$E_2 = 0.06\, (7.9/4.1)^{0.7}$$
$$= 9.5 \text{ per cent.}$$

However, blower horsepower to supply reduced air input for the deeper tank does not decrease since higher pressure is required.

	Case I	*Case II*
Diffuser depth, m	4.1	7.9
Static head, g cm^{-2}	1374	1828
Diffuser and pipe loss, g cm^{-2}	1149	1149
Blower discharge, g cm^{-2}	2523	2977

Average oxygen uptake is expected to be about 1 kg kg^{-1} BOD removed, or approximately 54.4 kg O_2 min^{-1}. At 100 per cent transfer efficiency, the equivalent air input is:

$$\frac{54.4 \text{ kg } O_2}{\text{min}} \times \frac{0.453 \text{ kg air}}{0.10 \text{ kg } O_2} \times \frac{2.8317 \times 10^{-2} \text{ m}^3}{0.03 \text{ kg}} = 189 \text{ m}^3 \text{ air @ } 1033 \text{ g cm}^{-2}, 10°\text{C}.$$

Blower output required for Case I is: 189/0.06 = 3115 inlet m^3 min^{-1} at: 1033 + 2523 = 3556 g cm^{-2}.

Blower output required for Case II is: 189/0.095 = 1982 inlet m^3 min^{-1} at: 1033 + 2977 = 4005 g cm^{-2}.

Blower horsepower can be approximated from the following formula:

$$\text{HP} = 0.00436 \times P_1 \times V_1 \times \frac{n}{n-1} \times \left[\left(\frac{P_2}{P_1}\right)^{n-1/n} - 1\right]$$

where: P_1 = inlet pressure; P_2 = discharge pressure; V_1 = inlet volume; n = 1.4, air constant.

Case I

$$\text{HP} = 0.00436 \times 14.7 \times 110{,}000 \times \frac{1.4}{0.4} \times \left[\left(\frac{22.2}{14.7}\right)^{0.286} - 1\right] = 3080 \text{ HP}.$$

At 65 per cent blower efficiency, power input would be 4750 HP

Case II

$$\text{HP} = 0.00436 \times 14.7 \times 70{,}000 \times \frac{1.4}{0.4} \times \left[\left(\frac{27.7}{14.7}\right)^{0.286} - 1\right] = 3110 \text{ HP}.$$

At 65 per cent efficiency, power input would be 4785 HP.

For aeration systems other than submerged diffusers the relationship between tank depth and oxygen transfer is not readily demonstrated by formula. However, successful installations of deep tank gas dispersion mass transfer operations using submerged turbines are found in the pharmaceutical and chemical processes industries. Also, successful waste treatment aeration operations in deep tanks have been found as follows: submerged turbine aeration of paper mill wastes at 6.7 m (22 ft); and mechanical draft tube aerators in old Imhoff tanks converted to aeration tanks of depths in the order of 9.1 m (30 ft).

(b) *Mixing*

For the submerged diffuser system, driving force for liquid circulation is a function of the static pressure difference between a point away from the diffuser and a point near the diffuser as shown in FIG. 1.

Even if air input to the deeper tank is reduced in accordance with the expected improvement in oxygen transfer efficiency, air input per 0.09 m² of tank surface (and average surface escape velocity) will be much greater in the deeper tank.

Case I

$$\frac{3115 \text{ m}^3 \text{ min}^{-1}}{13.8 \times 10^4 \text{ m}^3 \text{ } 4.6 \text{ m}^{-2}} \qquad \left[\frac{110{,}000 \text{ ft}^3 \text{ min}^{-1}}{4.88 \times 10^6 \text{ ft}^3 \text{ } 15 \text{ ft}^{-1}}\right.$$

$$= 0.10 \text{ m}^3 \text{ min m}^{-2} \qquad \left. = 0.34 \text{ ft}^3 \text{ min}^{-1} \text{ ft}^{-2}\right].$$

Case II

$$\frac{1982 \text{ m}^3 \text{ min}^{-1}}{13.8 \times 10^4 \text{ m}^3 \text{ } 8.5 \text{ m}^{-1}} \qquad \left[\frac{70{,}000 \text{ ft}^3 \text{ min}^{-1}}{4.88 \times 10^6 \text{ ft}^3 \text{ } 28 \text{ ft}^{-1}}\right.$$

$$= 0.12 \text{ m}^3 \text{ min}^{-1} \text{ m}^{-2} \qquad \left. = 0.4 \text{ ft}^3 \text{ min}^{-1} \text{ ft}^2\right].$$

The mean density of air liquid mixture in the column above the diffuser could be assumed to be lower (less dense) for the deeper tank as a function of the higher average

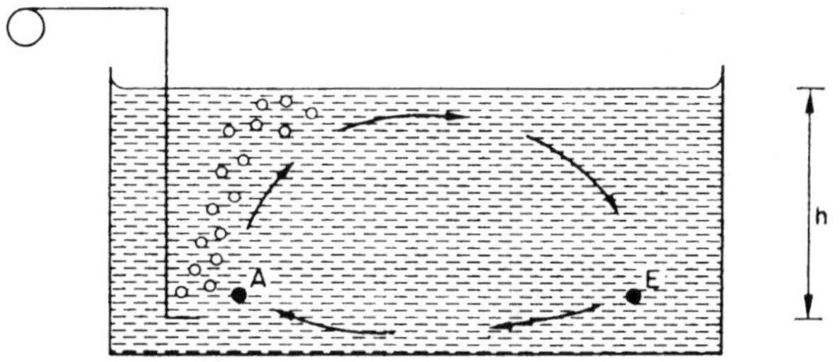

Driving force, $dP = P_e - P_a$

$= h(D_e - D_a)$

where D_e = Density of liquid

D_a = Mean density of air-liquid mixture above point a.

FIG. 1.

surface escape velocity. However, for present purposes, it is safer to assume that this mean density is the same for both tanks and that circulating driving force is proportional only to depth of diffuser. Thus it is rationalized that circulating driving force for Case II 8.5 m (28 ft) deep tanks with diffusers at 7.9 m (26 ft) deep is approximately twice as great as for Case I, 4.6 m (15 ft) deep tanks with diffusers at 4.1 m (13.5 ft). It is believed that the 8.5 m (28 ft) deep tanks present no problem for designing to achieve circulating transverse velocities required to maintain solids in suspension. Also, diffuser arrangement should provide "wide band" aeration in order to prevent excessive transverse velocity which could cause formation of an undesirable quiescent core.

For other aeration systems, including submerged turbine and mechanical draft tube types, reputable manufacturers have indicated that there are reliable scale up factors and sufficient experience in mixing of deep tanks such that mixing of the proposed 8.5 m (28 ft) deep aerators could be approached with confidence.

(c) *Higher pressure environment*

No reason has been found to suspect that the maximum pressure, 18.77 g cm^{-2} (26.7 psia) in the proposed 8.5 m (28 ft) deep tanks will adversely affect the biological sludge. Maximum pressure in a conventional 4.6 m (15 ft) deep tank is 1490 g cm^{-2} (21.2 psia) but such tanks are only a portion of a system in which the solids are subjected to pressures of 1736 to 4548 g cm^{-2} (24.7–64.7 psia) at return sludge pumps. Also, research projects involving pressurized activated sludge type systems have been reported on in the literature, with no indication of adverse affects on the sludge solids.

(d) *Availability of equipment*

The proposed 8.5 m (28 ft) depth does not in itself restrict or limit the type of aeration equipment that can be used.

Systems that could be applied, and for which it has been found that equipment is available in the market are as follows: submerged diffuser system with air blowers at higher discharge pressures than are normally used in sewage treatment; relatively shallow diffused air system incorporating draft tube like assemblies and in which system, blowers at "normal" discharge pressures could be used; submerged turbine system consisting of mechanical mixers for mixing and for dispersal of oxygen from relatively small amounts of air supplied from blowers; and mechanical draft tube aerators which do not require process air blowers.

PROCESS AIR BLOWERS

There would be no problem in getting blowers for discharge pressures required for deep tank diffused aeration. Two manufacturers have each provided preliminary proposals for machines fully in conformance with City specifications, except that higher pressure requirement necessitates four-stage machines rather than the three-stage type normally used by the City.

Gas handled	Air
Inlet volume	708 m^3/min^{-1} (25,000 acfm)
Inlet temperature	38°C (100°F)
Inlet pressure	1033 g cm^{-2} (14.7 psia)
Discharge pressure	2018 g cm^{-2} (28.7 psia)
Speed	3580 rpm

Impeller tip speed	181 m s^{-1} (595 ft s^{-1})
Brake horsepower required	1642 bhp
Discharge temperature	128°C (262°F)

AERATION SYSTEM

Four aeration systems have been considered for the proposed 8.5 m (28 ft) tanks. These are: deep diffuser; shallow diffuser; submerged turbine; and mechanical draft tube.

It is believed that each of these systems has its own merits and could be successfully applied with no great disparity of energy consumption between one system and another. On balance, it appears that deep diffusers would be the best system for North River for reasons as follow:

It offers the greatest similarity to other City plants both in terms of operation and equipment. Thus it would create the least problem in transferring personnel from plant to plant.

Rotating machinery is concentrated both in location and number of units. This permits better maintenance and also makes it more reasonable to install standby spare equipment. Therefore, it provides greater security for continuous process performance. (This advantage also applies to shallow diffusers.)

Special structural features such as bridges and supports, needed for both turbine and mechanical draft aerators, are not required.

It offers simplest and widest process control since air supply can be varied greatly by the number of blowers in service and by regulating output of individual units.

It does not require extensive high capacity hoisting equipment and procedures needed to remove gear reducers and motors for either turbine or mechanical draft tube systems.

The electrical system for large, centrally located motors is much simpler and less costly than would be required for the great number of smaller motors dispersed over a wide area, as needed for other systems.

Noise level in the large aeration tank area will be much lower than would be the case for systems with a great many motors and gear reducers spread throughout the area.

It appears likely that the deep diffuser system could be least costly to install. This because of the less extensive electrical system, absence of special structural features, fewer motors, absence of gear reducers and simplicity of process control.

A summary of major equipment and estimates of average power consumption for each of the four systems is presented below. The values shown are approximations based on calculations presented in preceding pages and on some base values prepared

	Deep diffusers	Shallow diffusers	Submerged turbine	Mechanical draft tube
Installed equipment				
Mixers				
HP/units			150	100
No. units	None	None	24	48
Blowers				
HP/unit	3000	2000	1000	
No. units	6	6	6	None
Total installed				
HP	18,000	12,000	9600	4800
Average operating conditions				
BHP				
Mixers	0	0	3430	4000
Blowers	4785	4750	2040	0
Total	4785	4750	5470	4000

by manufacturers for this study. The figures listed should be considered with limitations as follows:

A shallow diffuser system using very large draft tube-like appurtenances is being tested in Detroit. By including estimates for power and blowers, it is assumed that this system would be suitable for North River. Validity of this assumption has yet to be demonstrated.

For the mechanical draft tube system, it is questionable whether the listed horsepower for installed units is comparable to the other systems in terms of capability for peak loads of 150 per cent of average.

For the deep diffuser system, values listed for installed blower HP do not reflect anticipated improvement in transfer efficiency. However, this improvement is credited in average operating HP.

SUMMARY

The required treatment will be achieved within the limited area by using 8.5 m (28 ft) liquid depth aeration tanks; together with final tanks double decked over hypochlorite contact units and secondary pump rooms, and again preliminary tanks double decked over pump and service areas.

The aeration system for the deep tanks will be fine bubble, porous media diffusers placed at a depth of 7.9 m (26 ft).

REFERENCES

ASSOCIATE CONSULTING ENGINEERS, Tippetts-Abbett-McCarthy-Stratton, Feld, Kaminetzky & Cohen, Gibbs & Hill, Inc.

ENVIRONMENTAL PROTECTION ADMINISTRATION, Department of Water Resources, Martin Lang, Commissioner and Director, Joseph Cunetta, Deputy Director, Jospeh Miller, Chief Engineer.

Water Research Pergamon Press 1972. Vol. 6, pp. 413–416. Printed in Great Britain

FLOW VELOCITIES IN AERATION TANKS WITH MECHANICAL AERATORS

K.-H. KALBSKOPF

Emschergenossenschaft, D-43 Essen, FRG

WITH mechanical aerators, the pumping capacity of the aeration system produces, along the surface of the aeration tanks, a radial or also somewhat spiral water flow towards the wall of the tanks. This flow is conducted along these walls to the floor and then, aided by the suction effect of the aerator, to the center of the tank. The flow velocity reached on the tank floor must be sufficiently high to ensure that sludge will not settle out. In our experience, flow velocities of 15–20 cm s^{-1} are sufficient for this purpose. The flow pattern caused by the mechanical aerator and the level of the flow velocity depend on the power consumption of the aerator as well as on the shape and volume of the aeration tank. FIGURE 1 shows the flow pattern determined in model tanks of different geometrical shapes. The flow patterns are similar for full size tanks.

The flow velocities, however, increase with the same power density per tank volume (W m^{-3}) with increasing tank volume. As FIG. 2 shows, Froude's Law largely applies. According to measurements by the Emschergenossenschaft, the power consumptions

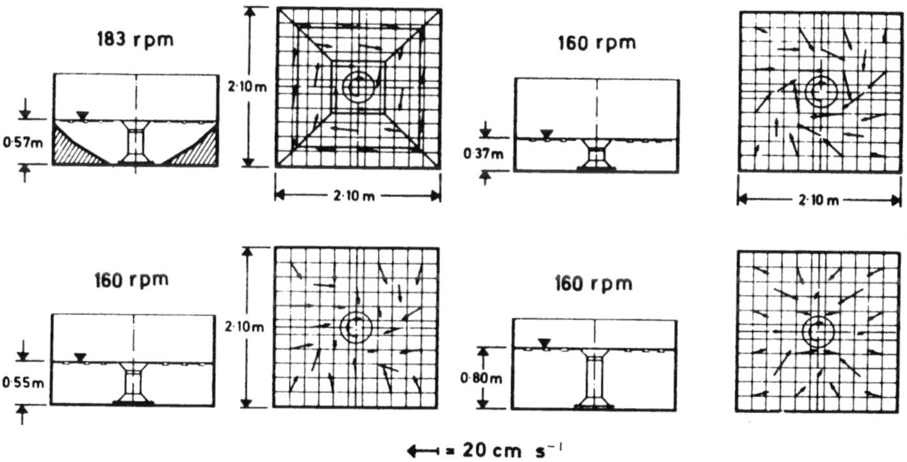

FIG. 1. Flow pattern across the bottom of aeration tanks

given in FIG. 3 are necessary for sufficient mixing and circulation of the water in the aeration tank, depending on the shape and volume of the tanks; flow velocities of 25–30 cm s^{-1} are reached on the tank floor.

With waste water constituents that reduce the surface tension of the water, the flow conditions in aeration tanks with mechanical aerators can give negative effects. An example of the decrease of the surface tension (dyn cm^{-1}) with increasing detergent content (mg l^{-1}) is given in FIG. 4. With reduced surface tension of the water, more

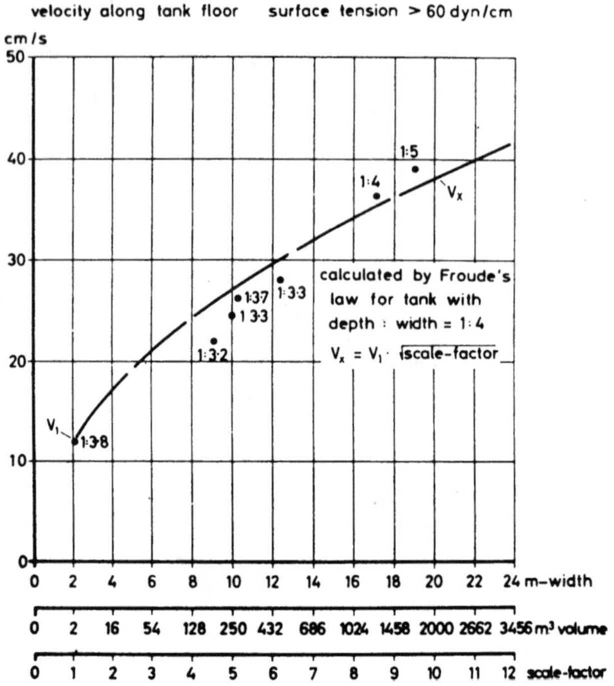

FIG. 2. Velocity along the tank floor

and finer air bubbles are introduced into the water thus checking the spiral flow and possibly causing density currents.

FIGURE 5 shows the decrease of the flow velocity along the floor of a 400 m^3 tank measured as a function of detergent content. As activated sludge tanks with mechanical aerators are complete mixing tanks, the detergent content in such tanks is between 1 and 2 mg l^{-1} for domestic waste waters. With regard to the decrease of flow velocity

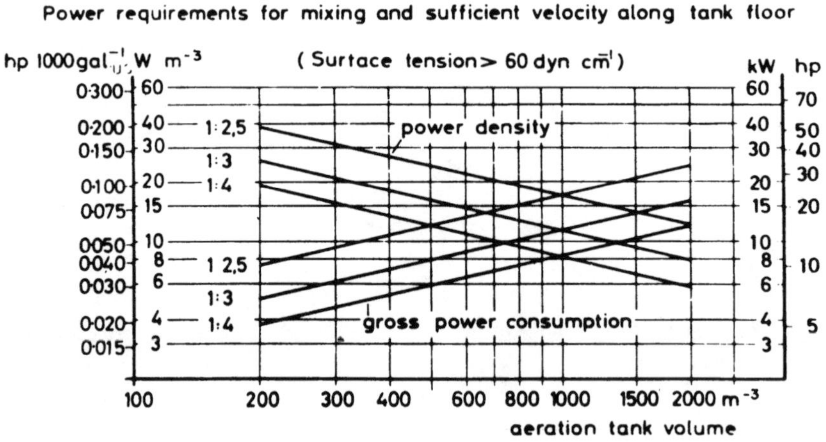

FIG. 3. Power requirements for aeration tank mixing

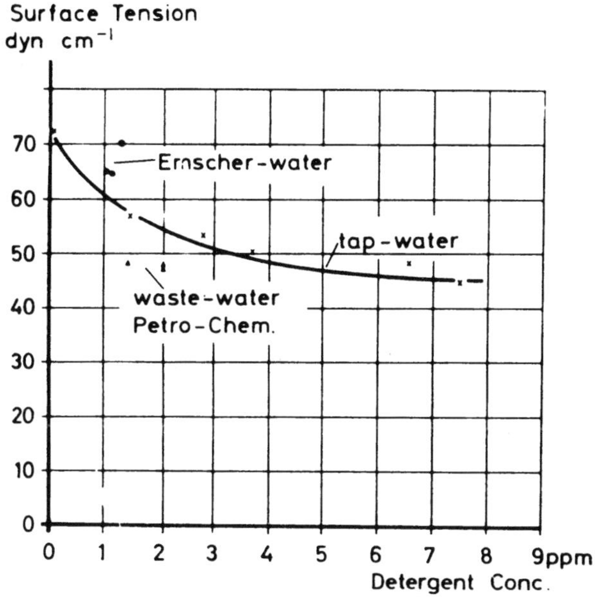

FIG. 4. Surface tension–detergent concentration

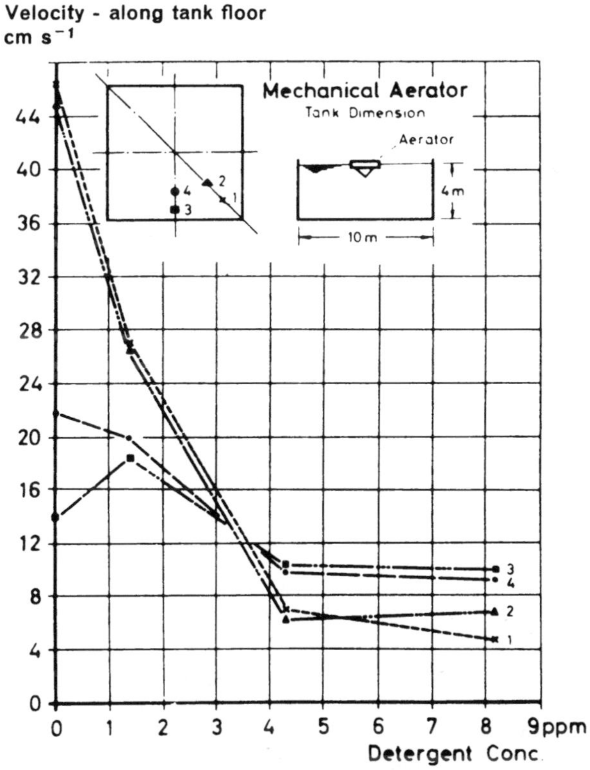

FIG. 5. Influence of detergent conc. on velocity

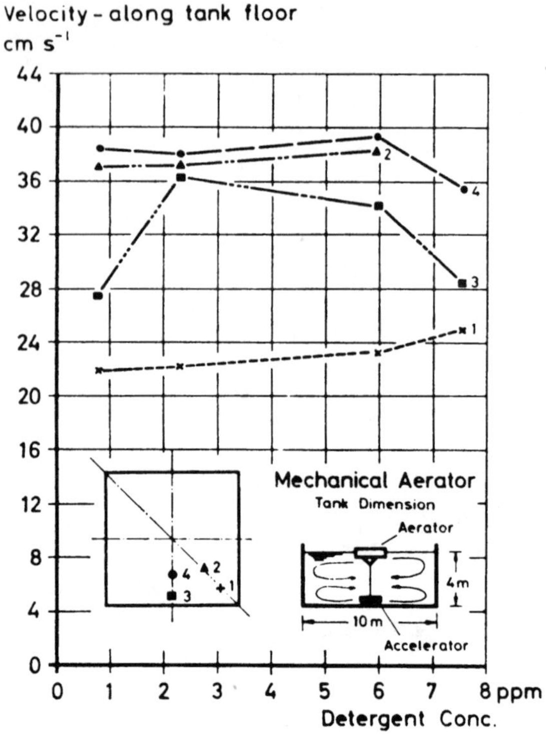

FIG. 6. Velocity detergent conc. with aerator and accelerator

along the tank floor, this amounts to about 30 per cent compared with pure water, and the flow velocities are reduced from 25 to 30 cm s^{-1} to between 17 and 21 cm s^{-1}. At such flow velocities sludge sedimentation does still not occur. With industrial wastes, however, there may be very low surface tensions due to different constituents and even critical flow velocities below 15 cm s^{-1}, with the result that sludge settles out. According to tests run by the Emschergenossenschaft, a sufficient flow velocity can be obtained along the tank floor, as the results proved, by using the aeration system shown in FIG. 6.

REFERENCE

KNOP E. and KALBSKOPF K.-H. (1969) Energetische und hydraulische Untersuchungen an mechanischen Belüftungssystemen. (Energetic and hydraulic tests with mechanical aerating systems). *Gas- und Wasserfach* **110,** S. 198–201 and 266–269.

Water Research Pergamon Press 1972. Vol. 6, pp. 417–421. Printed in Great Britain

TANKS WITH TURBULENT FLOW GENERATED BY MAMMOTH ROTORS

W. STALZER and W. VON DER EMDE

Institut für Wasserversorgung, Abwasserreinigung, und Gewässerschutz, Technische Hochschule Wien, Karlsplatz 13, A-1040 Wien, Austria

(1) TANKS with Mammoth rotors where the mixed liquor is circulated can be regarded as a combination and further development of Haworth-basins and oxidation-ditches with Kessener-brushes.

(2) Tanks with Mammoth rotors have been chosen for the treatment plant Wien-Blumental for the following reasons:

2.1 As primary clarification is omitted fibrous material that may form lumps is kept in circulation. Clogging of the aerators is avoided.

2.2 There is hardly any limit to the size of the tanks. For 300,000 population equivalent, two tanks with 6000 m^3 (1·585 U.S. MG) each, 150 m (492 ft) long, 17 m (56ft) wide, 2·5 m (8 ft 3 in.) deep, are sufficient (FIG. 1). By applying large units and simple methods for the civil engineering works, the construction expenses will be low.

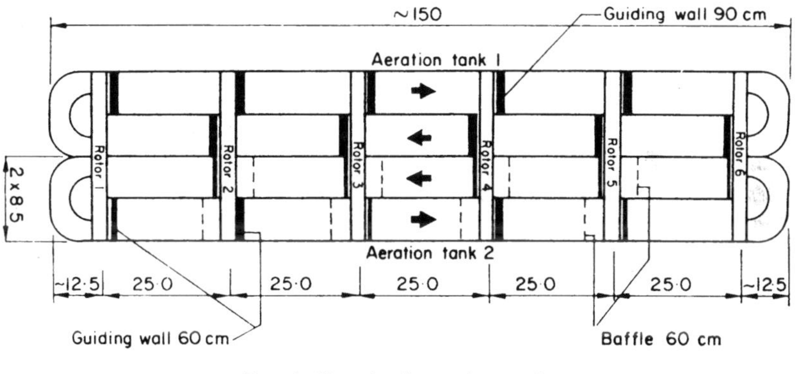

FIG. 1. Sketch of aeration tanks.

2.3 Mammoth rotors can be applied to basins with small water depths [e.g. 1·5 m (4 ft 10 in.) to 3·0 m (9 ft 8 in.)]. In these cases the floor slab of the tanks can be erected above the ground water level.

2.4 The tanks need not to be emptied if maintenance work is necessary with the Mammoth rotors.

2.5 A large range of the energy input actually applied (10–60 W m^{-3}, equivalent to 50–300 HP 10^{-6} gal) ensures adaptation to uncertain future conditions of operation.

2.6 A simple adaptation of the oxygenation capacity on the oxygen uptake can be achieved by either varying the submergence of the aerators or the numbers of rotors in operation or both.

2.7 Two tanks with Mammoth brushes where the mixed liquor is circulating bring about a number of different combinations for operation (both tanks in series with the same or a different intensity of aeration in each tank; contact-stabilization; separated aerobic stabilization of the excess sludge).

(3) Before the treatment plant was put into operation, the flow distribution and velocities, the oxygen transfer and the energy consumption were studied in clear water (water from the adjacent rivulet Liesing):

3.1 The energy input increases linearly with the depth of submergence (FIG. 2).

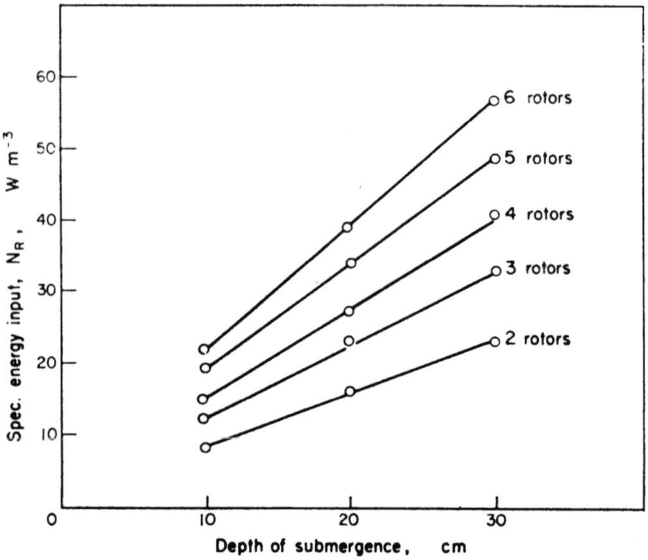

FIG. 2. Spec. energy input and depth of submergence for guiding walls.

3.2 The mean flow velocity in the channel is dependent on the specific energy input *Nr* (W m^{-3} or HP 10^{-6} gal) and nearly independent from the combination of

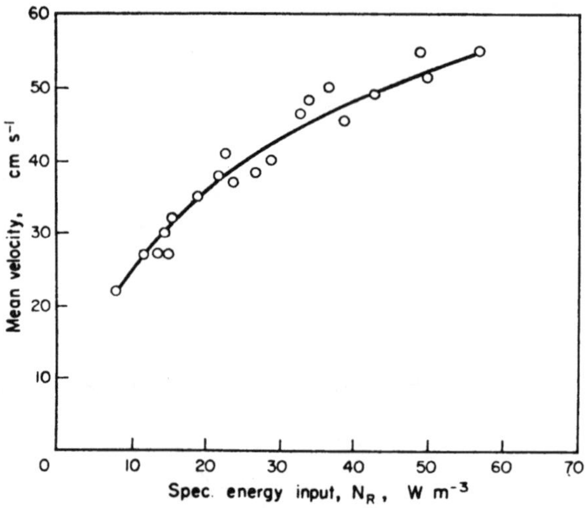

FIG. 3. Spec. energy input and mean velocity.

rotors chosen (FIG. 3). With an energy input of 10 W m^{-3} (50 HP 10^{-6} U.S. gal), a mean velocity 25 cm s^{-1} (10 inches s^{-1}) can be obtained.

3.3 Guiding walls (in the direction of flow after the rotor) or guiding walls and baffles (in direction of flow before the rotor), respectively, are useful in order to slow down the flow velocity in the upper water layer. Where these baffles and guiding walls are not applied, the surface velocity of the circulating mixed liquor will be large and come close to the velocity of the rotors (3·6 m s^{-1} equivalent to 12 ft s^{-1}). Thus, the difference of these two velocities will become small and so will the oxygen transfer rate. In cases where only guiding walls are applied after the rotors, the submergence is 30 cm (12 in.) and all six rotors are driven, the surface velocity is 1·2 m s^{-1} (4 ft s^{-1}) in front of the next rotor (FIG. 4).

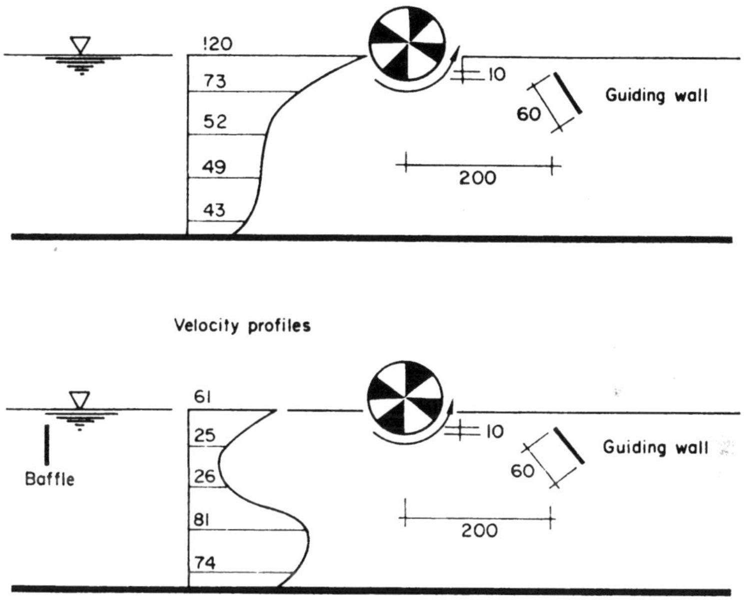

FIG. 4. Velocity profiles.

3.4 The smallest flow velocities in the tanks occur at the base of the tanks, below the rotors. Thus a driven rotor disperses a much larger quantity of water than the volume of the upper part of the circulating mixed liquor. Where a rotor is not driven, the guiding walls will provide equalization and a distribution of the velocity profile, but only after the rotor. Where an additional baffle is applied in front of the rotor, the distribution of the velocity profile will occur directly in front of the rotor. An increase in the velocity at the base of the tank is the result (FIG. 5).

(4) In connection with experiments that were carried out to test the specified performance, the following mean values were obtained for the oxygen transfer (clean water from the rivulet Liesing):

No. of rotors	Submergence (in.)	Energy kW	HP 10^{-6} U.S. gal	Oxygen-transfer lb ft^{-1} h^{-1}	lb/HP^{-1} h^{-1}	MO
4	11	235	185	5·50	3·46	G
4	10	264	210	5·44	3·14	GB

MO = mode of operation, G = with guiding wall, GB = with baffle and guiding wall.

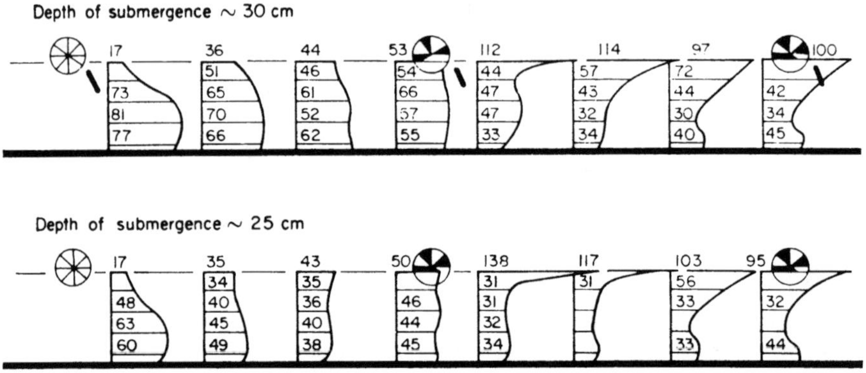

FIG. 5. Velocity profiles after driven and undriven rotors.

(5) When the plant was in operation, the oxygen transfer was determined under operating conditions, according to the method of KAYSER (1967, 1968). With a mixed liquor suspended solids concentration of 10 kg m^{-3}, the following mean values were measured:

No. of rotors	Submergence (in.)	Energy kW	HP 10^{-6} U.S. gal	Oxygen-transfer lb ft^{-1} h^{-1}	lb HP^{-1} h^{-1}	MO
3	9	180	145	5·24	3·14	G
4	8	193	160	4·50	3·46	G
6	7	265	215	3·02	2·48	G

TABLE 1.

No. of rotors	Submergence (cm)	Energy kW	W m^{-3}	O_2-Transfer kg m^{-1} h^{-1}	kg kWh^{-1}	MO
4	27	235	37	8·2	2·1	G
4	26	264	42	8·1	1·9	GB

TABLE 2.

No. of rotors	Submergence (cm)	Energy kW	Energy $W\ m^{-3}$	O_2-Transfer $kg\ m^{-1}\ h^{-1}$	O_2-Transfer $kg\ kWh^{-1}$	MO
3	25	180	29	7·8	1·9	G
4	21	193	32	6·7	2·1	G
6	19	265	43	4·5	1·5	G

(6) After 6 months of operation, both tanks were emptied. The deposits were found to be 2 per cent of the net volume. These deposits were relatively large below the rotors but were not found in the turns. They did not affect the efficiency of treatment.

Up to now only guiding walls after the rotors have been installed. We do not know yet whether the application of additional baffles (transformation of the velocity profile from the surface to the base of the tank) would prevent or hinder the formation of sediments.

(7) During 2 years of operation, good results have been obtained at the treatment-plant Wien–Blumental, with tanks where the mixed liquor is circulating and which are equipped with Mammoth rotors.

This system can be applied for plants with low and medium organic loads, where the energy input necessary for oxygen transfer is in the range of 10–50 $W\ m^{-3}$ (50–250 HP 10^{-6} U.S. gal^{-1}). Advantages, too, may exist when nitrification and denitrification is a design goal (two tanks in series) with this system without primary clarification.

REFERENCES

EMDE W. V. D. (1969) Entwurf Belüftungssysteme, Wiener Mitteilungen, Band 4 Technische Hochschule Wien.

EMDE W. V. D. (1971) Die Kläranlage Wien-Blumental, Österr. Wasserwirtschaft, Heft 1/2, Springer, Wien.

KAYSER R. (1967) Ermittlung der Sauerstoffzufuhr von Abwasserbelüftern unter Betriebsbedingungen. Veröffentlichung des Institutes für Stadtbauwesen der TU Braunschweig, Heft 1.

KAYSER R. (1969) Comparison of Aeration Efficiency under Process Conditions. *Advances in Water Pollution Research*, IAWPR Prague.

STALZER W. (1971) Strömungsverhältnisse in langgestreckten Umlaufbecken mit Oberflächenbelüftung Tagung für Abwasserreinigung, Budapest.

Water Research Pergamon Press 1972. Vol. 6, pp. 423–427. Printed in Great Britain

THE ECONOMY OF MECHANICAL AERATORS

H. WEINRICH

Klein, Schanzlin und Becker AG, Johann-Klein-Strasse 9,
D-6710 Frankenthal/Pfalz, FRG

INTRODUCTION

EXISTING publications do not allow an exact consideration of the various factors of importance regarding mechanical aerators. Therefore, for new designs one has to depend on actual tests. Below, the equation for the oxygen capacity is interpreted in a way that allows predictions for new designs to be made from known results. Furthermore, the economy, i.e. the oxygen capacity per kilowatt-hour, can be calculated.

1. GENERAL RELATIONS

1.1 *The oxygen capacity*

The oxygen capacity is the *velocity of the oxygen feed.* It is expressed by the increase of the saturation capacity as a function of the time by means of the following equation:

$$O_C = \frac{26{\cdot}1}{1000}\,\frac{1}{t_1 - t_0} \cdot \log \frac{c_s - c_0}{c_s - c_t} \sqrt{\frac{k_{10}}{k_t}} \; [\text{kp m}^{-3}\,\text{h}^{-1}] \tag{1}$$

$26{\cdot}1$ = $2{\cdot}303 \times 11{\cdot}33$
($2{\cdot}303$ = recalculation of natural logarithms into common logarithms,
$11{\cdot}33$ = saturation value of O_2 in water at 10°C and 760 mmHg in pm^{-3},
c_s = saturation value of O_2 in water at test temperature,
c_0 = O_2-concentration at test begin,
c_t = O_2-concentration at test end,
$t_1 - t_0$ = duration of test in hours,
k = diffusion constant,
$\sqrt{\frac{k_{10}}{k_t}}$ = recalculation value for the diffusion velocity from the test temperature to 10°C.

1.2 *The stream of liquid to be revolved*

Every aeration device feeds the liquid with air at one point in the tank. The air admitted is transported together with the flowing liquid. In this way renewal of the contact surfaces takes place according to the degree of turbulence of the stream and at the same time oxygen is absorbed. An important increase in concentration is possible only when liquid again comes into contact with the aeration device. Thus, the circulation velocity is a factor which is of influence to the oxygen capacity.

We assume the ideal case where the path of circulation for all the parts of the liquid is of the same length. This can also be considered as the *most probable* length of the path during circulation. The period for a *single circulation* of the tank volume will be $\Delta t = t_1 - t_0$[h]. The reciprocal value of Δt results in the *number of circulations per hour.*

$$Z = \frac{1}{\Delta t} \; [\text{h}^{-1}]. \tag{2}$$

By multiplying through m^3 the above dimension becomes

$$[m^3\ m^{-3}\ h^{-1}].$$

Thus, the value calculated from equation (2) represents the stream of water in $m^3\ h^{-1}$ to be revolved in order to feed 1 m^3 of sewage with the necessary oxygen. If the *stream of water revolved is given the designation* Q, then

$$Q = \frac{1000 \times O_C}{26{\cdot}1 \times \log \dfrac{c_s - c_0}{c_s - c_t} \sqrt{\dfrac{k_{10}}{k_t}}}\ [m^3\ m^{-3}\ h^{-1}]. \quad (3)$$

During a single circulation the oxygen concentration increases from c_0 to c_t.

1.3 *The economy*

The economy is the relation of the oxygen capacity O_C to the power input N.

$$W = \frac{O_C}{N}\ [kp\ O_2\ kW\ h^{-1}] \quad (4)$$

2. MECHANICAL AERATORS

2.1 *Oxygen capacity*

Mechanical aerators always effect a definite circulation. As the oxygen feed is effected directly by the mechanical aerator or near it, all parts of the liquid in the aeration tank should logically stream through the aeration device. There are parts of course which do not stream through the aerator, but get the oxygen from the turbulence. On an average during a single circulation of the tank volume, the concentration increase c_t is reached.

Mechanical aerators split up the surface of the liquid in more or less big drops. For these a very quick oxygen saturation at the surface is reached. The concentration increase of the mixture is a function of the "throwing range", i.e. of the energy value being transferred from the aeration device to the liquid.

Once again we consider the concentration increase during a single circulation, and that as a function of the energy value H [mkg kg^{-1}] transferred to the liquid.

It is: $c_t = f(H)$.

The oxygen capacity and the stream of liquid revolved were taken from KNOP *et al.*, (1964), KAELIN and TOFAUTE (1968, 1969), BAARS and MUSKAT (1962), SCHERB (1965), and v. d. EMDE (1961). Then the concentration increase c_t could be calculated by means of equation (3). The energy value H [mkg kg^{-1}] transferred from the aerator to the liquid can be calculated from the power input according to the relation used for centrifugal pumps:

$$N = \frac{\gamma Q H}{367{,}000\ \eta_p}\ [kW\ m^{-3}]. \quad (5)$$

Thus

$$H = \frac{367{,}000 \eta_p N}{\gamma Q}\ [mkg\ kg^{-1}\ resp.\ m]. \quad (6)$$

Accordina to FIG, 1. the concentration increase rises as a function of the energy value transferred and it already approximates to the saturation value at $H = 1{\cdot}5$ to $1{\cdot}6$ [mkp kp^{-1}]. The curve drawn shows the average values of various designs of mechanical aerators.

2.2 *The economy*

By using equation (3) and (5)

$$W = \frac{O_C}{N} = \frac{26{\cdot}1\ Q\ 367{,}000\ \eta_p}{\gamma\ Q\ H\ 1000} \times \log \frac{11{\cdot}33}{11{\cdot}33 - c_t}$$

respectively

$$W = \frac{9{\cdot}4\ \eta_P}{H} \times \log \frac{11{\cdot}33}{11{\cdot}33 - c_t}. \tag{7}$$

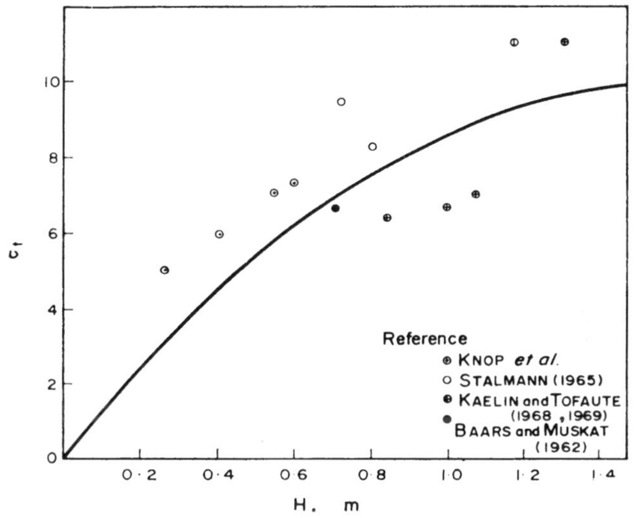

FIG 1. Concentration increase during a single circulation.

FIGURE 1 shows the relation between c_t and H *as a result of the tests*. Thus, the only factor still to be determined is the efficiency of the energy transfer to the liquid stream. For this factor there are values that have been obtained with centrifugal pumps. Figures can be derived as a function of the energy value H as shown in FIG. 2.

Determination of all factors for the "economy" equation (7) gives results shown in the following table and in FIG. 3.

H	0·2	0·4	0·7	1·0	1·4	1·8
c_t	2·4	4·5	6·9	8·55	9·8	10·0
η_p	0·43	0·58	0·65	0·69	0·72	0·72
W	2·10	3·0	3·53	3·90	4·2	3·52

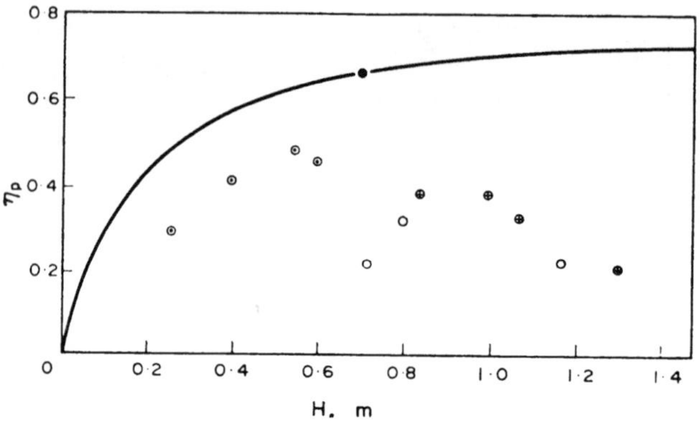

FIG. 2. Efficiency of the energy transfer to the liquid.

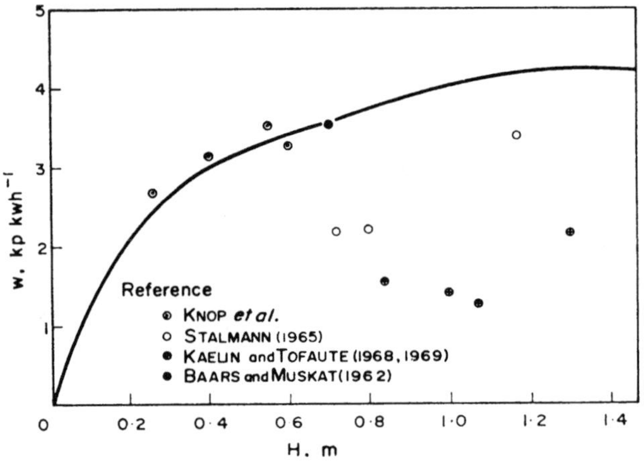

FIG. 3. The economy of mechanical aerators.

2.3 *Conclusions*

Up to an energy transfer of 0·8 mkp kp^{-1} the calculated curve of economy coincides with test results for various types of aerators. For higher energy values the calculated curve of the economy shows that an oxygen capacity of more than 4 kp kWh^{-1} seems to be possible. Tests carried out up to now do not give such favourable values. According to FIG. 2 the reasons for this are that the efficiency of the energy transfer decreased as the energy value increased. This is a matter for future development, i.e., the efficiency of the energy transfer needs to be increased, but at the same time the increase in the oxygen concentration should not fall below the values now attainable.

REFERENCES

BAARS and MUSKAT (1962) Der Sauerstoffeintrag in Wasser mit Hilfe rotierender Bürsten. *Gesundheitsingenieur*, Heft 7.

EMDE W. VON DER (1961) Belüftung, Arten und Systeme. *ATV-Berichte*, Heft **13.**

EMDE W. VON DER (1968) Belüftungssysteme und Beckenformen. Münchener Beitrage zur Abwasserbiologie, Bd. **5,** 2. Aufl.

KAELIN und TOFAUTE (1968) Leistungsfähigkeit von BSK-Turbinen in großen Belebungsbecken. *Wasser, Luft und Betrieb* **12,** Nr. 12 sowie; *ibid* (1969) **13,** Nr. 1.

KNOP, BISCHOFSBERGER and STALMANN (1900) Versuche mit verschiedenen Belüftungssystemen im technischen Maßstab, Teil 1. Vulkan-Verlag, Essen.

KNOP, BISCHOFSBERGER and STALMANN (1900) Versuche mit verschiedenen Belüftungsysstemen im technischen Maßstab, Teil 2. Vulkan-Verlag, Essen.

SCHERB (1965) Vergleichende Untersuchung über das Sauerstoffeintragungsvermögen verschiedener Belüftungssysteme. *Münchener Beiträge zur Abwasserbiologie*, Bd. **12.**

STALMANN (1965) Die BSK-Turbine–ein neues Hochleistungs-Belüftungssystem der Abwassertechnik. *Das Gas- u. Wasserfach*, **106**, Jhrg., Heft 22, Oldenburg-Verlag, München.

Water Research Pergamon Press 1972. Vol. 6, pp. 429–431. Printed in Great Britain

DISCUSSION OF THE DESIGN PARAMETER FOR SECONDARY SEDIMENTATION TANKS

K.-H. KALBSKOPF
Emschergenossenschaft, D-43 Essen, FRG

TESTS and observations on the sewage treatment plants on the Emschergenossenschaft and Lippeverband showed that the design data usual for primary settling tanks such as detention time (h), overflow rate q_F (m h^{-1}), depth of tank (m) and average flow-through velocity V (m s^{-1}) are not sufficient to explain fully the operational conditions of secondary settling tanks varying because of the different operation of secondary sedimentation and the characteristics of the activated sludge.

The performance of a final settling tank is determined, besides by the hydraulic load, also by the solids loading and the sludge volume index. The more activated sludge is added into the secondary sedimentation and the higher the index of this activated sludge, the more dumping area is necessary for the thickening process of the recycle sludge in order to keep the sludge level in the secondary settling tank so low that no activated sludge carry over in the effluent occurs.

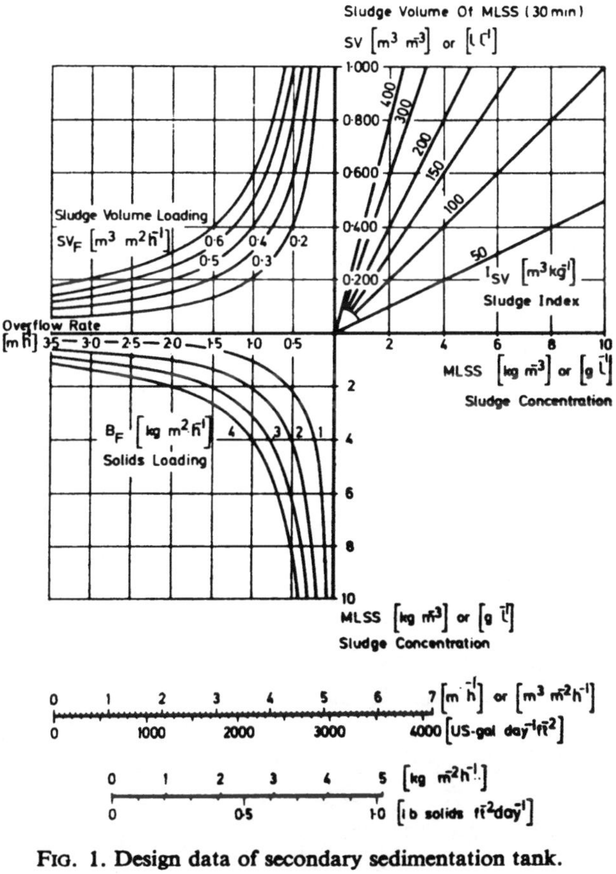

FIG. 1. Design data of secondary sedimentation tank.

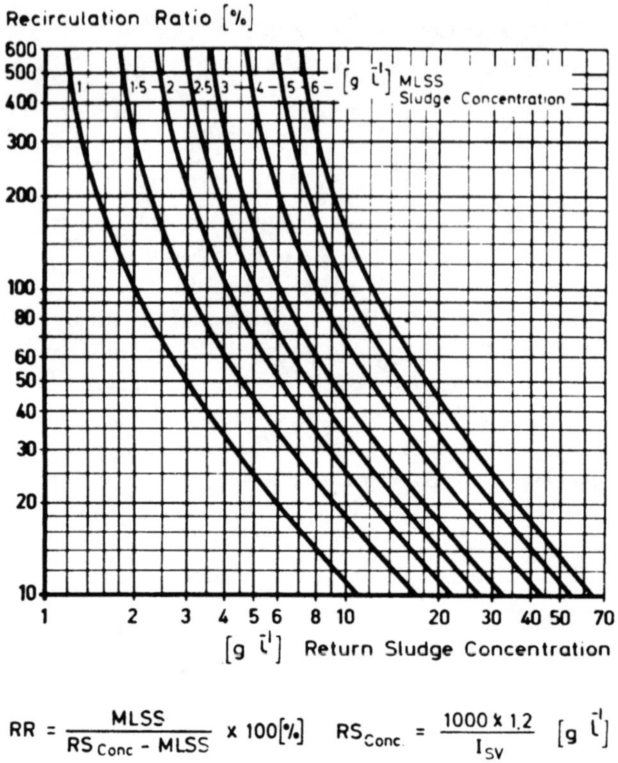

FIG. 2. Recirculation ratio–MLSS–return sludge concentration.

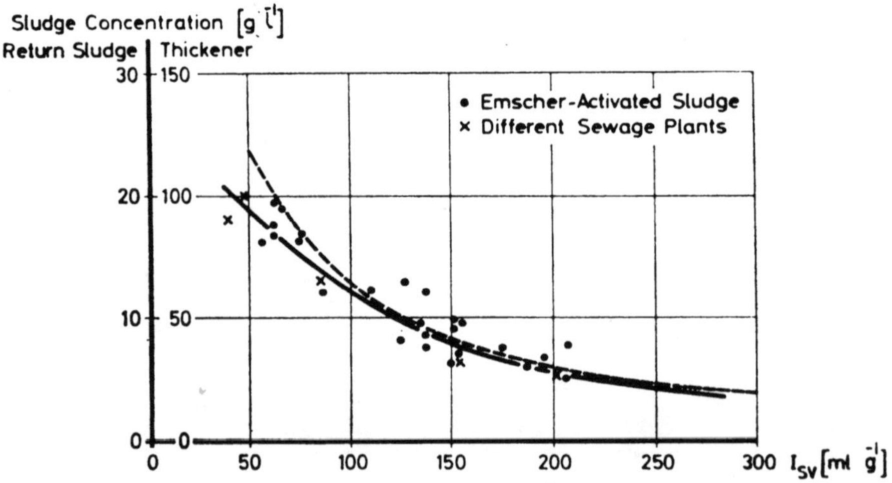

FIG. 3. Return sludge concentration–index SV relationship.

In order to reach a satisfactory removal effect in the secondary sedimentation, design parameters must thus be used which take the solids loading and the sludge characteristics into consideration. As a practical design parameter, the sludge volume loading ($m^3\ m^{-2}\ h^{-1}$) can be taken.

The functional relationship of this parameter to the known design parameters is given by the following equations:

sludge volume loading SV_F ($m^3\ m^{-2}\ h^{-1}$) = overflow rate q_F ($m\ h^{-1}$) . sludge volume index I_{sv} ($m^3\ kg^{-1}$) . solids concentration TS_R ($kg\ m^{-3}$),

the product of sludge volume index and solids concentration being equal to the sludge volume SV of the activated sludge tank after 30 min settling time and the product of overflow rate and solids concentration giving the solids loading of the secondary sedimentation B_F ($kg\ m^{-2}\ h^{-1}$).

As FIG. 1 shows, this functional relationship can also be plotted.

If admissible limit values are introduced for the sludge volume loading, (SV_F=0·3), the overflow rate of the secondary sedimentation is limited, too, depending on the characteristics and mass of the activated sludge. This shows clearly the mutual influences of the operational conditions in the activated sludge and final settling tanks, which must be taken into regard especially when fixing the recycle sludge ratio. Whether the desired sludge concentration can be attained in the aeration tank with a definite return sludge ratio, depends in the end on the thickening of the activated sludge in the secondary sedimentation, as shown in FIG. 2.

The solids concentration to be expected of the recycle sludge is again influenced by the compaction characteristics of the activated sludge, the sludge volume index being a suitable parameter. The known formula for the calculation of the return sludge concentration as a function of the sludge volume index gives values which have the same proportion to the attainable underflow concentrations with sludge thickening (FIG. 3).

Considering all three graphical representations, it will be seen that the design data for the secondary settling can be selected with sufficient security. It will be subject to discussion, however, what sludge volume loading and which depths of the tanks as a function of the type of tank, the type of the sludge removal equipment and the return sludge ratio should be taken as limiting rates in order to secure still an economical design of secondary sedimentation tanks.

REFERENCES

PLANZ P. (1966) Über das Absetzen des belebten Schlammes in horizontal durchströmten Nachklärbecken. (About the settling of activated sludge in secondary sedimentation tanks with horizontal flow-through). *Veröffentlichungen des Institutes für Siedlungswasserwirtschaft der TH. Hannover*, Heft **25**.

KALBSKOPF K.-H. (1969) *European Practices in Sedimentation, Water Quality Improvement by Physical and Chemical Processes*, The University of Texas at Austin No. 3.

Water Research Pergamon Press 1972. Vol. 6, pp. 433–435. Printed in Great Britain

TWO-STOREYED FINAL SEDIMENTATION TANKS

I. KLEFFNER

Baubehörde Hamburg—Stadtentwässerung, Neuer Wall 72, D-2 Hamburg 36, FRG

SOMETIMES the design engineer is confronted with the problem of having to construct a sizeable sedimentation tank on an inadequate small area. In the case of primary sedimentation tanks he may get over the difficulty by increasing the depth to a certain extent. But the same is not possible where long rectangular final sedimentation tanks are concerned because of the limiting factor of tank surface loading. Generally it is uneconomic to allow a depth of more than 2·5 m in final sedimentation tanks. In such a case the two-storeyed tank may provide an answer. The construction costs per unit volume are found to be lower than that of the normal sedimentation tanks of comparable performance. One disadvantage with the two-storeyed sedimentation tank is the compulsion to install band scrapers on endless chains for sludge removal. Also scum should be removed before the inlet in order to avoid putrefying conditions at the ceiling of the lower tank.

In 1963 two storeyed final sedimentation tanks were constructed at sewage treatment plant Stellinger Moor. The design was based on Swedish prototypes. These final tanks are part of the activated sludge plant shown in FIG. 1. The complete biological plant consists of:

activated sludge basins: diameter 16 m, 6 m deep, covered, with improved Dorr–Oliver turbines;
operational data:
organic load
$B_R = 3{\cdot}0\text{–}3{\cdot}5$ kg BOD_5 m^{-3} day^{-1}
sludge loading
$B_{TS} = 0{\cdot}7\text{–}0{\cdot}8$ kg BOD_5 kg^{-1} $MLSS^{-1}$ day^{-1}.

Paddle basins: cylindrical, parallel flow; for removal of scum, fats and air bubbles prior to the final tanks and for formation of well settling flocs.
Detention time t_{18} = 20 min.

Final sedimentation: two-storeyed tanks with band scrapers.
Volume = 1940 m^3 for two storeys (FIG. 2).

Design data:	DWF	SWF
detention time t_{18}	3·2	16·4
overflow rate q_F	0·75	1·5 m h^{-1}

Operational results

The treatment plant was commissioned in 1965. The results obtained with the two-storeyed tanks are given in the attached table (Table 1). It is of interest to note that

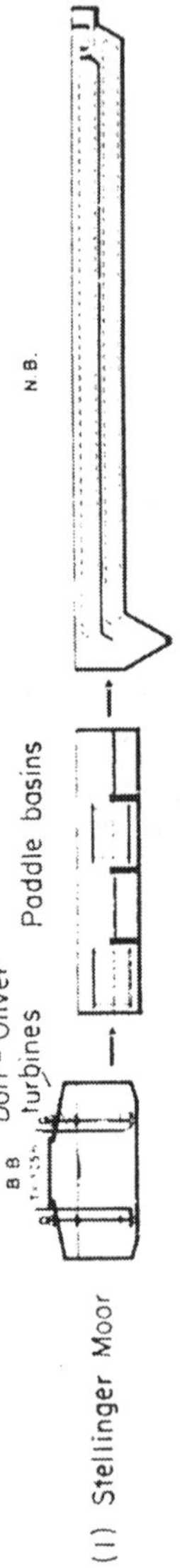
(1) Stellinger Moor
B B
Dorr - Oliver turbines
Paddle basins
N B.

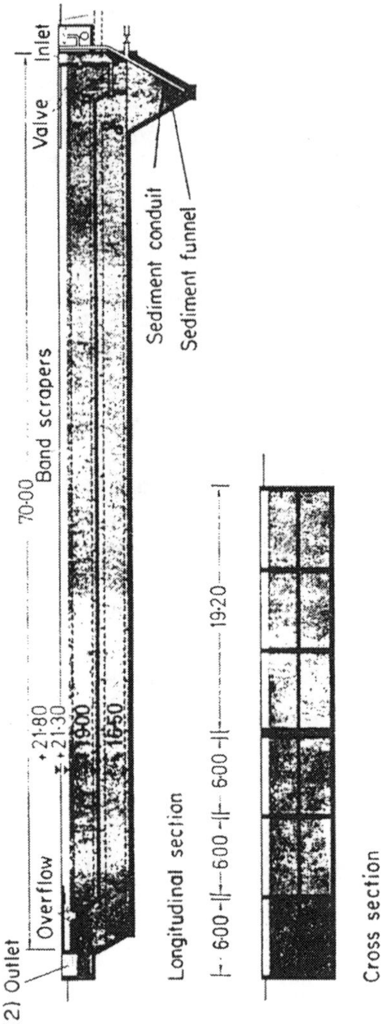
(2) Outlet
Overflow
+21·80
+21·30
1900
1650
70·00
Band scrapers
Valve
Inlet
Sediment conduit
Sediment funnel
Longitudinal section
6·00
6·00
6·00
19·20
Cross section

FIG. 1.

FIG. 2.

constantly low dry solid effluent contents of 30 mg l^{-1} and below were observed under the following operational conditions:

detention time 2·0–4·7 h;

hydraulic overflow rate 0·5–1·1 m h^{-1}

mass loading 1·3–4·8 kg dry solids m^{-2} h^{-1} and 125–400 l. m^{-2} h^{-1}.

TABLE 1. OPERATIONAL RESULTS FROM TWO-STOREYED FINAL SEDIMENTATION TANKS

Influent			Loadings				Effluent	
Q/18 (m^{-3} h^{-1})	TS_R MLSS (kg m^{-3})	J_{SV} (l. kg^{-1})	t (h)	q_F (m h^{-1})	B_F (kg m^{-2} h^{-1})	(l. m^{-2} h^{-1})	TS Dry solids (mg l^{-1})	BSB$_5$ BOD$_5$ (mg l^{-1})
1170	2·6	99	4·7	0·49	1·27	125	16	36
1360	3·6	83	4·0	0·57	2·06	170	13	34
1670	3·9	76	3·3	0·70	2·70	206	16	25
1940	4·0	84	2·8	0·82	3·30	280	10	40
2200	4·3	82	2·5	0·93	4·00	330	13	24
2500	4·5	83	2·2	1·05	4·75	395	30	25
2670	3·8	75	2·0	1·12	4·25	320	20	34

Return sludge constant ~ 300 m^3 h^{-1} for 2 stages.

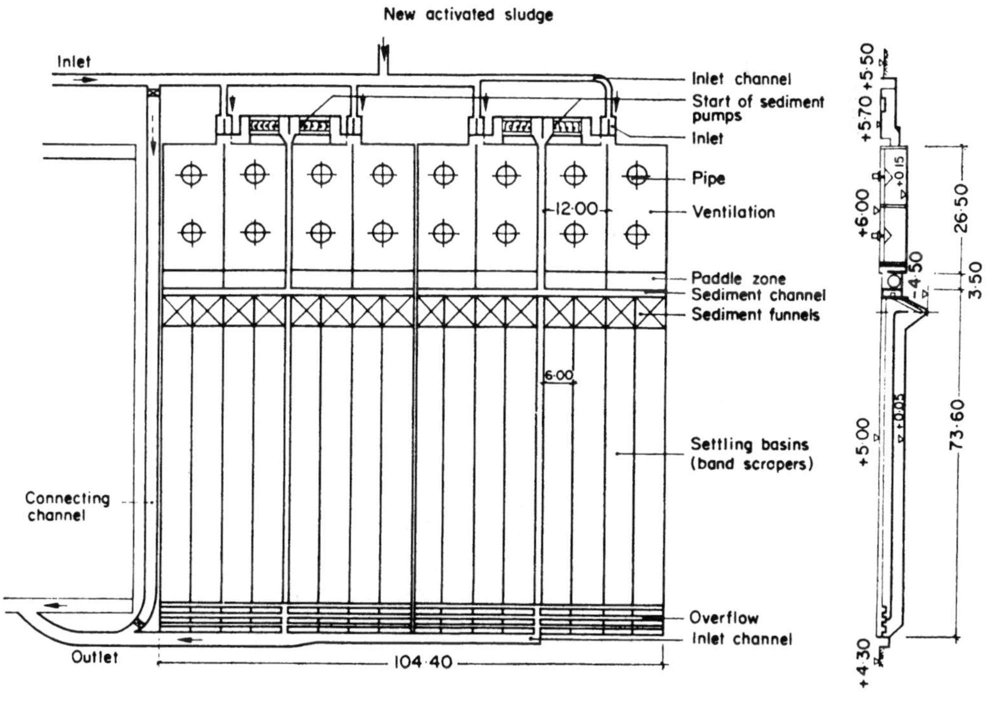

FIG. 3.

Since 1965 the band scrapers were operated either simultaneously or alternatively at regular intervals of 1 h. Alternatively means that either the scraper at the top or bottom was in operation. The quality of effluent was observed to be unaffected by the mode of operation. The concentration of return sludge was increased from 8 to 10 g l^{-1} in the case of simultaneous scraper operation to 9–12 g l^{-1} in the case of alternative scraper operation. Since alternative scraper operation diminishes the working time by 50 per cent compared to the simultaneous operation it actually means a doubling of the scraper's useful life.

Further tests showed that equal loading of both storeys depend only on the equal discharges at the outlet weirs. The head losses at the inlet devices are equalized within the two basins.

Encouraged by the satisfactory operation of the two storied final sedimentation tanks at Stellinger Moor it was decided to construct similar tanks at the Köhlbrandhöft sewage treatment plant expansion scheme as well (FIG. 3). The design features of the proposed tanks are as follows:

detention time $t = 1{\cdot}9$ h, $q_F = 1{\cdot}1$ m h^{-1}.

In this case the hydraulic weir loading will be decreased by the two additional outlet channels.

Water Research Pergamon Press 1972. Vol. 6, pp. 437–440. Printed in Great Britain

CONTROL OF ACTIVATED SLUDGE PLANTS

R. Wood

The West Hertfordshire Main Drainage Authority, Maple Lodge, Denham Way, Rickmansworth, Hertfordshire, England

The activated sludge plant at the Maple Lodge Works of the West Hertfordshire Main Drainage Authority, England, serves a population of, currently, 521,700. The capacity of the diffused air units (partly ridge and furrow, partly "high density" dome arrangement) allows for 10 h detention based on the dry weather flow of 127,288 m^3 (33·5 U.S. mg d). Tables 1 and 2 show the performance of the plant during 1970/71.

It will be seen that the average BOD loading was 23,358 kg day^{-1} (51,481 lb day^{-1}), but in view of the almost complete nitrification of the plant it is suggested that this figure is misleading. It might be more useful to consider the load as being applied as McGowan Strength of the settled sewage, i.e. (6·5 × N/8 permanganate value or PV) + (4·5 × Total N) which was 743, the nitrogen exerting one-third of the load.

Table 1. Maple Lodge works—activated sludge plant 1970–71

Month	Sewage Flow		Air Flow	
	10^3 m^3 day^{-1}	10^3 U.S. g day^{-1}	10^3 m^3 day^{-1}	10^6 ft^3 day^{-1}
April	137·3	36·19	1694·0	59·82
May	128·9	34·03	1546·9	54·62
June	129·0	34·06	1314·2	46·40
July	128·1	33·82	1352·3	47·75
August	127·5	33·66	1237·9	43·71
September	133·4	35·22	1389·2	49·05
October	130·2	34·37	1571·1	55·48
November	158·6	41·87	1565·8	55·29
December	136·6	36·06	1717·4	60·64
January	156·2	41·24	1941·1	68·54
February	140·9	37·20	1841·6	65·03
March	143·8	37·96	1976·5	69·79
Average	137·5	36·31	1596·2	56·34

The flow through the 19 units is controlled by Transdata equipment, measuring the flows by magnetic flowmeters and adjusting the outlets by pneumatically-operated plug valves, to give approximate equal flow through each unit at all times. This is the only automatic control installed.

It could be that similar automatic air control, to control the air flow equally to each unit, might be advantageous should the impurity load to be treated rise to a critical level. In view of other developments, involving the construction of another treatment plant, it is not expected that this need will arise.

The total air flow is controlled manually to maintain adequate dissolved oxygen conditions in the mixed liquor, to influence the Sludge Density Index, and to ensure less than one mg l^{-1} ammoniacal N in the final effluent.

TABLE 2. MAPLE LODGE WORKS—ACTIVATED SLUDGE PLANT 1970–71

Month	S.S.		Nitrogen							BOD$_5$		Anionic surface active agents	
			Total	Ammoniacal		Albuminoid		Nitrous	Nitric				
	Settled sewage	Final effluent	Settled sewage	Settled sewage	Final effluent	Settled sewage	Final effluent	Final effluent	Final effluent	Settled sewage	Final effluent	Settled sewage	Final effluent
April	145	9·4	57	37	0·15	5·9	0·73	0·21	17·7	218	7·9	12·7	0·41
May	157	5·8	54	34	0·17	6·2	0·62	0·05	18·9	196	5·2	13·8	0·40
June	124	7·1	50	34	1·09	4·8	0·65	0·32	19·0	150	7·2	12·1	0·43
July	114	7·2	50	32	0·30	4·5	0·62	0·10	20·6	141	5·1	11·9	0·40
August	96	6·7	48	34	0·40	4·5	0·69	0·12	19·8	134	5·4	12·1	0·45
September	108	6·0	49	32	0·29	4·8	0·70	0·12	17·2	154	4·3	11·4	0·42
October	110	5·5	57	41	1·44	5·5	0·83	0·29	20·6	176	6·2	10·6	0·40
November	108	9·4	47	32	0·31	5·5	0·77	0·12	20·7	155	7·2	9·3	0·35
December	128	8·5	52	39	0·24	5·6	0·79	0·17	22·6	188	6·5	10·5	0·38
January	124	12·0	44	30	0·21	5·1	0·90	0·19	20·3	168	8·8	9·1	0·32
February	131	8·6	52	37	0·25	6·3	0·82	0·17	22·8	183	7·2	11·0	0·37
March	135	8·5	44	33	0·21	5·4	0·80	0·21	18·1	184	6·6	11·5	0·39
Average	122	8·0	51	34	0·43	5·3	0·74	0·21	19·6	170	6·5	11·2	0·39

Air is supplied from up to 8 compressors, so that control steps are in units of 12·5 per cent of maximum. It will be seen that the variations in the quantity of air supplied are very considerable. In August 1970, the air supplied was only 62·6 per cent of that supplied in March 1971 (maximum). The load applied to the plant, measured as BOD, was, in August, 64·6 of the March figure; measured as McGowan strength it was 88·7 per cent, being exactly in proportion to the flow.

There was at no time any air control required to correct D.O. or ammoniacal N figures: all control was used in an endeavour to control the Sludge Density Index. The variations in this index are shown in FIG. 1. In June, when the Density Index reached 2·4, the average sludge percentage in the mixed liquor dropped to 10·8 per cent; in March it reached 56 per cent. The MLSS are permitted to rise in winter to compensate for the reduced activity of the sludge under lower temperature conditions, but careful study of FIG. 1 will show that alteration of density precedes alteration of

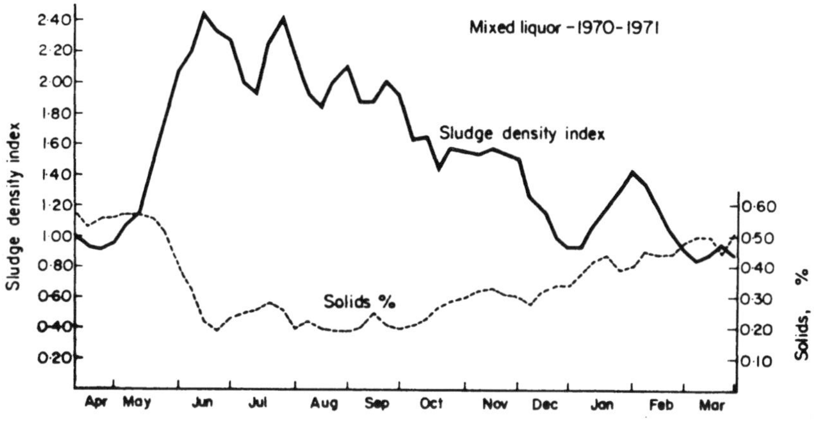

FIG. 1.

the level of MLSS carried in the plant. During a period of reducing Sludge Density Index the two factors influence each other, for a reduction in sludge density causes greater difficulty in sludge wastage because of its greater volume, with the result that the level of MLSS tends to rise further than is desired.

The cause of the wide variations in Sludge Density Index is not known: the variation are seasonal, but very imprecise in their timing. Air control, coupled with adequate sludge withdrawal techniques can be used to adjust it, but it is a very blunt instrument.

There is, however, clearly no need to automate, or even mechanise, the control of total air flow to the plant. Adjustments are not required frequently: at the most, with the size of compressor units installed, adjustment every 12 h is required; at the least, control once per month is needed. The use of smaller compressor units would be no advantage: rather would, in general, larger units with facilities for blowing off surplus air be more economical. In the new treatment plant being constructed to serve a future population of 250,000, only 3 units are being installed, and air will be blown off: the reduction in capital cost of equipment and size of building to hold it more than offset the cost of additional power used.

The problems of control of Sludge Density Index in a fully nitrifying plant should not be confused with sludge bulking, which is associated with low D.O., and growth of filamentous organisms. Under the conditions described there is no discernable difference in the quality of the sludge (numbers and types of organisms present) during periods of low or high density index.

Rate of sludge return is constant at approximately 118,000 m^3 day^{-1} (26 mgd), and although it is not considered that this has any adverse effect upon plant performance, at the new plant provision is being made for the rate of return to be automatically controlled by the rate of sewage flow received. This is considered necessary in view of the requirement to pass 4·5 DWF through the activated sludge plant in times of storm, compared with less than 3 DWF at the Maple Lodge plant.

To sum up: under the conditions prevailing in a fully-nitrifying plant (in fact, one removing 47 per cent of the nitrogen fed to it) automatic flow distribution and probably automatic air flow distribution are required. The variation in the rate of sludge return may not be necessary except to control hydraulic conditions in final settling tanks, and air supply can be manually controlled to the requirements shown by routine daily laboratory checks on free ammonia in the effluent and D.O. in the mixed liquor. With a knowledge of the factors causing variations in Sludge Density Index more sophisticated controls may be required, but for plants having a high sludge age (18 days was the average figure for Maple Lodge in 1970/1971) it is unlikely that frequent changes in MLSS or air supply would be necessary.

Water Research Pergamon Press 1972. Vol. 6, pp. 441–445. Printed in Great Britain.

DISSOLVED OXYGEN CONTROL OF ACTIVATED SLUDGE AERATION

R. A. RYDER

Kennedy Engineers, 657 Howard Street, San Francisco, California 94105, U.S.A.

THE DEVELOPMENT of electrochemical sensing elements for the measurement of dissolved oxygen has made possible the automatic control of this important factor of the activated sludge process. The methods used are polarographic (BUSCH and SAWYER, 1952), electrode sensing (CLARK, WELD and TAYLOR, 1953) including temperature-compensating thermistors (CARRITT and KANWISHER, 1959). Process oxygen monitoring and control systems then become possible.

DISSOLVED OXYGEN CONTROL

Our initial use of dissolved oxygen control systems for activated sludge dates to 1963 when it was for the activated sludge and post-aeration process control at the 120,000 m^3 day^{-1} (32 mgd) capacity wastewater treatment facility at Reno-Sparks, Nevada. Following this initial success, we have designed dissolved oxygen control systems at activated sludge treatment plants for the Oro Loma Sanitary District in San Lorenzo, California, the University of California at Davis, Carmel, California, and the San Francisco International Airport.

At the Reno-Sparks plant there was a requirement for dissolved oxygen post-aeration control to economically assure that the discharge standards would be met. Also, the stringent effluent BOD and solids standards required that the activated sludge system selected for secondary wastewater treatment operate at nearly maximum efficiency all of the time.

CONTROL SYSTEMS

A typical dissolved oxygen control system is illustrated in FIG. 1. The minimum elements of this system include: A. DO sensor. B. DO amplifier. C. DO recorder and automatic controller. D. Aeration air input controller. E. Controlled element (such as flower or turbine speed, throttled air valve, or turbine submergence).

A solid electrode DO sensing element is illustrated. An important requirement is that the flow velocity of the liquid passing the electrode be sufficient to prevent inaccuracies due to depletion of oxygen concentration at the interface of the membrane, and electrode polarization. The velocity required is usually in excess of 0·3 ms^{-1} (1 ft s^{-1}) and can be provided by a small pump unit on the electrode assembly as illustrated, or in closed pipe systems by sizing the piping to maintain this minimum velocity with insertion of the electrode into the stream of flow.

DIFFUSED AERATION CONTROL

The earliest examples of aeration control were applied to diffused aeration plants utilizing centrifugal blowers. In one instance, a multiple system of electrodes were monitored.

A different approach to control of centrifugal blowers was utilized at Reno-Sparks. In this case there was control of both activated sludge aeration tanks and a post-aeration tank following the secondary sedimentation tanks using the same group of

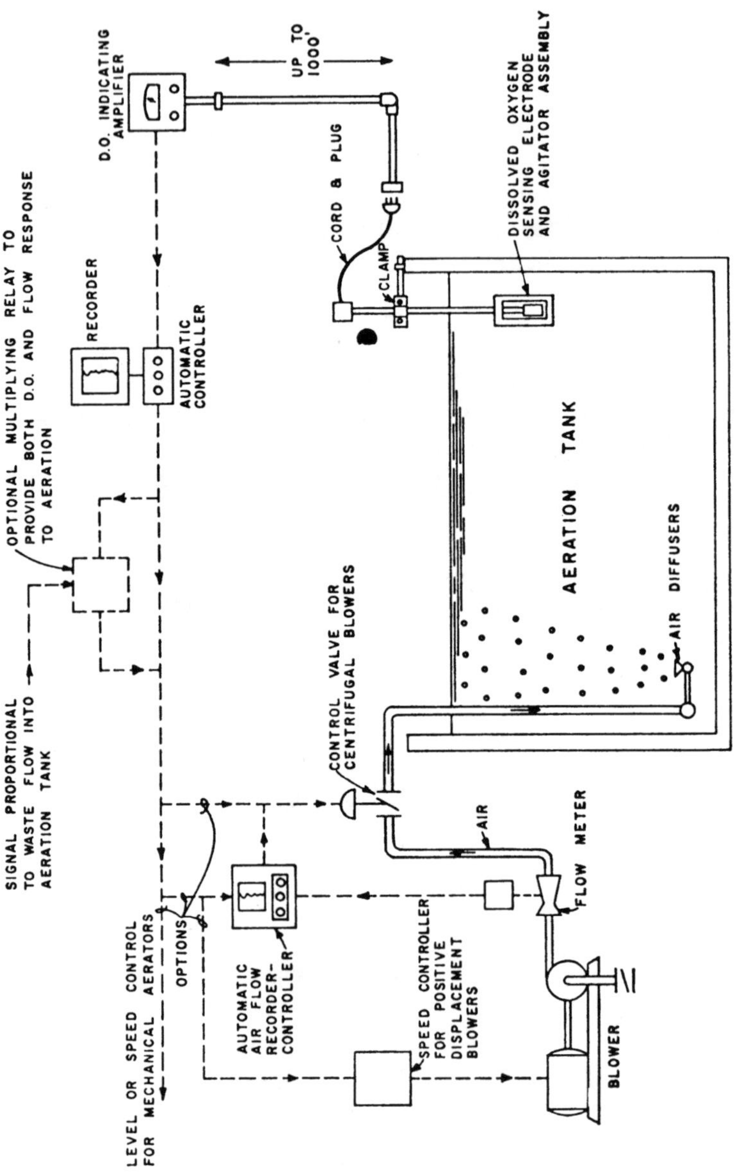

FIG. 1. Typical dissolved oxygen automatic control systems.

aeration blowers. The desired range of dissolved oxygen could vary from 0·5 to 3 mg l^{-1} in the mixed liquor and above a minimum of 6 mg l^{-1} in the post-aeration tanks. A schematic diagram of the control system utilized at Reno-Sparks is shown in FIG. 2. Two centrifugal blowers of 8·5 m^3 s^{-1} (18,000 CFM) capacity were utilized to supply all of the air. Regulation of the amount of air to each of three aeration tanks and the post-aeration tank was accomplished by modulating butterfly valves responsive to both the wastewater flow and dissolved oxygen concentration. The aeration tanks provide a triple pass and can flexibly be operated in a number of modifications of the activated sludge process depending upon loading conditions and operational choice. Three alternative locations for dissolved oxygen monitoring were provided at the third points of aeration tank.

Swing pipe headers with coarse bubble diffusers attached to pipe laterals were utilized. The number of diffusers on each swing header could be varied by plugging or attaching diffusers to pipe laterals to provide a tapered aeration pattern to match the desired aeration profile. A portable dissolved oxygen analyzer was utilized to determine oxygen profiles whenever a change was made in either the step feed inflow pattern or when a portion of the first pass was utilized exclusively for sludge reaeration. The number of diffusers placed throughout the tank would then be adjusted to maintain a relatively constant concentration of dissolved oxygen throughout.

The combined use of flow and dissolved oxygen cascade control has produced a quickly responsible and highly stable control system. Experience in over two years of operation has indicated that it is possible to maintain a desired concentration of dissolved oxygen within $\pm 0{\cdot}2$ mg l^{-1} of the set point over periods of time as long as several weeks.

MECHANICAL AERATOR CONTROL

The dissolved oxygen may be controlled with equal facility in mechanical aeration systems as readily as those with diffused air. For example, in the Oro Loma Water Pollution Control Plant there are three aeration tanks, each with eight 18·6 kW (25 h.p.) surface aerators installed in pairs in each of the aeration tanks. The first quarter of the tanks are utilized for returned sludge reaeration. The remaining three-quarters of the tanks can optionally be operated as plug flow or continuous mixing aeration patterns.

The dissolved oxygen signal is utilized in a cascade control system to adjust the set point of the submergence controller to maintain a predetermined concentration of dissolved oxygen in the aeration tanks, usually 1 or 1·5 mg l^{-1}. Only one of the surface aerators in each tank is positioned from the DO control system. The other units follow the leader to maintain the same submergence setting. The two aerators in the reaeration zone of each tank are controlled by a similar but independent system.

The cost of providing dissolved oxygen control is approximately $5000 for the sensor, amplifier, transducer, controller, and positioner. At the Oro Loma Wastewater Treatment Plant there are six control loops of this kind, representing a cost in the order of $30,000. Computed power cost saving of aerator operation by continuous dissolved oxygen control at Oro Loma is about $15,000 per year. It is evident that considerable economic savings will be provided by this control system.

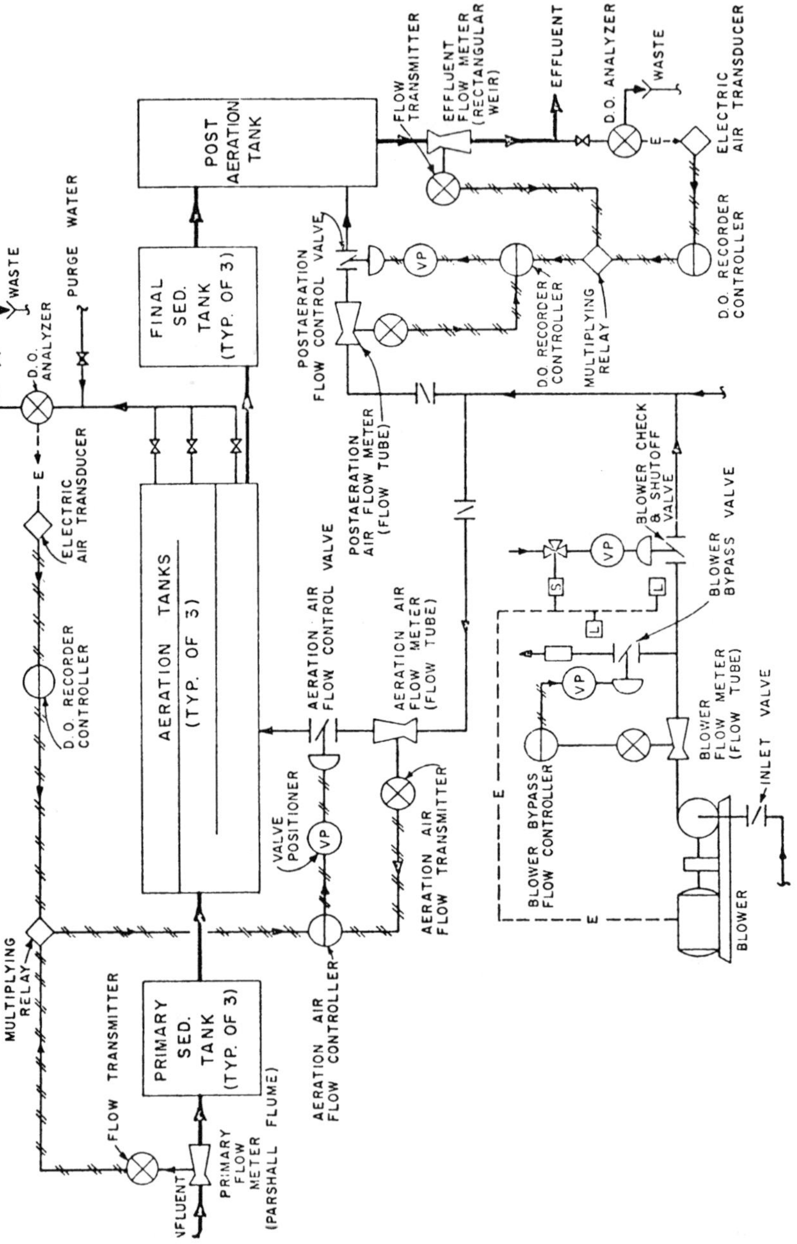

FIG. 2. Dissolved oxygen control of activated sludge aeration and effluent postaeration.

Several of our more recent installations utilize control of variable speed motors on turbine aerators in response to dissolved oxygen concentration. Both saturated reactor induction motors and hydraulic coupling drives have been utilized.

DISSOLVED OXYGEN LEVELS

With methods of continuously maintaining a selected concentration of dissolved oxygen readily available, plant operation must choose at what level that concentration should be. From the standpoint of economy in operation related to power consumption, it is the minimum oxygen concentration that will maintain an effective activated sludge. The larger the gradient between saturation and maintained oxygen concentrations, the greater the oxygen transfer for an equivalent amount of aeration and power consumption. There are lower limits imposed by the period when the sludge is in the secondary sedimentation tank and the dissolved oxygen is depleted by the metabolic activity of the organisms.

Higher dissolved oxygen concentrations can result in excessive nitrification, proliferation of filamentous organisms, and/or by mechanical shear result in poorly settling activated sludge. Experience in the initial startup of the Reno-Sparks, Oro Loma and Davis plants clearly defined these problems.

It has been found that a dissolved oxygen concentration of mixed liquor in the aeration tanks in the range of 0·75 to 1 mg l^{-1} in the aeration tank, and permitting zero dissolved oxygen in the final sedimentation tanks impose an unfavourable environment for filamentous organisms, and control sludge bulking.

The practical implication is that optimum conditions of an efficient and rapidly settling activated sludge can be obtained by controlling dissolved oxygen within close limits of from 0·5 to 1·5 mg l^{-1}. This is particularly important when due to either overloading or high carbohydrate feed there are conditions that enhance the growth of filamentous organisms.

REFERENCES

BUSCH A. W. and SAWYER C. N. (1952) Determination of the B.O.D. and sewage and industrial wastes with the polarograph. *J. Analyt. Chem.* **24,** 12, 1887.

CARRITT D. E. and KANWISHER J. W. (1959) An electrode system for measuring dissolved oxygen. *J. Analyt. Chem.* **31,** 5.

CLARK L. C., WELD R. C. and TAYLOR Z. (1953) Continuous recording of blood oxygen tension. *J. appl. Physiol.* **6,** 189.

Water Research Pergamon Press 1972. Vol. 6, pp. 447–449. Printed in Great Britain

CONTROL OF ACTIVATED SLUDGE PLANTS

W. VON DER EMDE and U. SCHOPPER
Technische Hochschule Wien, Karlsplatz 13, Wien 1V, Austria

(1) THE PURIFICATION effected by an activated sludge plant can be influenced by

(a) dissolved oxygen content (c_x)

(b) sludge concentration (TS_R)

in the aeration tanks.

(2) The DO concentration depends on the respiration rate of the micro-organisms (OV), the oxygenation capacity (OC), and the DO saturation value of the mixed liquor (c_s).

(3) The oxygen consumption depends on the removal of organic matter and on the sludge content of the mixed liquor in the aeration tank.

The decomposition of organic matter (e.g. TOC or COD) may not be used for the control of the oxygen supply, as the removal of organic matter and the oxygen consumption progress at different rates in waste waters containing suspended matter and colloids.

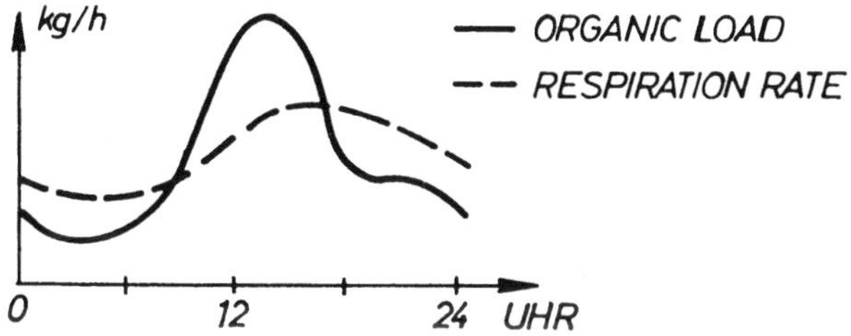

FIG. 1.

(4) The control of the oxygenation capacity

$$OC = \frac{c_s}{c_s - c_x} \times OV$$

is therefore based on:

(a) the dissolved oxygen content c_x or

(b) the oxygen consumption OV.

(5) The dissolved oxygen content based control offers many advantages. Difficulties may be experienced however with the dissolved oxygen based control at low oxygen contents in aeration tanks (e.g. lightly loaded plants) and when localized oxygen input is applied (surface aerators).

(6) The dissolved oxygen concentration in the aeration tanks is derived from the relation between the respiration rate and the oxygenation capacity. The oxygenation capacity for a given dissolved oxygen content can be found directly from the respiration rate. The respiration rate can be determined by occasional respiration tests

(oxygen consumption under simulated load in a closed container) or continuously in a model of a large plant (e.g. oxygen content reduction from a given amount of added air in a closed container).

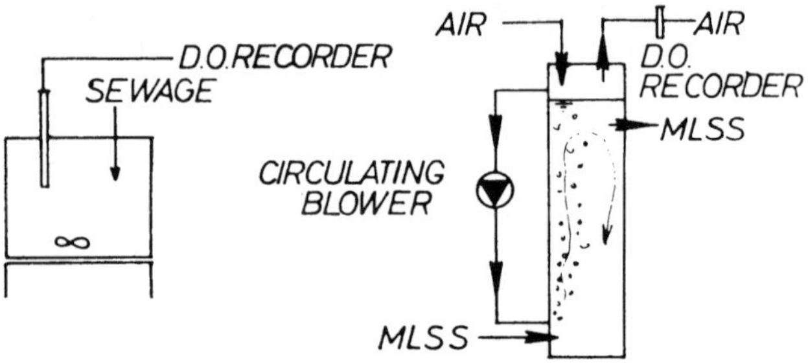

FIG. 2.

Since there exists, for various aerators, a relationship between oxygenation capacity and energy consumption, the oxygen input may be controlled through the energy consumption.

$$OC = 9/9\text{–}1{\cdot}5 \times OV; \; OC/N = 1{\cdot}8 \text{ kg}O_2 \text{ kWh}^{-1}; \; N = 0{\cdot}67 \times OV$$

e.g. OV = 300 kg O_2 h^{-1} demands N = 200 kW.

The dissolved oxygen in the aeration tank has to be recorded for control purposes.

(7) The sludge content of the aeration tanks is controlled by the withdrawal of excess sludge (assuming constant recirculating sludge quantities). In contrast to the oxygen supply and the respiration rate, excess sludge removal and increase in sludge quantity need not balance at all times since little variations from the desired MLSS (e.g. $TS_R = 3{\cdot}5$ kg m^{-3}) do not influence the efficiency.

(8) The withdrawal of excess sludge should therefore be coordinated with other plant operations. Excess sludge is known to thicken well when mixed with the sludge from primary sedimentation basins. It would appear to be expedient to convey to the thickeners excess sludge with proportional amounts of primary sedimentation sludge.

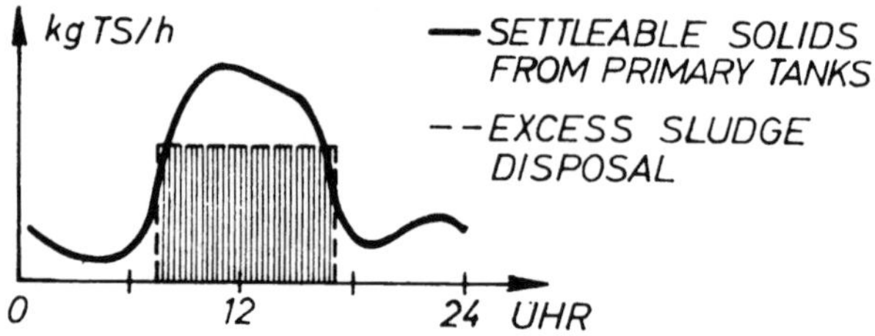

FIG. 3.

(9) The quantity of excess sludge to be removed daily can be determined from the MLSS of the sludge in the aeration basins (return sludge). A quick test or a continuous determination of MLSS could be possible using the TOC (diluted and homogenised sample) as there is a nearly constant ration between the TOC and the mixed liquor volatile solids.

(10) The excess sludge yield can be approximately assessed by tracing the TOC load.

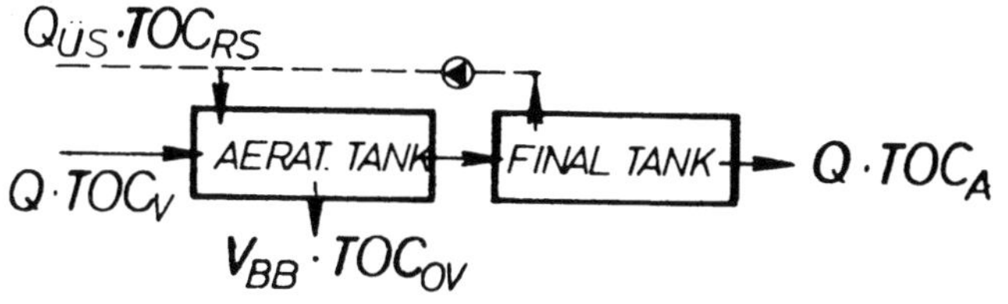

FIG. 4.

$Q_{\ddot{u}s} \times TOC_{RS} = Q(TOC_V - TOC_A) - V_{BB} \times TOC_{OV}$

$Q_{\ddot{u}s}$ = excess sludge (m^3 day^{-1})

Q = waste water quantity (m^3 day^{-1})

TOC_{RS} = TOC return sludge

$TOC_{V,A}$ = TOC effluent primary sedimentation, effluent final sedimentation, respectively.

The organic carbon (TOC_{OV}) which is oxidized to carbon dioxide can be calculated from the respiration rate. If the respiration rate is not measured, a factor g can be determined (from operational data) for excess sludge yield in relation to TOC decomposition.

$$Q_{\ddot{u}s} \times TOC_{RS} = g \times Q(TOC_V - TOC_A) \text{ or}$$

the excess sludge quantity to be withdrawn

$$Q_{\ddot{U}S} = g \times Q \frac{(TOC_V - TOC_A)}{TOC_{RS}}.$$

The TOC of the return sludge and the sludge blanket in the final sedimentation tanks (particularly suitable for circular tanks) serve as control.

(11) It is suggested that the maximum inflow should be assumed as the design basis for the return sludge quantity, and the plant operated with a constant volume of return sludge. The difference in elevation between the aeration and final sedimentation tanks should remain reasonably small (0·5 m). The energy consumption for 1 m^3 return sludge is of the order of 5 W, while about 100 W are required to purify 1 m^3 waste water. It would therefore be unreasonable to vary the quantity of return sludge to save energy costs.

Water Research Pergamon Press 1972. Vol. 6, pp. 451–454. Printed in Great Britain

CONTROL OF ACTIVATED SLUDGE PROCESS: APPLICATIONS

P. Brouzes

C.R.O.D.A., 9 Bd Louis Seguin, 92 Colombes, France

This paper summarizes a technique for the retrieval of data on the process of cellular growth in activated sludge, using the respiration of the culture. The growth rate is computed and controlled by this technique. With small plants one may control oxygenation and wastage of excess sludge, and obtain an indication of daily pollution data. For industrial plants, such equipment may be used for the continuous control of oxygenation and growth rate, recording the quantity of organic pollution removed, calculating the mineral salt dosage proportional to pollution removed, and, if required, regulating the admission of polluting matter to the treatment plant.

Digital computers are used for performing these operations in real time in large plants, where they can also be used for centralising and processing a large number of other parameters. Aeration costs are reduced by the use of this technique.

REMINDER OF THEORY

The factors which influence oxygen consumption in a culture of activated sludge are the requirements of new cells in the presence of polluting matter, decrease of consumption owing to auto-oxidation and decrease due to loss of activated sludge.

EXAMPLE OF APPLICATION FOR A PAPER PULP MILL

Factory: Cellulose des Ardennes Belges (2nd stage of installation).
Water flow treated: 2150 $m^3\ h^{-1}$.
BOD to be removed: 15,600 kg day^{-1}.
Primary sedimentation: 138 m^2.

Biological treatment

BOD to be removed biologically: 11,000 kg day^{-1}.
Volume of aeration tanks: 11,700 m^3.
Detention time: 5·50 h.
Aeration by 12 55-kW surface aerators at constant speed and variable immersion.
Regulation of aeration by modification of the levels in the aeration tanks by numerically controlled siphons.
Reactivation of activated sludge: volume 3500 m^3.
Aeration by 4 30-kW surface aerators.

Secondary sedimentation

Area: 1970 m^2.
Return of activated sludge: 80 per cent.

Centralized measurements

Water flow treated.
Temperature of activated sludge.

CONTROL PANEL

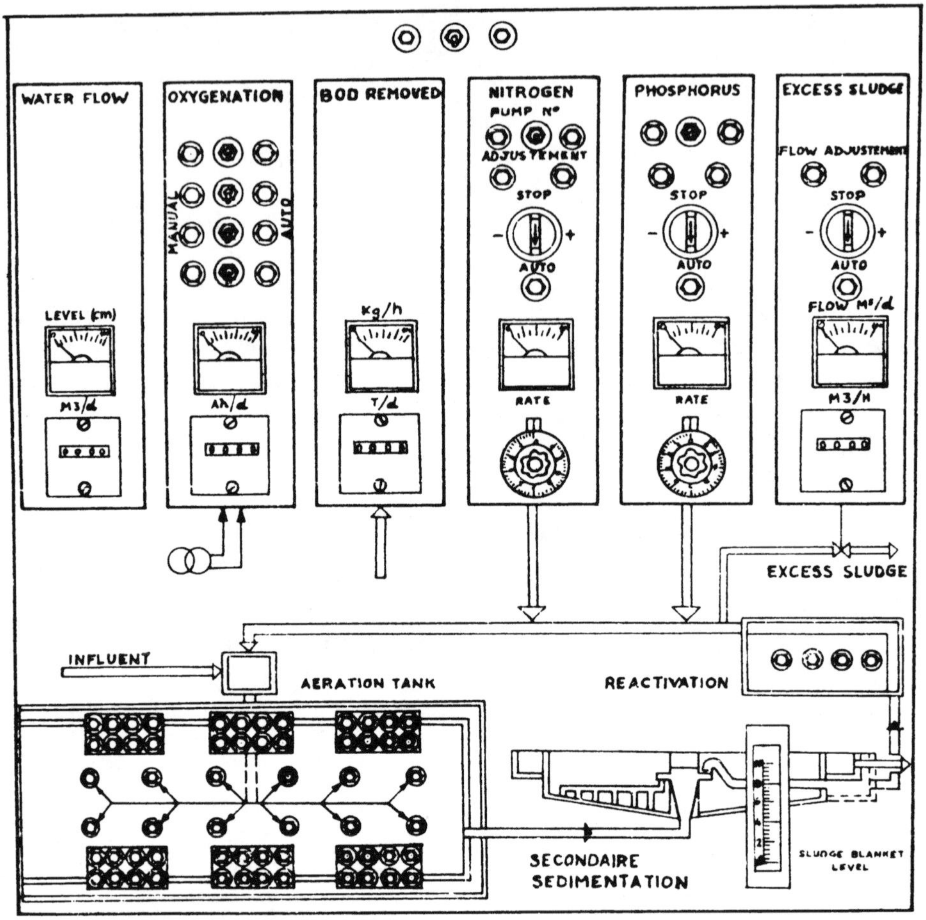

FIG. 1.

Dissolved oxygen in aeration tanks.
Water level in aeration tanks.
Current used by surface aerators.
Flow of excess sludge withdrawn.
Flow of nitrogen solution.
Flow of phosphorus solution.

Automatic controls

Numerical control of siphons for regulating oxygenation.

Automatic control of wastage of excess of activated sludge to maintain correct growth rate.

Automatic addition of nitrogen solution required.

Automatic addition of phosphorus solution required.

Continuous calculation of BOD removed.

EXAMPLE OF THE USE OF A DIGITAL COMPUTER IN REAL TIME PURIFICATION STATION AT PARIS ACHERES (3rd STAGE)

Flow treated: 900,000 m^3 day^{-1}.

Population served: 3 million inhabitants.

Primary sedimentation: 8 50-m dia. tanks.

Biological treatment: activated sludge at high load–sludge digestion.

Contractors: Service Technique de la Ville de Paris, Monsieur Olivesi, Directeur des services industriels et commerciaux. Monsieur Feuillade, Ingénieur général des services techniques.

Research and development for the centralisation of measurements and for control: Centre de Recherche de l'Omnium d'Assainissement (CRODA).

Operational by: end of 1971.

CENTRALIZED CONTINUOUS MEASUREMENTS

Ambient conditions	
Water level of Seine	1
Ambient temperature	1
Sewers	
Liquid level	5
Positions of gates	14
Pre-treatment	
Number of grit chambers in operation	8
pH	3
Raw sludge flow	2
Primary sedimentation	
Water flow treated	2
Sludge level	8
Current used by scrapers	8
Concentration of crude sludge	8
Crude sludge flow	8
Mass of crude sludge removed	8
Biological treatment	
Flow treated	4
Compressed air flow	4
Compressed air pressure	1
Return sludge flow	3
Excess sludge flow	4
Activated sludge temperature	1
Dissolved oxygen	8
Optical density of activated and return sludge	8
Secondary sedimentation	
Sludge level	9
Current used by scrapers	9
Turbidity of treated water	12
Digestion	62
Electric power used	1

AUTOMATIC ADJUSTMENTS

Eleven main controls shall be performed automatically. The primary decantation and biological treatment flows received are to be automatically limited to adjusted values.

The oxygenation of activated sludge, which requires power of the order of 2400 metric h.p., is to be controlled by the effective respiration of the micro-organisms. The total production of compressed air and its distribution in the various parts of the plant must be controlled automatically. The regulation of the growth rate which maintains steady purification conditions is computed according to theoretical requirements. Calculation of the excess sludge flow to be removed will be performed in real time by the computer.

	Centralized adjustments	Automatic control by computer
Flow of settled liquid	2	2
Flow of purified liquid	4	4
Distribution of air supply	1	1
Total air supply	1	1
Flow of excess sludge	3	3
Air inlet gates	3	
Standard raw sludge extraction	8	
Distribution of settled liquid	30	
Control of gates on sewers	14	
	66	11

CALCULATIONS

The calculations performed by the computer are concerned with:

The clock function for starting the various sub-programmes and for keeping track of the time.

Scaling.

Linearization of the signals received (extraction of square roots and special functions).

Supervision of alarm points set (96 different channels checked).

Additions, integrations, ratios, derivatives, etc.

Water Research Pergamon Press 1972. Vol. 6, pp. 455–456. Printed in Great Britain

DISCUSSION
INVESTIGATION OF THE BENEFITS FROM AUTOMATIC CONTROL OF DISSOLVED OXYGEN IN THE AERATION CHANNELS OF A LARGE ACTIVATED SLUDGE PLANT ACHIEVING FULL NITRIFICATION

A. B. WHEATLAND

Water Pollution Research Laboratory of the Department of the Environment, Stevenage, England

1. INTRODUCTION

THE INVESTIGATION was carried out at the Rye Meads sewage purification works of the Middle Lee Regional Drainage Scheme following pilot-scale work at the Laboratory, and was a collaborative effort of the staff of the Water Pollution Research Laboratory and of the Middle Lee Regional Drainage Scheme. The works had 8 flat-bottomed aeration channels each of 3800 m^3 capacity and each designed to treat approximately 0·1 m^3 sewage s^{-1} (2,400,000 U.S. gal day^{-1}). Each aeration channel was divided into four passes and was aerated by means of 7-in. dome diffusers. Channel No. 6 had 4584 domes uniformly spaced on the base of the tank and served as a control. Channel No. 8 had a similar total number of domes but they were more closely spaced near to the inlet. The first pass contained 2016 diffusion domes whereas the third and fourth passes contained only 600 each. Mackereth-type dissolved oxygen electrodes were provided at 8 points in each aeration channel so that the concentration of dissolved oxygen present in the mixed liquor could be monitored. In addition provision was made for the signals from four of the electrodes in Channel 8 to be used to control automatically the flow of air to the diffusers so that the concentration of dissolved oxygen in the mixed liquor could be maintained at about 2 mg l^{-1}. The conditions of operation and the performance of the plants have been described in detail by JONES *et. al.* (1969).

The conditions of operation of the units and the composition of the sewage treated were approximately as follows: BOD loading on aeration tanks, 0·4 kg m^{-3} day^{-1}; Mean retention period based on flow of sewage, 9·4 h; Return-sludge flow, 100 per cent dry weather flow of sewage; Concentration of solids in mixed liquor, 5000–6000 mg l^{-1}; Air flow, 15 volumes air, per volume of sewage treated.

Average composition of settled sewage fed to the aeration plant; BOD, 165 mg l^{-1}; Suspended solids, 125 mg l^{-1}; Ammoniacal nitrogen, 34 mg l^{-1}; Anionic detergent (as Manoxol OT) 16 mg l^{-1}.

2. PERFORMANCE OF PLANT

The performance of aeration Channel No. 8 equipped with automatic control of dissolved oxygen was compared with that of Channel No. 6 (control) over a period of 6 months from July to December 1967 when so far as possible both channels were operated under similar conditions in respect of loading.

Both channels produced effluents of similar high quality which were fully nitrified for the whole period. Channel No. 8 used 22 per cent less air than the control Channel No. 6. At times the concentration of dissolved oxygen in Channel No. 8 exceeded

2 mg l^{-1} because a minimum air flow was needed to prevent diffuser blockage, and the rate of introduction of dissolved oxygen at certain times exceeded that necessary to satisfy the BOD.

The sludge from Channel No. 8 had a lower density index and a higher capillary suction time than sludge from the control plant. It is thought that this was due in large measure to the difficulties of modifying a full-scale plant which necessitated the use of a smaller sludge-return pump which gave rise to higher shearing forces than that used for the control channels.

3. PROSPECTS FOR CONTROLLED AERATION

The extent to which controlled aeration will result in a direct reduction in the operating costs of a large sewage-treatment works will depend on a number of factors including:

(a) Whether a 1-point or a multi-point control system is used. One-point control is more easily engineered than a multi-point system such as used at the Rye Meads works. A completely-mixed plant is ideally suited to 1 point control.

(b) Whether the works is already built or at the design stage. The maximum saving with a piston-flow plant would be achieved if all of the first passes could be supplied from an air main regulated by a single oxygen probe in one of the first passes and all the second passes could be supplied with air from a second air main regulated from an electrode in one of the second passes and so on. The increased flexibility of control would permit maximum savings in air.

(c) Whether the aeration is by means of diffused air or by a mechanical method. It is possible that mechanically aerated plants would lend themselves more easily than diffused-air plants to automatic control of aeration.

(d) Source and cost of energy. The tariff arrangements applying to the purchase of electrical energy will have a major effect on the magnitude of any financial saving that might result from reductions in the energy requirements of aeration equipment.

(e) Possible increase in treatment capacity.

(f) Size of works. The cost of control equipment does not increase in proportion to the size of the works and the capital cost would be likely to be recovered more rapidly the larger the works.

REFERENCE

JONES K., BRIGGS R., CARR J. G. and POTTEN A. H. (1969) *Instn publ. Hlth Engrs J.* **68,** 271.

Water Research Pergamon Press 1972. Vol. 6, pp. 457–463. Printed in Great Britain

STUDY OF THE METABOLIZATION OF POLLUTANT PRODUCTS

P. BROUZES

C.R.O.D.A. 9 Bd Louis Sequin, 92 Colombes, France

BIOLOGICAL purification is mainly a result of the growth of micro-organisms. Other processes which take part in purification, such as adsorption, filtration by the flocs of activated sludge, and sedimentation, are also dependent on biochemical reactions or on the ecology of the culture. Respiration is one of the most characteristic indications of micro-organism metabolism and thus allows one to follow the growth.

By feeding the culture in discrete quantities, various phases of respiration corresponding to phases of growth, can be analysed, while the processes which occur in reality are reproduced.

The combination of control of respiratory metabolism and incremental feeding give results which are characteristic, and which can be obtained very rapidly.

By mathematical analysis the relationships between the oxygen used and the pollution removed can be defined, and the toxicity and biodegradability of various substrates characterised.

CULTURE FEEDING METHODS

It is known that there are significant differences in ecology and kinetics between continuous and batch cultures. We have adopted a method of feeding by discrete quantities, intermediate between these two techniques, but closer to reality. In practice in treatment plants, the contact each floc of activated sludge has with the water to be treated depends on the method of feeding, and on rate of rotation in spiral flow. With complete mixing tanks with surface aerators, this period is longer than 10 min.

In the laboratory, feeding by discrete quantities, which has the advantage of reproducing reality, also allows one to analyse the phases of feeding and synthesis without modifying the adaptation process.

RESPIROMETERS

Up till now much research has been conducted using WARBURG type respirometers. The interesting nature of the results obtained with this apparatus should not make one forget its limits in use. Fast processes are almost impossible to analyse. For rapid respiration cycles the removal of CO_2 is the main limiting factor if forced circulation of gas across the adsorber is not used. Nor does scanning by an operator allow the detection of fast processes.

The kinetics observed from dissolved oxygen measurements are very fast, making one think that certain phenomena may have passed unnoticed with conventional methods. The synthesis phase can end in a few minutes.

During this period, respiration often reaches several times the endogenous level.

Furthermore, the details obtained are highly characteristic. A computer presents these results directly (FIG. 1).

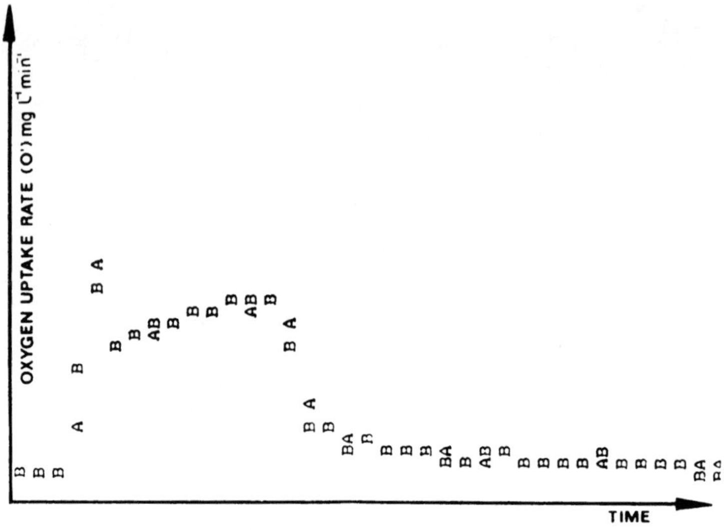

FIG. 1. Variation of oxygen consumption.

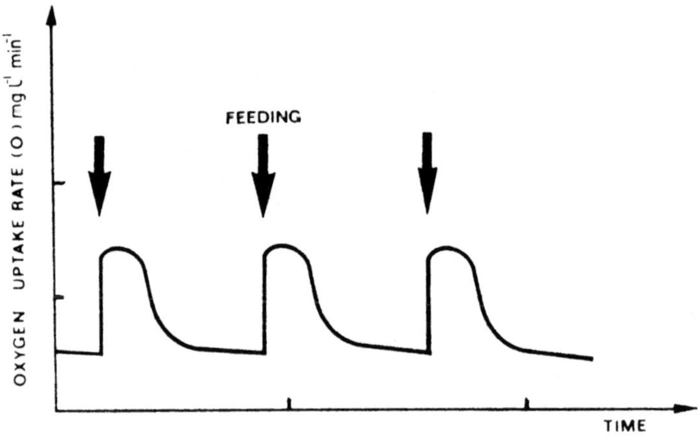

FIG. 2. Usual case of consumption graph.

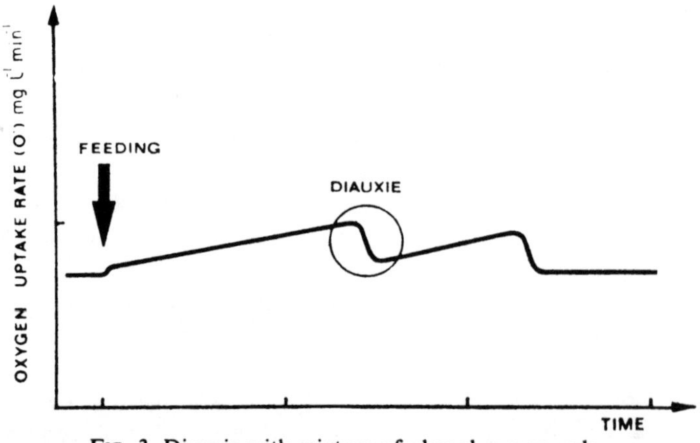

FIG. 3. Diauxie with mixture of phenol compounds.

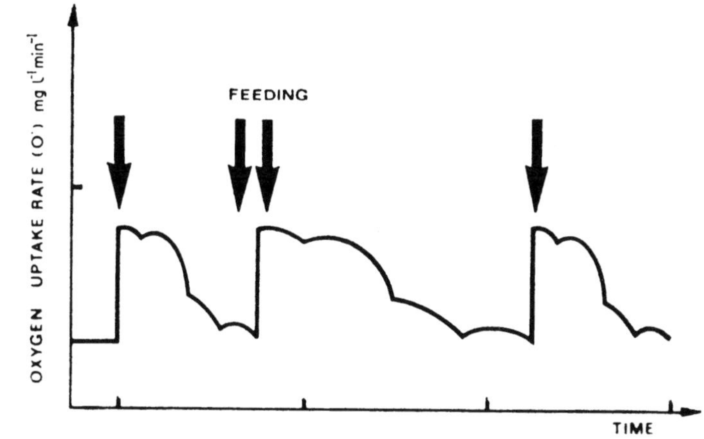

FIG. 4. Predominance of various species for different feed quantities.

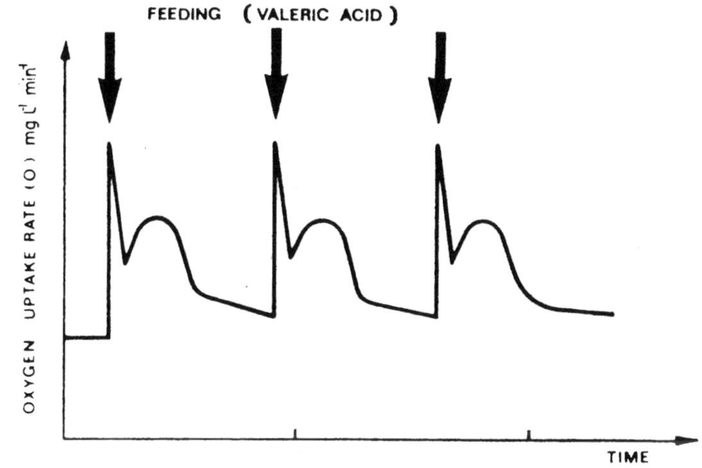

FIG. 5. Valeric acid.

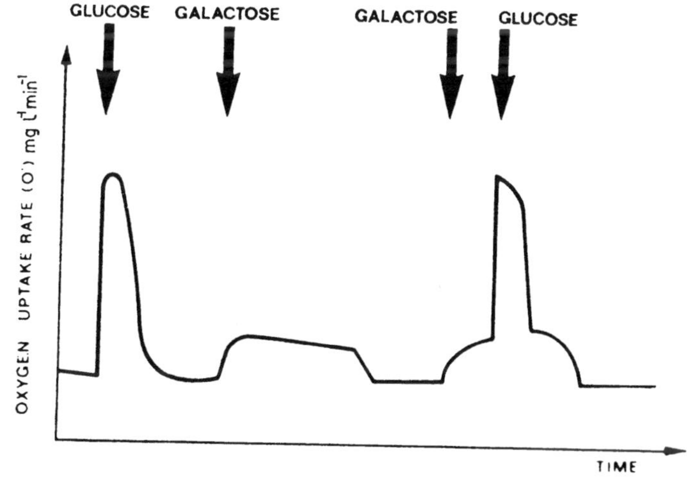

FIG. 6. Glucose and galactose.

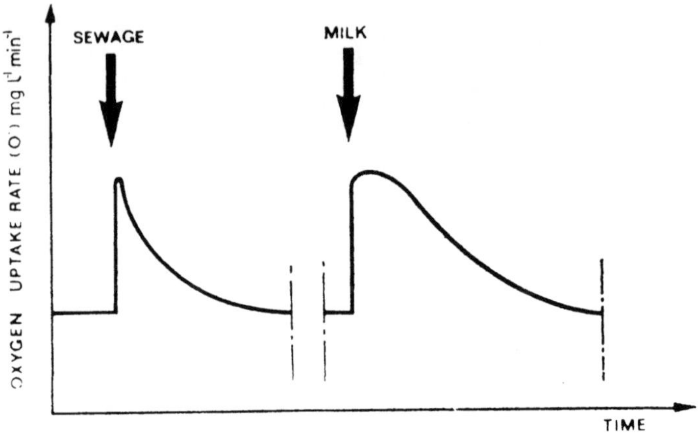

FIG. 7. Domestic effluent and milk.

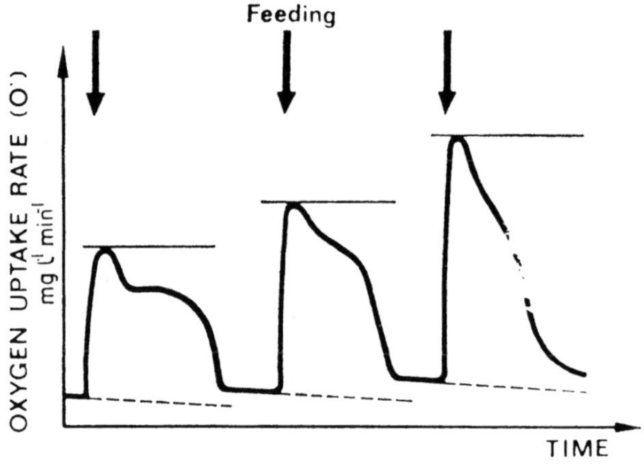

FIG. 8. Adaptation of a culture to a substrate.

CHARACTERISTIC RESULTS

The results have the general form shown in FIG. 2. As soon as the substrate is introduced, respiration increases. In a fraction of a second, consumption shoots up. In certain cases, it can reach 15 times endogenous respiration. The step corresponding to synthesis is more or less even, and can last from a fraction of a minute to several dozen minutes.

At the end of the synthesis, consumption falls back in an approximately progressive manner to about its original value. If the culture is not further fed, consumption decreases exponentially.

Exceptionally, with extremely specialized cultures, one can observe phases of diauxie,* corresponding to conversion of the enzyme system (FIG. 3). If several cul-

* Diauxie: Monod's double growth.

tures, each adapted to some substrate constituents, are mixed, the influence of various species of micro-organisms can be recognized (FIG. 4).

With a fixed culture, it is possible to obtain characteristic graphs (FIGS. 5–8):

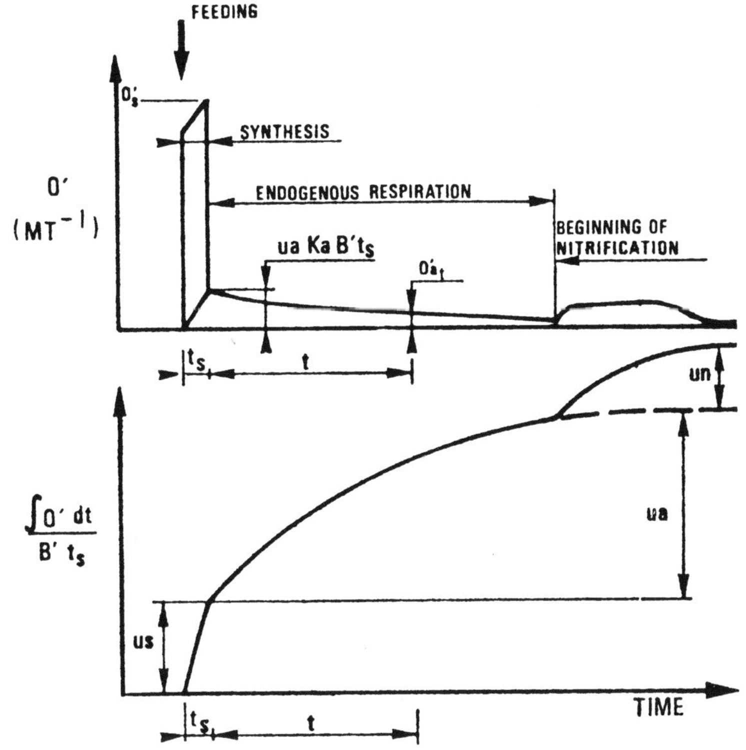

FIG. 9. Utilization of oxygen for synthesis and endogenous respiration.

RELATIONSHIP BETWEEN OXYGEN USED AND THE SYNTHESIS OF NEW CELLS

The endogenous respiration $O'a$ corresponding to cells formed previously decreases at a constant rate Ka. If a fraction of the cells is continuously removed, the dilution rate Kd reduces further the respiration of the sludges previously formed, so that:

$$O'a_{t_1} = O'a_{t_0} \exp([Ka + Kd)] (t_0 - t_1). \qquad (1)$$

During the formation of new cells, two processes are superposed (FIG. 9): one the use of oxygen in reactions providing life energy and oxygen for incorporation in the cells synthesized: the other, an increase of endogenous respiration due to these new cells formed.

The increase of endogenous respiration due to the new cells formed is proportional to BOD removed, B'. This respiration starts to fall at a constant rate by auto-oxidation, and by dilution if any, from the time that the cells are formed, so that:

$$O'a_{t_1} = ua\ Ka \int_{t_0}^{t_1} B'_t \exp[(Ka + Kd)(t - t_1)]\, dt. \qquad (2)$$

The oxygen used for synthesis, O′*s*, is directly proportional to pollution removed per time unit B'. Because of the speed of the process all these requirements have to be met during the detention time in the aeration tanks:

$$O's = B'\,us. \tag{3}$$

Thus the expression for the respiration becomes:

$$O'_{t_1} = O'a_{t_0} \exp\left[(Ka + Kd)(t_0 - t_1)\right] + ua\,Ka \int_{t_0}^{t_1} B'_t \exp\left[(Ka + Kd)(t - t_1)\right] \mathrm{d}t + us\,B'_{t_1} \tag{4}$$

in which the first part corresponds to the decrease in respiration of previously formed cells, the second to the increase in endogenous respiration due to synthesized cells, and the last to the requirements of the synthesis.

In steady state, the exponential of the first part would tend to zero, and in the second expression the integral would tend to $B'/(Ka + Kd)$ so that:

$$\frac{O'}{B'} = \frac{ua\,Ka}{Ka + Kd} + us. \tag{5}$$

To sum up, the effect of the principal parameters on respiration can be characterized by the following expression:

$$\frac{\mathrm{d}O'}{\mathrm{d}t} = -\,O'a\,(Ka + Kd) + B'\,ua\,Ka + \frac{\mathrm{d}B'}{\mathrm{d}t}\,us. \tag{6}$$

MATHEMATICAL EXPRESSION OF RESULTS (FIG. 10)

The broad outline and specific details of the metabolization of a substrate can be defined by a few expressions:

rate of decrease by auto-oxidation (*Ka*) observed during the periods of no feeding; mean growth rate of the culture (*Kc*) determined by the increase of endogenous respiration in a period corresponding to a whole number of feeding cycles

$$Kc = \frac{1}{t}\log\frac{O'a_t}{O'a} + Kd; \tag{7}$$

oxygen used in synthesis per unit of BOD removed

$$us = \frac{1}{B't}\int O's\,\mathrm{d}t; \tag{8}$$

increase of endogenous respiration per unit mass of oxygen used in the synthesis (characterizes the capability of a substrate to form new cells)

$$Ks = \frac{\mathrm{d}O'a}{O's\,\mathrm{d}t}. \tag{9}$$

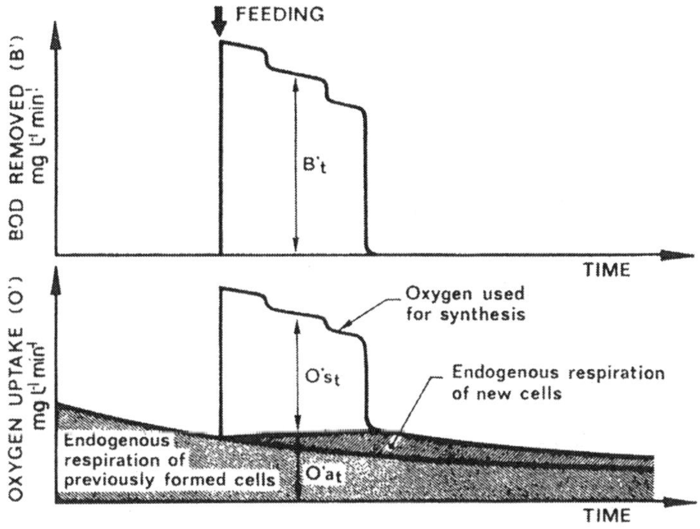

FIG. 10.

SYMBOLS

Mass transferred (mt^{-1}).

B' BOD removed per unit of time.

O' mass of oxygen used biologically per unit of time.

Rates of variation (t^{-1}).

Ka mean rate of decrease of oxygen consumption by auto-oxidation.

Kc culture growth rate.

Kd rate of dilution (rate of decrease of the oxygen consumption due to loss or removal of activated sludge).

Ks increase of endogenous respiration per unit mass of oxygen used for the synthesis.

Dimensionless

us quantity of oxygen used in synthesis per unit of BOD removed: $us = O's/B'$.

ua quantity of oxygen theoretically required for the total oxidation of sludge formed per unit of BOD removed (in the presence of a nitrification inhibitor).

e base of natural logarithms: e = 2·718.

exp ( ) = $e^{()}$.

Water Research Pergamon Press 1972. Vol. 6, pp. 465–474. Printed in Great Britain

THE USE OF MICROSTRAINERS AND SAND FILTERS FOR TERTIARY TREATMENT

PETER C. G. ISAAC and R. L. HIBBERD

Professor of Civil and Public Health Engineering and Head of Department, and
Lecturer in Public Health Engineering, Department of Civil Engineering,
University of Newcastle upon Tyne, England

TERTIARY treatment is increasingly frequently applied to normal secondary effluents to remove solids and to reduce BOD. The two processes most frequently used in Britain for large works are microstrainers and rapid sand filters. This paper reviews these processes and discusses their effectiveness and cost. The cost data are partly culled from the papers quoted in this review and partly from questionnaires which we sent to the managers of all those sewage works where we knew tertiary treatment was in operation.

MICROSTRAINERS

Microstraining was originally developed for water treatment and it was first tried for tertiary treatment at Luton in the late 1940s (PETTET *et al.*, 1949). The woven stainless steel fabric, which is made with nominal aperture sizes of 23, 35 and 65 μm, is carried around the periphery of a slowly revolving drum. The water to be treated passes from the inside to the outside of the drum, the head through the straining medium being limited to 15 cm. The mat of material which collects on the inside of the drum is washed away by high-pressure jets of clean water or strained effluent. When used for sewage treatment a zoogloeal mat is likely to form and grow on the fabric; this is prevented by a high-intensity ultra-violet lamp. [TRUESDALE and BIRKBECK (1967) showed that the lamp was effective in preventing the accumulation of biological growths, no increase in flow being observed after treatment of the fabric with sodium hypochlorite.]

The microstrainer will perform well over a very wide range of influent SS concentrations (TRUESDALE *et al.*, 1964, FIG. 7). The influent SS concentration has a small but significant influence on the concentration appearing in the final effluent. These investigators showed that in the range of influent concentrations from 25 to 120 mg l^{-1} the SS concentration in the effluent was usually less than 20 mg l^{-1} even when the flow through the drum was at a maximum. It is, however, highly relevant to the purpose of tertiary treatment that the Harpenden pilot experiment on unsettled filter effluent showed that a small residue of suspended solids (between 5 and 10 mg l^{-1}) could not be removed and always passed through the fabric.

In operation it is necessary to limit the flow so that the head lost through the fabric does not exceed about 15 cm. The concentration of suspended solids in the feed to the strainer may be expected, therefore, to affect the rate of treatment, and TRUESDALE *et al.* (1964) showed a highly significant correlation between SS content of feed and time taken to treat a given volume of liquid, the rate of treatment falling with an increase in influent solids.

Some typical results of treatment by microstrainers are given in TABLE 1. TRUESDALE *et al.* (1964) showed that SS concentration less than 10 mg l^{-1} were obtained

TABLE 1. PERFORMANCE AND COSTS OF MICROSTRAINERS

Works	Process description†	Performance‡ SS	PV	BOD	NH_3-N	Oxidized N	Surface-active matter¶	Size of installation: Max flow Av. flow 10^3 m^3 d^{-1} (m.g.d.)	Surface area m^2 (ft^2)	Cost§ (£): 1970 capital (year of construction)	Operating (p.a.)	Capital m^{-2} area	Total per unit flow £ m^{-3} (£ 1000 gal^{-1})
Harpenden (TRUESDALE *et al.*, 1964) Aug. 1962–Jan. 1963	I 330 (290) (max)							1·5 (0·3) (max)	4·5 (47)	6700 (1962)	—	1500	—
Effluent from percolating filters		39·9	16·4	16·2	9·2	33·3							
Microstrainer effluent		10·6	11·0	9·9	9·1	34·0							
Removal		73	33	40	1	−20							
Humus-tank effluent**		18·5	12·2	13·4	9·9	31·1							
Essex R.B. Works (FISH, 1966)	II (?) 300–430 (260–390)							13 (2·8) 9 (2·0)	30 (310)	40,000 (1961 ?)	1000	1300	0·0018 (0·008)
Humus-tank effluent		26·0	10·4	11·5	1·8	28·3‖	3·7						
Microstrainer effluent		14·5	8·6	8·8	1·8	28·6‖	3·7						
Removal		44	17	24		−1							
Luton (PETTET *et al.*, 1949) (June 1949)	IIA 400 (350) (max)							13·5 (3) (max)	33 (354)	42,000 (1950)	Not now in use	1300	—
Settled effluent		31	19	18									
Microstrainer effluent		7	14	9									
Removal		77	26	50									
*Luton** (1961–1962)													
Settled effluent		14	9	10									
Microstrainer effluent		8	8	8									
Removal		45	10	20									
Bracknell (CASSIDY, 1960) (Nov. 1956–Mar. 1959)	I 120–320 (100–270)							7 (1·5) 3 (0·6)	22 (235)				
Humus-tank effluent		16·2	10·5	18·5	4·1	21·2	5·9						
Microstrainer effluent		6·9	9.4	12·3	3·9	21·6	5·9						
Removal		57	10	34	5	−2							
*Bracknell** (1970–1971)													
Humus-tank effluent	150–440	20		13				29·0 (6·3)	65	117,000	1300	1800	0·045
Microstrainer effluent	(130–370)	11		9				10·0 (2·2)	(700)	(1956 and 1967)			(0·021)
Removal		45		30									
Letchworth (TRUESDALE and BIRKBECK, 1968) Apr. 1966–Nov. 1967	0 231–302									Pilot			
Act.-sludge final tank effluent	(197–257)	17·2	12·6	19·2	12·4	20·6	1·1						
Microstrainer effluent		6·6	10·6	9·5	11·6	20·8	1·1						
Removal		62	16	50	15	−1							
*G.L.C. Redbridge**	I 310–440 (270–380)							12·5 (2·7) 9·0 (2·0)	29 (310)	68,000 (1961)	1050	2300	0·003 (0·013)
Humus-tank effluent		30	9	13	1·0	3	3						
Microstrainer effluent		15	7	8	1·0	26	3						
Removal		50	20	40	—	—	—						

Hambledon R.D.C.	110–520						11·5 (2·5)	22	21,500	550	1000	0·003
*Elmbridge**	(100–450)						2·5 (0·6)	(236)	(1966)			(0·014)
Humus-tank effluent	1	14		6·7								
Microstrainer effluent		8		6·4								
Removal		45		5								
Microstrainer effluent	0	7·7		6·0								
Removal		45		10								
*Leighton–Linslade U.D.C.**	I						9·0 (2·0)	29	31,000	900	1050	0·002
	190–310						5·5 (1·2)	(310)	(1967)			(0·010
Humus-tank effluent	(160–270)	29		28	4·7							
Microstrainer effluent		11		16	4·3							
Removal		60		40	—							
Hatfield R.D.C.	I						10·0 (2·25)	33	33,500	3200	1000	0·0045
*Mill Green**	140–300						4·5 (1·0)	(354)	(1966)	(High mech.		
Humus-tank effluent	(120–260)	25/9††		19/6††						replacement		
Microstrainer effluent		12/2††		10/2††						costs)		
Removal		50		30								
*Rugby B.C.**	0						45·5 (10)	146	150,000	1600	1000	0·003
	120–310	Under construction					18·0 (4)	(1,570)	(1971)	Estimated by		(0·013)
	(100–270)									Rugby)		
*Battle R.D.C.**	I						3·5 (0·75)	10	19,000	325	1900	0·008
	100–350						1·0 (0·25)	(118)	(1964)			(0·033)
Humus-tank effluent	(90–300)	9/13	7/6·5	3/4	0·1/1·1	20/32						
Microstrainer effluent		12/7	7·5/6	2·4/3	0·1/0·8	21/31						
Removal		25/50		25	—	—						
				(2 grab samples)								
*Basingstoke**	0						14·5 (3·2)	146	138,000	1600	950	0·005
	70–100						10·0 (2·2)	(1570)	1966 and 1969)			(0·023)
Act.-sludge final tank	(60–90)											
effluent		13·1	9·8	10·4	1·6	24·4						
Microstrainer effluent		3·9	8·6	4·5	1·7	24·8						
Removal		70	10	60	—	—						
*Chertsey U.D.C.**	I						?	62	70,000	1000	1100	—
								(670)	(1967)			
Humus-tank effluent		15	16									
Microstrainer effluent		7	7									
Removal		50	55									
*Fleet U.D.C.**	I						16·5 (3·6)	57	59,000	900	1000	0·004
	100–290						5·5 (1·2)	(620)	(1967)			(0·018)
Humus-tank effluent	(90–250)	15	10									
Microstrainer effluent		6	8									
Removal		60	20									

* Data obtained from the authors' survey.
† Process description: Microstrainer fabric mark
Treatment rate $m^3\ m^{-2}\ d^{-1}$ (gal $ft^{-2}\ h^{-1}$). Average and maximum values are calculated on the total fabric area.
‡ All analyses are mg l^{-1}. Removals are % long-term average values.
§ See Appendix for details of cost analysis.
¶ As Manoxol OT.
|| NO_3-N
** The humus-tank effluent analyses are shown for comparison; the effluent was passed directly to the microstrainers without humus-tank settlement.
†† Max/min values.

about 40 per cent of the time, the influent concentration corresponding with this frequency being about 40 mg l^{-1}. On the other hand, a concentration below 5 mg l^{-1} was obtained less than 1 per cent of the time. The Harpenden works were producing a well oxidized effluent from two-stage biological filtration. In order to determine the performance of microstrainers on a poor biological effluent the order of the filters was reversed, the load being 0·39 kg BOD m^{-3} day^{-1} (0·65 lb BOD yd^{-3} day^{-1}); the BOD of the filter effluent was about 50 mg l^{-1}. As a result of slimy growths on the fabric the rate of flow rapidly dropped. As the SS concentration of the filter effluent increased from 25 to 200 mg l^{-1} the microstrained effluent correspondingly increased from 10 to 45 mg l^{-1} (TRUESDALE and BIRKBECK, 1967).

Field experiments (April 1966–March 1967) at Letchworth showed that the microstrainer was very effective in polishing well-nitrified and settled activated-sludge effluent. Throughout the year the effluent from the activated-sludge plant was of consistently high quality, the average suspended solids varying from 10 to 34 mg l^{-1}. It will be seen from TABLE 1 that the microstrainer produced a marked improvement in SS and BOD, the SS concentration exceeding 10 mg l^{-1} on only 7 per cent of occasions and being below 5 mg l^{-1} about one-quarter of the time.

The washwater will consume about 5 per cent of the flow (TRUESDALE and BIRKBECK, 1967 and 1968). Effective cleaning is important to maintain a high rate of flow through the fabric. In the field investigation at Letchworth new high-pressure cleaning jets were fitted to the microstrainer and increased the average rate of flow from 231 to 302 m^3 m^{-2} day^{-1} (197 to 257 gal* ft^{-2} h^{-1}).

The capital and operating costs of microstraining are also shown in TABLE 1. While there is considerable variation in the capital cost per unit area of filter fabric, the cost for all seven plants constructed after 1965 fall in the range £950 to £1100 m^{-2} (with average flows of 2·5–18·0 $\times$ 10^3 m^3 day^{-1}). The wide variation in the total costs per unit of flow is caused mainly by the variations in the average rates of flow per unit area of fabric at the different works.

The maximum rate of flow possible is limited by the loss of head through the fabric. This, in turn, depends on the nature of the effluent being treated, the fabric used, the efficiency of the fabric-cleaning process, and the rate of rotation of the drum. A typical maximum rate of flow is 400 m^3 m^{-2} day^{-1} (350 gal ft^{-2} h^{-1}).

The annual operating costs of a microstraining unit are of the order of 2 per cent of the capital cost. This gives a minimum total cost of £0·0009 m^{-3} (£0·004 1000 gal^{-1}). Where the microstrainer has to deal with varying flows the cost per unit average flow increases accordingly. A large decrease in cost with increase in scale cannot be expected with this kind of treatment unit, which is increased in capacity by replication of smaller units.

Where there is a high suspended-solids load (e.g. while biological filters are sloughing) it is probable that the microstrainer would not be able to pass the high flow mentioned above.

RAPID SAND FILTERS

Rapid sand filters have also been used in Britain for the last 20 yr for the tertiary treatment of sewage. Such filters are especially suitable at the larger sewage-treatment works, where skilled operation and regular maintenance can be provided. The earlier

* All gallons in this paper are Imperial gallons.

filters were downflow of the same kind as have been in use for water treatment for many years. In more recent years upward-flow (*Immedium*) filters have also been used, and anthracite/sand filters have been tested on the pilot scale. The main feature of these innovations is that the flow passes through a coarser medium before reaching the fine sand.

At Luton one of the earliest downflow sand-filter plants had sand graded 10–18 BS mesh (0·85–1·67 mm) in a bed 3 ft (1 m) deep (EVANS and ROBERTS, 1952; NAYLOR *et al.*, 1967). The settled final effluent is filtered at the rate of about 120–240 m^3 m^{-2} d^{-1} (100–200 gal ft^{-2} h^{-1}) (TRUESDALE and BIRKBECK, 1967). The grade of sand and the design rate of filtration will be governed by the size and nature of the solids to be removed, and to a lesser extent by the permissible head loss through the bed. In the early pilot investigations at Luton and Finham anthracite (8–16 mesh) was also used as filtering medium and performed satisfactorily (PETTET *et al.*, 1949), but its use has not become general. Whilst the use of anthracite as a coarse medium in the dual-media filters is increasing in water-treatment practice, it appears to have no advantage over sand as a filter medium, and is used only because of its low density which causes it to remain above a finer sand after backwashing (TEBBUTT, 1971).

The filter must be backwashed with filtrate; this may be expected to amount to some 2·5 per cent of the volume filtered and must be returned to the works inlet for further treatment (TRUESDALE and BIRKBECK, 1967). Some trouble has been experienced in keeping the sand uniformly clean after backwashing; the "mud balls" which form can gradually spread until disruption occurs with considerable loss of sand (NAYLOR *et al.*, 1967).

The first full-scale *Immedium* upward-flow filter was commissioned at the East-Hyde works of Luton B.C. in mid-1969. It was designed to treat up to 68,000 m^3 day^{-1} (15 m.g.d.) to achieve an effluent standard of BOD 7, suspended solids 7 mg l^{-1}. Pilot trials at Luton (NAYLOR *et al.*, 1967) had shown that the solids in the feed to the filter were glutinous and were not easily removed by normal backwashing, even with air scour. Pulsating air scour and backwashing was, therefore, adopted. This plant removes 1 ton day^{-1} of dry solids (BARRETT, 1971). Additional differences between the upflow and downflow filters are that the upflow filter often has a somewhat greater depth of sand than the downflow [5 ft (1·6 m) at Luton], that the sand may be "retained" during normal filtration by a grid close to the surface of the sand, and by the useful fact that unfiltered water is used for backwashing. It is also claimed that passage of the plant effluent to be treated through the underbed induces mechanical flocculation of the solids, the washwater settling very rapidly (BOBY and ALPE, 1967). NAYLOR *et al.* (1967) suggest that—presumably with a well-nitrified effluent (see, for example, discussion by C. D. FURNESS on WOOD *et al.*, 1968)—it might be possible to dispense with humus tanks before treatment in upflow filters. Results of rapid sand filtration are given in TABLE 2.

A continuously backwashed, radial-flow filter has been developed in Hungary and is marketed in the U.K. as the *Simater* filter by Simon–Hartley Ltd. of Stoke-on-Trent. A type-110 *Simater*, with an output of 520–640 m^3 d^{-1} (0·11–0·14 m.g.d.) and the production of 52–64 m^3 day^{-1} (0·011–0·014 m.g.d.) sludge, was investigated at Derby between May and October 1968. In the first series of tests the sand was quite fine, being graded 0·5–1 mm; in the second series a coarser sand (1–2 mm) was used. The filter is very effective in removing suspended solids, removals of 70–90 per cent being

TABLE 2. PERFORMANCE AND COSTS OF RAPID SAND FILTERS

		Performance‡						Size of installation		Cost§ (£)			
Works	Process description†	SS	PV	BOD	NH_3-N	Oxidized N	Surface-active matter¶	Max flow Av. flow 10^3 m^3 d^{-1} (m.g.d.)	Surface area m^2 (ft^2)	1970 capital (year of construction)	Operating (p.a.)	Capital m^{-2} area	Total per unit flow £ m^{-3} (£ 1000 gal^{-1})
Luton (EVANS and ROBERTS, 1952) (May 1951–Mar. 1952)	DS 0·85–1·7 mm (1 m) 115–250 (100–200)							45·5 (10) 20·0 4·3	180 (1900)	240,000 (1950 and 1953)	4000	1300 (Includes pumping main flow)	0·004 (0·020) (Prototype filters)
Humus-tank effluent		15·4	9·8	13·0									
Sand-filter effluent		4·2	7·7	6·7									
Removal		73	21	48									
*Luton** (Sept. 1969–Apr. 1970)	US 0·85–1·7 mm (1·5 m) 225–300 (100–260)							70·0 (15·5) 29·0 (6·3	230 (2500)	217,000 (1969)	4000	940 (Includes pumping main flow)	0·0025 (0·012)
Humas-tank effluent		30	12	25	4·0	18							
Sand-filter effluent		5	8	10	3·0	18							
Removal		80	25	60	25	0							
Ashton-under-Lyne (MICHAELSON, 1970)	US Pilot 350–370 (300–320)							——	——	Pilot	——	——	——
Humus-tank effluent		34·3											
Sand-filter effluent		8·6											
Removal		75	25	30									
Letchworth (TRUESDALE and BIRKBECK, 1968) (Apr. 1966–Mar. 1967)	US 1–2 mm 310 (264)							——	——	Pilot	——	——	——
Act.-sludge final tank effluent		17·2	12·6	19·2	12·4	20·6	1·1						
Sand-filter effluent		6·9	9·6	9·0	10·6	20·3	1·0						
Removal		60	24	53	15	2	9						
*Upper Stour Main Drainage** *Freehold Works*	DS 0·5–2·5 mm 70–140 (60–120)							27·0 (6·0) 13·5 (3·0)	190 (2050)	92,000 (1966)	3000	480 (No pumping)	0·0025 (0·012) (Trouble with fine sand in washwater)
Act.-sludge final tank effluent		12											
Sand-filter effluent		5											
Removal		60											

*Cheltenham**	DS 70–120 (60–180)		Only recently commissioned—no results to date					55·0 (12) 18·0 (4)	260 (2800)	210,000 (1970)	Not yet known	810 (Includes pumping main flow)	0·0045 (0·020) (Based on est. operating cost of £6000 p.a.)
*Harpenden U.D.C.**	DS 1·1 mm							14·5 (3·2) 5·0 (1·1)	85 (900)	39,000 (1967	1500	460 (No pumping)	0·003 (0·015)
Humus-tank effluent	60–170	20		5	0·3	38							
Sand-filter effluent	(50–150)	5		5	0·3	34							
Removal		75		0		—							
*Swindon** *Rodbourne Works*	DS 1·5–3 mm 90–180 (80–160)							66·0 (14·5) 33·0 (7·2)	350 3800)	210,000 (1968)	8000 (Estimated by Swindon)	600 (Includes pumping main flow)	0·0025 (Biological filters produce poor winter effluent–flow by-passed)
Humus-tank effluent		21	10	18	9	14	—						
Sand-filter effluent		5	7	6	8	12	—						
Removal		75	30	65	10	15	—						
*Oxford**	DS 130–190 (110–165)		Only recently commissioned—no results to date					34·0 (7·5) 23·0 (5·0)	180 (1900)	90,000 (1970)	Not yet known	500 (No pumping)	0·0015 (0·007) (Based on estimated operating cost of £3000 per annum)
*Derby** (JOSLIN and GREENE, 1970)	DS *Pilot*							136·0 (30) 91·0 (20)	360 (4000)	135,000 (1970) (Excluding pumping) 300,000 (1970) (Including pumping)	Not yet known	400 (No pumping) 830 (Includes pumping main flow)	0·0015 (0·006) (Based on estimated operating cost of £9000 per annum)
Humus-tank effluent	1–2 mm	22		20									
Sand-filter effluent	230–350	9		9									
Removal	(200–300)	60		55									

* Data obtained from the authors' survey.
† Process description: D, downflow S, sand
U, upflow A, anthracite
Sand grading (depth).
Treatment rate $m^3\ m^{-2}\ d^{-1}$ (gal $ft^{-2}\ hr^{-1}$).
‡ All analyses are mg l^{-1}. Removals are % long-term averages.
§ See Appendix for details of cost analysis.
¶ As Manoxol O.T.

obtained. Unfortunately the rates of filtration are relatively low, being equivalent to about 94 m^3 m^{-2} day^{-1} (80 gal ft^{-2} h^{-1}) with the finer sand. The coarser sand gave an output of 115 m^3 m^{-2} day^{-1} (100 gal ft^{-2} h^{-1}) without any marked deterioration in the quality of the filtrate. In addition the washwater sludge represents about 10 per cent of the throughput, which is very high. However, this was probably a feature of the particular unit which had an underloaded separation zone. A value of 5 per cent is usual for this filter. The capacity of the largest unit at present commercially available is only 1400 m^3 day^{-1} (0·3 m.g.d.). It was decided, therefore, that this filter was not economically feasible for the scale of operation envisaged at Derby (JOSLIN and GREENE, 1970). The 1400 m^3 day^{-1} (0·3 m.g.d.) filter unit costs £5500.

The data from TABLE 2 suggest that conventional rapid sand filters, working at a filtration rate of 200 m^3 m^{-2} day^{-1} (260 gal ft^{-2} h^{-1}) and costing £500 m^{-2} would cost £3500 neglecting the effect of scale. The continuous-flow filter provides considerable scope for development and it is likely that the production of the 4550 m^3 day^{-1} (1 m.g.d.) unit at Stoke-on-Trent will reduce the difference in costs especially for small installations. A continuous plant is currently being compared with upflow filters on Thames side and the results of this investigation should be of considerable value.

There is evidently some biological action in a rapid sand filter, since the water usually loses oxygen in passage through the sand bed (WOOD *et al.*, 1968). (In the Monsanto plant, however, serious deoxygenation did not occur, possibly owing to bactericidal or bacteriostatic material in the chemical effluent—WILSON, 1960.) That such biological action has little effect on other quality parameters is evident from a comparison of the Letchworth results in TABLES 1 and 2; the removal of ammoniacal nitrogen is no more in the rapid filter than in the microstrainer and the BOD removal is closely similar.

The evidence from the pilot investigation of WOOD *et al.* (1968), however, points much more strongly to the effectiveness of biological treatment at relatively low filtration rates in removing BOD in upflow filters. The pilot *Immedium* filter was run at four filtration rates, 130–360 m^3 m^{-2} day^{-1} (108–300 gal ft^{-2} h^{-1}). By comparison of TABLES 1 and 2 in these investigators' paper it can be seen that the ratio of BOD:SS removed is high at low filtration rates and rapidly drops above 230 m^3 m^{-2} day^{-1}, as shown below:

Filtration rate (m^3 m^{-2} day^{-1})	Ratio $\frac{\text{BOD removed}}{\text{SS removed}}$	BOD removed (%)
130	1·27	93
230	1·44	91
290	0·75	87
350	0·73	58

This effect is probably due to a combination of improved flocculation and removal conditions at the lower rates enabling the more "biological" solids to be removed and an increased effect of any biological activity. This deoxygenation must be kept in mind and steps may be required to reoxygenate the filtered effluent. WILSON (1960) provided a Kessener brush for reaeration, which has proved unnecessary, and WOOD *et al.* (1968) constructed a small cascade aerator to reoxygenate the filtrate from their pilot plant.

A bacteriological investigation was carried out by ALLEN *et al.* (1949) in association with the early pilot tests on sand filters at Luton and Shenley. This investigation showed that as the filters became blocked, up to 89 per cent of the bacteria (plate counts at 20°C) could be removed by the sand filters and up to 87 per cent by the anthracite filters. Nevertheless, very large bacterial populations remained in the filtrate, the sand filtrate containing on average 400,000 coliforms 100 ml^{-1}, 366,000 faecal coliforms 100 ml^{-1} and 19,000 faecal streptococci 100 ml^{-1} (ALLEN *et al.*, 1949). It is clear, therefore, that even sand filtration cannot be relied upon to remove substantially all the bacteria from a sewage effluent.

The cost of sand filters is affected considerably by the need for pumping the main flow. A head loss of the order of 3 m (9 ft) occurs through the unit and not many plants have sufficient fall for this to be provided by gravity. The cost data summarized in TABLE 2, although limited, indicate a cost of £400–500 m^{-2} where pumping is not required, and £600–940 m^{-2} for recent plants requiring pumping. The filtration rates in current use are substantially less than the 600 $m^3\ m^{-2}\ day^{-1}$ used in the pilot tests at Minworth (biological filtration plant) (TEBBUTT, 1971). From the Minworth results it was concluded that the efficiency of the filter does not appear to be greatly affected by the filtration rate within the range 100–600 $m^3\ m^{-2}\ day^{-1}$. The results of the work of JOSLIN and GREENE (1970), however, do show a slight reduction in efficiency at the higher rates, but it would appear that significant reduction in costs could be obtained by working at these higher filtration rates. For a maximum filtration rate of 400 $m^3\ m^{-2}\ day^{-1}$ (350 gal $ft^{-2}\ h^{-1}$) and annual operating costs equal to 3 per cent of the total capital costs the minimum total cost is £0·0004 m^{-3} (£0·002 1000 gal^{-1}) where no pumping is required (£450 m^{-2}) and £0·0008 m^{-3} (£0·0035 1000 gal^{-1}) if pumping is necessary (£800 m^{-2}). These costs are significantly less than the microstrainer costs quoted earlier. It is appreciated that the size of the sand-filter plants studied were on average considerably larger than the microstrainer unit. The decrease in unit cost with increasing scale was not very marked and would not be expected to make much difference to the comparison. Costs of treating the washwater have not been compared, but again are unlikely to be significantly different.

In spite of the high costs of rapid gravity sand filters it is probable that, except where land is very plentiful for the construction of lagoons, rapid sand filtration is the only tertiary treatment that can be relied upon to achieve standards consistently more stringent than 10/10. As may be seen from TABLE 2 rapid filtration can remove 60–75 per cent SS and about one-half of the BOD; TCHOBANOGLOUS (1970) has shown that substantially higher removals (up to 95 per cent SS) may be achieved with the aid of polyelectrolytes.

Acknowledgements—The figures given in the tables were those received in answer to our questionnaire by the middle of 1970; they do not claim to cover all the plants in Britain. We should like to express our thanks to the Water Resources Board, which invited us to undertake a survey of the further treatment of sewage effluents. We also owe very real thanks to the many borough engineers and plant managers, consulting engineers and equipment manufacturers who were so helpful in providing us with the information upon which this paper is based.

REFERENCES

ALLEN L. A., BROOKS E. and WILLIAMS I. L. (1949) Mechanical filtration of sewage effluents—II. Effect on the bacterial population of treatment of sewage effluents in sand or anthracite filters. *J. Proc. Inst. Sew. Purif.* 1949, 411–15.

BARRETT A. D. (1971) Immedium filter in tertiary treatment. *Process Biochem.* **6,** 33–5.

BOBY W. M. T. and ALPE G. (1967) Practical experiences using upward flow filtration. *Proc. Soc. Wat. Treat. Exam.* **17,** 215–30.
CASSIDY J. E. (1960) The operation of the Hazelwood Lane works of the Bracknell Development Corporation. *J. Proc. Inst. Sew. Purif.* 1960, 276–93.
EVANS S. C. and ROBERTS F. W. (1952) Twelve months' operation of sand filtration and microstraining plant at Luton. *J. Proc. Inst. Sew. Purif.* 1952, 333–41.
FISH H. (1966) Some investigations of tertiary methods of treatment. *J. Instn pub. Hlth Engrs* **65,** 33–47.
JOSLIN J. R. and GREENE G. (1970) Sand filter experiments at Derby. *Paper to Northern Branch, Institute of Water Pollution Control,* 13 January 1970.
MICHAELSON A. P. (1970) Under the solids limit at Ashton-under-Lyne. *Paper to Northern Branch, Institute of Water Pollution Control.*
NAYLOR A. E., EVANS S. C. and DUNSCOMBE K. M. (1967) Recent developments on the rapid sand filters at Luton. *Wat. Pollut. Control* **66,** 309–20.
PETTET A. E. J., COLLETT W. F., and SUMMERS T. H. (1949) Mechanical filtration of sewage effluent. I. Removal of humus. *J. Proc. Inst. Sew. Purif.* 1949, 399–411.
TCHOBANOGLOUS G. (1970) Filtration techniques in tertiary treatment. *J. Wat. Pollut. Control. Fed.* **42,** 604–23.
TEBBUTT T. H. Y. (1971) An investigation into tertiary treatment by rapid filtration. *Water Research* **5,** 81–92.
TRUESDALE G. A. and BIRKBECK A. E. (1967) Tertiary treatment processes for sewage works effluent. *J. Instn pub. Hlth Engrs* **66,** 82–97.
TRUESDALE G. A. and BIRKBECK A. E. (1968) Tertiary treatment of activated-sludge effluents. *Wat. Pollut. Control* **67,** 483–92.
TRUESDALE G. A., BIRKBECK A. E. and SHAW D. (1964) A critical examination of some methods of further treatment of effluents from percolating filters. *J. Proc. Inst. Sew. Purif.* **63,** 81–95.
WILSON I. S. (1960) The treatment of chemical wastes. In *Waste Treatment* (Edited by ISAAC P. C. G.) Pergamon Press, Oxford, 206–25.
WOOD R., SMITH W. S. and MURRAY J. K. (1968) An investigation into upward-flow filtration. *Wat. Pollut. Control* **67,** 421–8.

APPENDIX

Basis of cost analysis

a. *Cost Index.* The 1970 cost of existing facilities was obtained by using a cost index obtained from published construction statistics prepared by the Department of the Environment. An extract is detailed below.

Year of construction	Cost Index
1960	156
1962	149
1964	136
1966	124
1968	113
1970	100

b. *Annual capital repayments and loan charges.* Equal annual charges on capital were assumed. Amortization periods of 30 years for civil engineering structures and 15 years for mechanical plant were used. Loan charges of 9 per cent on the total capital were used.

For the *rapid gravity sand filters,* analysis of the breakdown of costs indicated that about 50 per cent of the total capital cost could be classified as civil, giving an *annual total charge on capital of* 11 *per cent.*

For the *microstrainers* a similar analysis indicated civil costs of 25 per cent of the total capital cost, *giving an annual total charge on capital of* 12 *per cent.*

Water Research Pergamon Press 1972. Vol. 6, pp. 475–476. Printed in Great Britain

DESIGN OF TREATMENT FACILITIES FOR THE CONTROL OF PHOSPHORUS

F. M. MIDDLETON

Advanced Waste Treatment Research Laboratory, EPA, R.A. Taft Water Res. Center, 4676 Columbus Parkway Cincinnati, Ohio 45226, U.S.A.

IN THE United States regulatory bodies are setting standards for phosphorus in receiving water as low as 0.2 mg l^{-1} (as P) and quite often at 1 mg l^{-1}, levels. Conventional biological treatment will seldom remove phosphorus below a 4–7 mg l^{-1} concentration. Raw sewages in the United States contain 10–12 mg l^{-1} of phosphorus. Plant designers have several choices to treat for phosphorus removal and the choice depends upon the type of plant and quality of effluent to be achieved.

Biological removal capabilities of a plant should be used to the maximum extent but in nearly all cases chemical treatment will have to be used. Lime, alum and iron compounds are effective in precipitating phosphorus and are common chemicals that are widely available. Iron compounds in waste pickle liquor from steel works are finding a place in the treatment scheme.

Phosphorus can be controlled at a trickling filter plant by adding chemicals to the primary treatment units, to the final settler, or at multipoints in the system. Polymers for added aid in coagulation and settling are used to supplement the metal salts. Phosphorus removal concurrently enhances removal of colloidal solids and hence BOD and solids removal is improved. Facilities for phosphorus control are easily added to existing plants. Chemicals are the major costs associated with the process. A 30-s flash mix, followed by a 1–5 min high energy flocculation and a 5–20 min low energy flocculation achieves good results.

Important parameters for successful removal of phosphorus include:

1. Assess the phosphorus concentration in and out hourly.
2. Set the mole ratio of the chemical used at the proper dosage, usually 2:1 or 1.5:1.
3. Polymer dosages are usually less than 1 mg l^{-1}.
4. Supernatant streams should be treated separately with alum or lime.
5. The weight of sludge may double but not the volume. Digestion and gas production remain good.
6. Sludges can be handled and disposed of in normal ways.
7. Costs are about 5¢ 1000 gal^{-1} (80 per cent removal). Higher removal costs more.
8. In some cases the levels of iron remaining in the effluent are higher than desired when iron is added to the final tank.

In activated sludge plants chemicals can be added at the primary stage, in the aeration tank (preferrably at the end of the process) or the treatment can be applied following full secondary treatment. Schematic representations of these control systems are shown in FIG. 1. When chemicals are added in the process train several advantages are seen:

1. Mixing and residence times are inherent in the process.
2. The final settler serves a dual purpose.
3. Biological flocculation aids in settling and chemicals help in settling rate.
4. Supernatant phosphorus is held by chemicals.

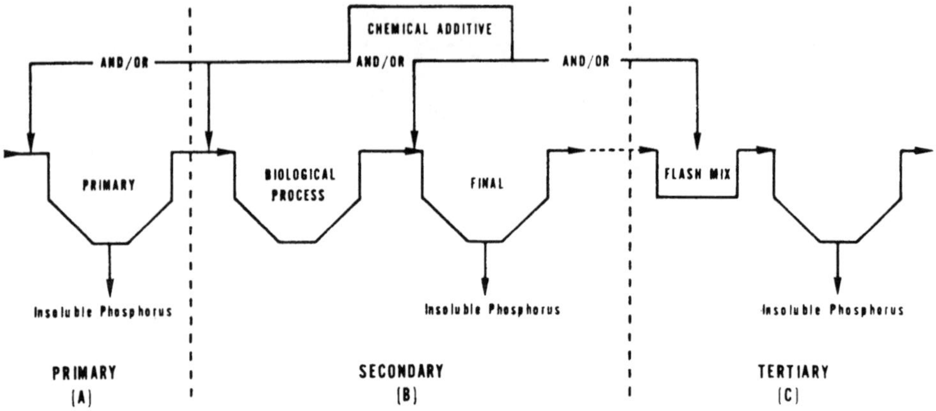

FIG. 1. Phosphorus removal systems.

Some disadvantages accrue from chemical addition in the form of added ions such as sulfate, chloride, or sodium. Lime used in the primary stage or after settling can be recycled and reused in part.

Phosphorus removal in conventional activated sludge plants will increase the cost as follows:

1 mgd plants	25 per cent
10 mgd plants	28 per cent
100 mgd plants	42 per cent

Several cities in the United States have phosphorus removal facilities either running or under design.

Water Research Pergamon Press 1972. Vol. 6, pp. 477–479. Printed in Great Britain

OPERATING EXPERIENCE AND RESULTS USING THE SIMULTANEOUS PRECIPITATION OF PHOSPHATES IN ACTIVATED SLUDGE PLANTS FOR 5000 TO 30,000 INHABITANTS IN THE CANTON OF ZURICH

P. WILDI

Gewässerschutzabteilung Kanton Zürich, Kantonale Baudirektion, CH-8090 Zürich, Switzerland

IN THE Canton of Zurich, all communities situated within the drainage area of a lake (Zürichsee, Greifensee and Pfäffikersee) have been obliged by the cantonal authorities since 1967, to equip the sewage treatment plants (apart from the primary and the biological treatment) with tertiary treatment in order to eliminate phosphates.

According to the Federal Standards of 1 September, 1966, the phosphate content in the effluent shall not exceed 2 mg l^{-1} PO_4^{3-}. In general elimination of 85 per cent will be required.

All of the 18 plants actually in operation with phosphate elimination in the Canton of Zurich have adopted the simultaneous precipitation process. This consists of adding a solution of ferric chloride (available with 11 per cent and with 14 per cent of Fe^{3+}) to the influent or the effluent channel of aeration tanks.

Preliminary tests and experiments have shown that, in general, a dosage of ferric chloride corresponding to 10 mg Fe^{3+} l^{-1} sewage should be sufficient to obtain the required effect.

The Chemische Fabrik Uetikon, producer of the ferric chloride solution has, since February 1971, tested the use of aluminium sulfate as precipitant with apparently the same effect as when ferric chloride was used. This firm gives the following relation of Fe to Al: 10 g Fe^{3+} corresponds to 71·5 g Fe Cl_3 with 14 per cent Fe or 46 ml solution or 53·6 g of aluminium sulfate with 17 per cent Al_2O_3 or 115 g of a 47 per cent solution of aluminium sulfate or 4·8 g aluminium. As we have no experience and no practical results of our own available concerning the recent use of aluminium we are unable to submit further information about it.

Simultaneous precipitation consists generally of a storage tank (glass fibre reinforced plastic or coated concrete) and a dosing device (FIGS. 1 and 2).

In smaller plants the dosing rate is continuous and is spread over a period of 24 h.

In other plants there are 2 stages (a higher rate during the daytime and a lower one at night). Some of the larger plants use 2 or 3 different specific dosing rates over a period of 24 h and the dosed quantity is automatically adapted to the dry weather flow.

In plants with a large volume of industrial wastes, the dosing rate is adapted to the fluctuating daily phosphate load over the week.

Exact adaptation to the sewage flow is not required. Continuous recirculation of return sludge has a compensating effect and peaks are equalized.

The installation costs vary from sFr. 10,000 up to sFr. 50,000 (storage tanks, dosing pumps, pipes, control systems etc.). This corresponds to 0·5–1·7 per cent of the total costs of the treatment plants.

The current consumption required for operation can be ignored, and maintenance and personnel requirements are not important.

Based on a dosing rate of 10 mg Fe l^{-1}, the theoretical costs of ferric chloride are nearly sFr. 12 1000 m^{-3} of sewage or sFr. 1·70–2·50 per inhabitant and year. For 7 plants (5000–30,000 inhabitants) the following effective costs for ferric chloride could be estimated for 1970: sFr. 9·60 1000 m^{-3} or 2·40 per inhabitant per year (varying from 1·40 to 3·10).

In the early stages the operation of precipitation was prejudiced for different reasons; for example: troubles with the dosing devices; clogging of valves and pipes due to impurities in the ferric chloride solution or due to low temperatures causing crystalization of the solution; the ferric chloride was not mixed intensively and rapidly enough with the sewage flow or, on the other hand, the initial flocculation was minimized by a too high turbulence in a centrifugal pump; industrial wastes which brought peak loads of phosphates (especially undesired are polyphosphates), or other harmful compounds (e.g. sulphide, giving ferri sulphide); and plant operators reduced and still tend to reduce the dosing rate to too low a level in order to save money.

The results for three municipal plants, one treating only municipal sewage, one municipal sewage and relatively harmless industrial wastes and one treating municipal sewage and a considerable volume of industrial wastes are given below:

TABLE 1. ACTIVATED SLUDGE PLANT, TREATING DOMESTIC SEWAGE OF NEARLY 7000 INHABITANTS

Results for 1970		Average	Variation
Raw sewage	mg l^{-1} PO_4^{3-}	20·6	6·0–33
Treated effluent	mg l^{-1} PO_4^{3-}	0·64	0·2– 2·0
Removal		97 per cent	
Dosing rate	mg Fe l^{-1}	7·7	6·8–10·2

This example shows that for domestic sewage, a dosage corresponding to 10 mg Fe^{3+} l^{-1} of sewage is more than sufficient.

TABLE 2. ACTIVATED SLUDGE PLANT TREATING THE SEWAGE OF 10,000 INHABITANTS AND A CERTAIN VOLUME OF (WITH REGARD TO RELATIVELY HARMLESS PHOSPHATE ELIMINATION) INDUSTRIAL WASTE

Results for 1970		Average	Variation
Raw sewage	mg l^{-1} PO_4^{3-}	16·1	6·0–34
Treated effluent	mg l^{-1} PO_4^{3-}	2·2	0·3–6·0
Removal		86 per cent	
Dosing rate	mg Fe l^{-1}	6·4	4·1–8·6

In this plant the dosing rate of only 6·4 mg Fe^{3+} l^{-1} of sewage was too little. The average rate ought to be increased to 10 mg Fe l^{-1}.

FIG. 1. Storage tanks (2·5 m^3 each) for ferric chloride and dosing pumps in an activated sludge plant for 5000 inhabitants.

FIG. 2. Storage tank (14 m^3), small tank (1 m^3) for one day's supply together with dosing pumps in an activated sludge plant for 20,000 inhabitants.

(*Facing p.* 478)

TABLE 3. ACTIVATED SLUDGE PLANT DEALING WITH THE SEWAGE OF 22,000 INHABITANTS AND A CONSIDERABLE VOLUME OF INDUSTRIAL WASTES. THE PRECIPITATION PROCESS IS PREJUDICED BY THE PRESENCE OF WASTES FROM TEXTILE INDUSTRIES, ESPECIALLY POLYPHOSPHATES. RESULTS FOR THE PERIOD MAY 1970 TO APRIL 1971

		Average	Variation
Effluent primary sedimentation	mg l^{-1} PO_4^{3-}	15·4	3·0–22
Treated effluent	mg l^{-1} PO_4^{3-}	3·1	1·3–6·0
Removal		80 per cent	
Dosing rate	mg Fe l^{-1}	11·8	7·7–15·6

In spite of the higher dosage, the results are not always satisfactory. Therefore the average rate needs to be increased to 15 mg l^{-1} Fe or to a dosing rate corresponding to 1 mg l^{-1} Fe per mg l^{-1} PO_4.

In the analysed effluent of 9 plants, the average Fe-content was 0·21 mg Fe^{3+} l^{-1}, varying from 0·08 to 0·56.

The simultaneous precipitation has no effect on the sludge treatment. Neither an increase of sludge quantity nor an influence on sludge digestion could be observed.

CONCLUSION

The simultaneous precipitation with ferric chloride has proved to be a suitable process for the elimination of phosphates with low installation costs. In order to obtain satisfactory and regular results the following factors have to be observed:

the dosing rate has to be sufficient and appropriate to the type of sewage (harmless domestic sewage or sewage with an important amount of industrial wastes), and to the daily phosphate load. This demands preliminary investigation and careful tests during the initial stage;

the right place for the addition of the precipitant to the influent or the effluent channel of the activated sludge tanks has to be tested in each case, particularly in regard to the essential intensive and rapid mixing with the sewage flow;

storage tanks and dosing devices should be protected from very cold weather and are therefore situated with advantage within a room. If not, the ferric chloride should be diluted down below 10 per cent Fe during the wintertime.

Water Research Pergamon Press 1972. Vol. 6, pp. 481–483. Printed in Great Britain

DESIGN OF TREATMENT FACILITIES FOR THE CONTROL OF NITROGENOUS MATERIALS

EDWIN F. BARTH

Environmental Protection Administration (EPA), R.A. Taft Water Res. Center, Advanced Waste Treatment Research Laboratory, 4676 Columbus Parkway, Cincinnati, Ohio 45226, U.S.A.

IN MANY receiving waters of the United States, it is becoming increasingly important to control either the qualitative or quantitative composition of effluents in regard to nitrogenous substances. Due to the high oxygen demand of ammonia nitrogen, it is desirable to produce a nitrified effluent, before discharge, in a great many watersheds. To a much lesser extent, it is necessary to provide for nitrogen removal in those cases where eutrophication or health hazards are a problem.

From pilot plant (BARTH *et al.*, 1968) and field evaluation studies (MULBARGER, 1970), a multiple-stage, combined chemical-biological process is being designed for several large United States facilities. The general design of such a process is shown on FIG. 1. The essential features of design are the stage concept, which essentially isolates each biological system so that optimum design conditions can be employed to maximize efficiency of each biological transformation. Secondly, experience has shown that

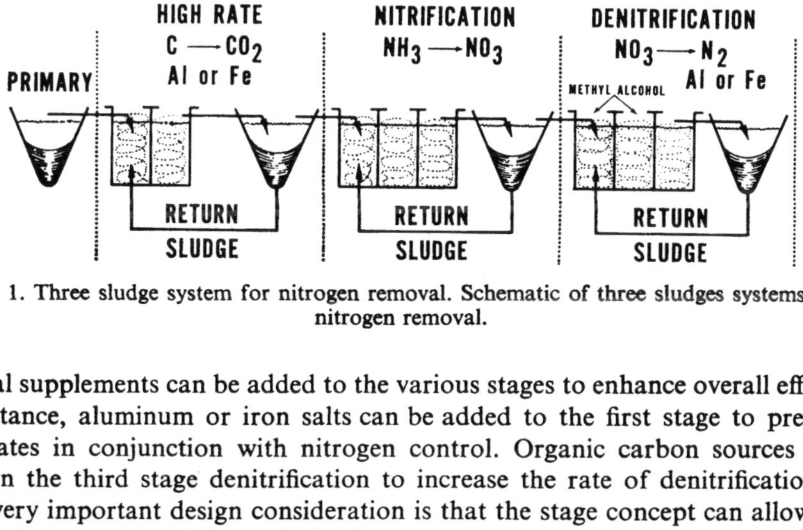

FIG. 1. Three sludge system for nitrogen removal. Schematic of three sludges systems for nitrogen removal.

chemical supplements can be added to the various stages to enhance overall efficiency. For instance, aluminum or iron salts can be added to the first stage to precipitate phosphates in conjunction with nitrogen control. Organic carbon sources can be added in the third stage denitrification to increase the rate of denitrification. And a last, very important design consideration is that the stage concept can allow initial construction to proceed through the first two stages for nitrification, and the third denitrification stage need not be added until a nitrogen removal requirement is definitely established.

This type design, employing suspended growth reactors, is now in the detailed design phase for Washington, D.C., 1,137,000 m^3 day^{-1} (300 mgd); Ft. Worth, Texas, 360,050 m^3 day^{-1} (95 mgd); Salt Creek, Illinois, 189,500 m^3 day^{-1} (50 mgd); Tampa, Florida, 189,500 m^3 day^{-1} (50 mgd); and Hobbs, New Mexico, 18,950 m^3 day^{-1} (5 mgd).

TABLE 1. PILOT PLANT RESULTS

Process stream	COD	S.S.	Total P	Org. N	NH₃ N	NO₂ N	NO₃ N	Total N
	mg l^{-1}							
Raw wastewater	320	157	12.6	10.3 +	11.3 +	—	— =	21.6
Primary effluent	218	90	11.9	5.9 +	13.7 +	—	— =	19.6
High-rate effluent	64	9	2.8	0.8 +	7.7 +	1.1 +	4.3 =	13.9
Nitrified effluent	43	7	2.6	0.4 +	0.6 +	0.3 +	11.5 =	12.8
Final (denitrified) effluent	44	7	1.5	0.4 +	0.3 +	0.3 +	0.9 =	1.9
Percentage removal	86	95	88					91

TABLE 2. RESULTS OF LARGE SCALE DEMONSTRATION MANASSAS, VIRGINIA

Final settler	Constituent	Muti-media effluent
	Effluent residuals, mg l^{-1}	
6	Suspended solids	1.3
6	BOD_5	1.1
22	COD	16
0.5	Total phosphorus	0.3
1.5	Organic-N	1.2
0	NO_2-N	0
0.9	NO_3-N	0.8

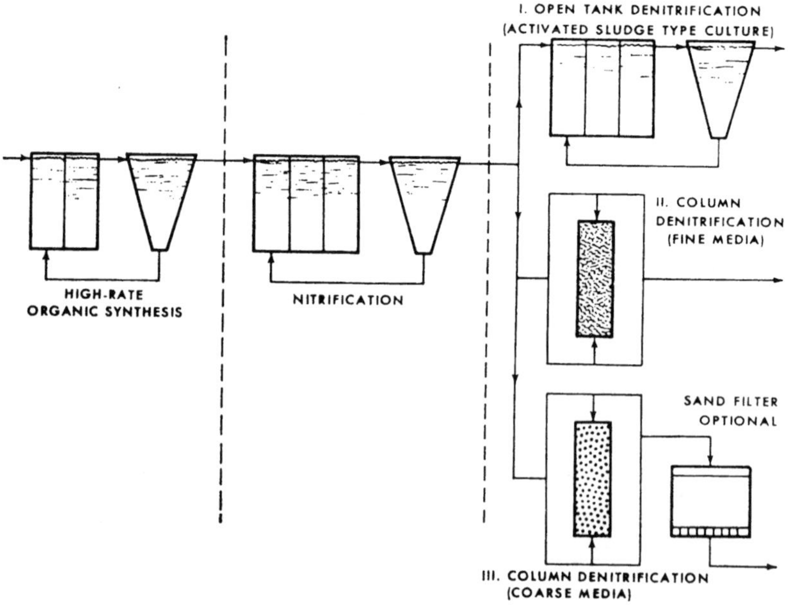

FIG. 2. Modifications of the dentrification process.

With a design of this type, modifications to suit specific needs are possible and we have an active research program investigating alternates. At Hatfield Township, Pennsylvania, 18,950 m^3 day^{-1} (5 mgd), the first-stage, high-rate biological carbon removal system is replaced by a lime coagulation in a solids contactor at pH 9.5. The second stage will be a suspended growth nitrification reactor.

At Midland, Michigan, 18,950 m^3 day^{-1} (5 mgd), a field-scale pilot plant is being used to evaluate the use of plastic trickling filter media as an attached growth process for second-stage nitrification. This is an important design consideration because treatment sites with limited land area could conserve space by use of vertical treatment units. Since the isolated second stage has a low net cell synthesis, the use of the attached growth process may eliminate the final settler in a two-stage nitrification design, if a filtration step is planned for a high quality effluent.

We have also investigated the use of pure oxygen, in place of atmospheric oxygen, in two-stage nitrification. Oxygen permits the use of shorter hydraulic detention times for efficient nitrification.

Alternate design modifications for the third-stage denitrification, if nitrogen removal is necessary, are shown on FIG. 2. Suspended growth reactors, fine-grain media columns (2–4 mm particles), and course media (2·5 cm rock) have all been shown to perform adequately. Each type of design has advantages and disadvantages which require engineering judgment for selection. In general, the suspended growth reactors would be favored for treatment facilities larger than 10,000 m^3 day^{-1} (3 mgd).

REFERENCES

BARTH E. F., BRENNER R. C. and LEWIS R. F. (1968) Chemical–biological control of nitrogen and phosphorus in wastewater effluent. *J. Wat. Pollut. Control Fed.* **40**, 2040.

MULBARGER M. C. (1970) Modifications of the activated sludge process for nitrification and denitrification. 43*rd Annual Conference, Water Pollution Control Federation*, Boston, Mass.

Water Research Pergamon Press 1972. Vol. 6, pp. 485–486. Printed in Great Britain

THE ELIMINATION OF NITROGEN IN THE TREATMENT PLANT OF VIENNA–BLUMENTAL

N. MATSCHÉ

Institut für Wasserversorgung, Abwasserreinigung und Gewässerschutz, Technische Hochschule Wien, Karlsplatz 13, A-1040 Wien, Austria

NITROGENOUS materials can be eliminated from sewage by physical, chemical or microbial processes. WUHRMANN (1964) and BRINGMANN (1962) have designed sewage treatment systems which take the microbial reaction into account: a heavily aerated nitrification basin is followed by a denitrification basin without aeration. BARTH (1971) refers to a similar system in which he uses methanol as the hydrogen donor. According to PASVEER (1964) nitrification and denitrification can occur simultaneously in the oxidation ditch.

Inspired by the paper by BARTH (1971) we obtained the following results in a 48-h experiment in the treatment plant of Blumental (2 aeration basins in series, first basin 4 rotors, second basin 2 rotors). The mean values of the loads are tabulated.

	Raw sewage ($mg\ l^{-1}$)	Effluent ($mg\ l^{-1}$)	Per cent removal
BOD_5	257	13	95
COD	475	50	90
TOC	153	14	91
Org.$^-$N	13·8	0·4	
NH_4^+N	21·4	3·8	
NO_2^-N	0·2	0	
NO_3^-N	0·3	0	
Total-N	35·7	4·2	88

The nitrogen loads of raw sewage, effluent and of the waste sludge are shown in the following balance.

Raw sewage − effluent − waste sludge = denitrified

1265 kg N day^{-1} − 144 kg N day^{-1} − 353 kg N day^{-1} = 768 kg N day^{-1}.

Transferred to a flow scheme of the treatment plant the elimination of nitrogen can be illustrated as follows (Fig. 1.).

HÜHNERBERG and SARFERT (1967) were not able to obtain a change of the nitrification-rate within oxygen concentrations of 2–6 mg l^{-1}. They could however get nitrification using mammoth-rotors at oxygen tensions as low as 1 mg l^{-1}. In our experiment we operated between 0·2–1·5 mg O_2 l^{-1} which seems to be high enough to get nitrification. The sludge load in the amount of approximately 0·1 kg kg^{-1} day^{-1} could be maintained in an optimal region as well as the temperature of 18–19°C.

According to HÜHNERBERG and SARFERT (1967) denitrification is mainly influenced by the oxygen concentration in the aeration basin. Because of the two aeration basins operated in series with 4 and 2 rotors in action, especially in some parts of the second basin the oxygen tension became very low which resulted in a complete denitrification of the effluent. The absence of the primary clarifier resulted in a relatively high oxygen consumption of the activated sludge which strongly influenced the oxygen balance.

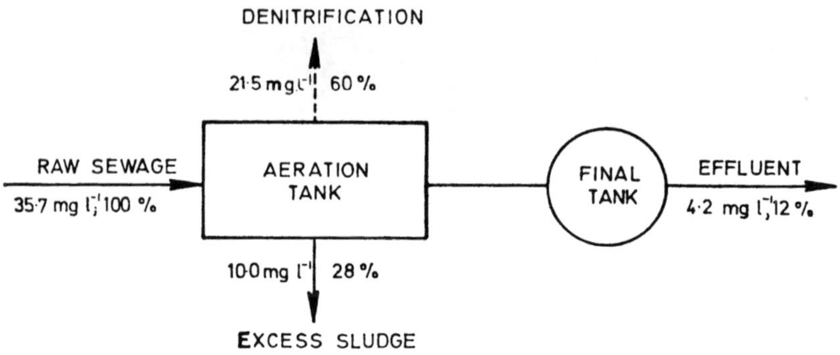

FIG. 1.

Unfortunately the flow of the waste sludge was kept too small which resulted in a slight accumulation of sludge in the plant. This might have changed the nitrogen balance a little towards waste sludge. The obtained elimination of nitrogen of 88 per cent however seems to be very satisfactory.

REFERENCES

BARTH E. F. (1971) Design of treatment facilities for the control of nitrogenous materials. *IAWPR Workshop*, Vienna 1971; *Water Research* **6**, 481 (1972).

BRINGMANN G. (1962) Erste Betriebserfahrungen an einer halbtechnischen Versuchsanlage zur biologischen Eliminierung von Stickstoff aus Abwässern. *Ges.-Ing.* **4**, 106.

HÜHNERBERG K. and SARFERT F. (1967) Versuche zur Stickstoffelimination aus dem Berliner Abwasser. *Gas-Wasserfach* **108**, 966.

PASVEER, A. (1964) Über Oxidationsgräben. *Schweiz. Z. Hydrol.* **26**, 466.

WUHRMANN, K. (1964) Stickstoff- und Phosphorelimination. *Schweiz. Z. Hydrol.* **26**, 520.

Water Research Pergamon Press 1972. Vol. 6, pp. 487–492. Printed in Great Britain

PHYSICAL–CHEMICAL TREATMENT FOR WASTEWATER

J. M. COHEN and I. J. KUGELMAN

EPA—R.A. Taft Water Res. Center, Advanced Waste Treatment Research Laboratory, 4676 Columbia Parkway, Cincinnati, Ohio 45226, U.S.A.

THE PHYSICAL–chemical treatment system which at present is most advanced in development, and which seems the best is the clarification–adsorption system schematically illustrated in FIG. 1. In this system chemical clarification and filtration are utilized to achieve almost complete suspended and colloidal solids removal. With a high enough dose of the proper chemicals essentially complete phosphorus removal can also be obtained. The function of the activated carbon step is the removal of soluble organics. As in any conventional system preliminary screening, grit removal and final disinfection are provided. The filtration step is shown as optional but conservative design dictates its use. The positioning of this step prior to or after carbon adsorption is dictated by the type of waste-carbon contacting system provided. The three steps of chemical clarification, filtration and carbon adsorption will be discussed in some detail below.

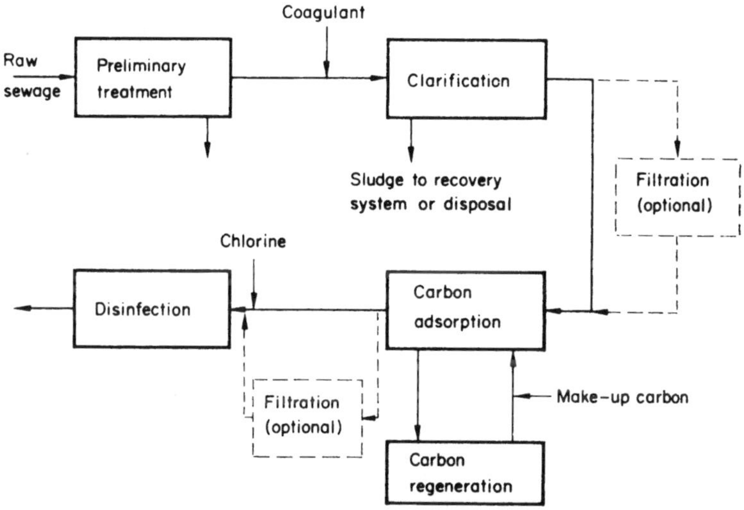

FIG. 1. Flow diagram of a physical–chemical treatment system.

This system does not provide ammonia or nitrate-nitrogen removal. Physical–chemical methods for soluble nitrogen removal are covered in another lecture of this symposium. Any of these nitrogen removal methods can be integrated into the clarification–adsorption system.

CHEMICAL CLARIFICATION

Chemical clarification provides the bulk of the pollutant removal achieved by this type of treatment system. Chemicals which have been successfully used to clarify raw sewage include organic polymers (RIZZO and SCHADE, 1967), iron salts (WEBER *et al.*, 1970), aluminium salts (HANNAH, 1971), and lime (VILLIERS *et al.*, 1970; BISHOP *et al.*, 1970). The inorganic coagulants have the advantage of providing for phosphorus removal. At present there is no rational method for predicting the dose of chemical required. For planning purposes, jar tests are suggested. Fortunately field control of the coagulant dose is possible. For all coagulants except lime the suggested method is monitoring of the clarifier effluent turbidity. Monitoring of phosphorus also appears promising, as good clarification is always obtained when sufficient chemical is added to provide good phosphorus removal. With lime as the coagulant excellent control is obtained by pH measurement. The pH required to achieve good clarification and phosphorus removal is dependent on the chemical characteristics of the sewage. In areas where the alkalinity and hardness are low the high lime process in which the pH is raised to approximately 11.5 is required. Generally, two-stage precipitation with intermediate recarbonation is utilized. In hard water, high alkalinity areas a low lime single-stage precipitation at pH 9.5–10 is sufficient.

TABLE 1 illustrates the treatment obtained with chemical clarification in pilot plants at a variety of locations.

TABLE 1. ACHIEVEMENTS OF CHEMICAL CLARIFICATION

Plant	Chemical	Organic removal (%)	SS removal (%)	P removal (%)
Ewing-Lawrence	170 mg l^{-1} $FeCl_3$	80	95	90
New Rochelle	Lime pH 11.5	80	98	98
Westgate, Va.	125 mg l^{-1} $FeCl_3$	70	—	—
Salt Lake City	80–100 mg l^{-1} $FeCl_3$	75	—	80
Blue Plains	Lime pH 11.5	80	90	95

Sludge disposal plays a dominant role in the economics of any chemical clarification system. Only limited data on the characteristics of sludges resulting from the chemical treatment of raw sewage are available. The data obtained indicate that with iron or aluminum salts as coagulants the resulting sludge volume sometimes exceeds and sometimes is less than primary sludge from the same sewage, and that the chemical sludge is more difficult to dewater in a vacuum filter than the corresponding primary sludge. However, with lime as the coagulation chemical dewatering is extremely rapid just by gravity thickening.

Lime is the only coagulation chemical which can be economically recovered. Recalcining the lime sludge in a furnace has been successfully conducted in a tertiary plant at South Lake Tahoe. The organic solids are incinerated thereby accomplishing ultimate disposal of the sludge. For these reasons it is anticipated that lime will be the coagulant of choice for most situations.

Design criteria for the coagulation equipment is similar to that used in water treatment plants—a flash mix of 1 min, flocculation for 15–30 min and sedimentation at upflow rates of 0.5 and 1.0 gpm^{-1} ft^{-2}.

CARBON ADSORPTION

The total organic removal achieved by the combination of clarification and carbon adsorption at several physical–chemical pilot plants shows that not only is the removal achieved quite high (95 per cent +) but equally important the residual organics after treatment are quite low. These effluents are superior to the usual quality of secondary effluent (TOC ≈ 20 mg l^{-1}. COD ≈ 40–50 mg l^{-1}).

The results reported from four works showed a removal of organic matter of 95–98 per cent giving rise to effluents containing 3–11 mg l^{-1} TOC. Granular carbon was used. In such systems the waste is passed either upward or downward through columns containing the carbon. Downflow columns function as packed beds and accomplish filtration of the wastewater. Flow rates of 2–8 gpm ft^{-2} have been employed. In this flow range essentially equivalent adsorption efficiency is obtained provided the same contact time is employed. At flow rates below 2 gpm ft^{-2} adsorption efficiency is reduced, while at flow rates above 8 gpm ft^{-2} excessive pressure drop takes place. Contact times employed are in the range of 30 to 60 min on an empty bed basis. In general increases in contact time up to about 30 min yield proportionate increases in organic removal. Beyond 30 min the rate of increase falls off with increases in contact time and at about 60 min contact becomes negligible. Carbon beds to be operated at the lower end of the flow rate range are generally designed for gravity flow. Those systems designed for the higher flow rates employ pressure vessels. A pressure vessel is more expensive to construct than a gravity flow vessel but it requires less land area, and provides greater ability to handle fluctuations in flow rate.

Provision must be made to periodically backwash downflow carbon beds because even if they are preceded by a filter they gradually collect suspended solids. In addition, biological growth takes place on the carbon granules and tends to clog the bed. It is advisable to include a surface wash and air scour to be assured of removal of the gelatinous biological growth. Backwash of the carbon beds satisfactorily relieves clogging but does not completely remove the biological growth.

The latest carbon contacting scheme to be applied to waste treatment utilizes powdered carbon (particle size below 200 mesh). This procedure provides for a mixture of a carbon slurry and the wastewater in a Reactor-Clarifier. Polymer addition is generally required to achieve a good gravity separation of the carbon from the wastewater following contact. The potential advantages of this system are the use of a cheaper carbon (10¢ per pound vs. 30¢ per pound for granular carbon) and a simpler type of contacting system.

In order to provide a good effluent and utilize most of the available capacity, countercurrent contact is required. An alternate arrangement which is used in some plants provides parallel flow through a number of identical contactors.

The factors which must be considered in designing any carbon removal system are the type of carbon contact time, flow rate, series or parallel operation, flow direction, hydraulic considerations and organic capacity. Pilot plant operation shows that the carbon capacity varies from 0.15 to 0.3 lb TOC removed per 1 lb of activated carbon, or from 0.4 to 0.6 lb COD removed per lb of activated carbon.

A carbon requirement at this level would be prohibitively expensive if regeneration and reuse of the exhausted carbon were not possible. At present a technically and economically feasible method is available for the regeneration of granular activated carbon. This method requires heating of the carbon in a multi-hearth furnace in the presence of steam to ≈1750°F. The overall loss expressed as percent of weight of virgin carbon required to restore the total original capacity of the batch ranges from 5 to 10 per cent.

DESCRIPTION AND PERFORMANCE OF SOME PHYSICAL–CHEMICAL PILOT PLANTS

Ewing–Lawrence (WEBER *et al.*, 1970)

For the last several years FMC Corporation has been conducting studies of physical–chemical treatment at the Ewing–Lawrence Treatment Plant near Princeton, N.J. Coagulant is added to the settled sewage, the coagulated solid removed by settlement, and the organic matter removed by adsorption. The results are shown in Fig. 2.

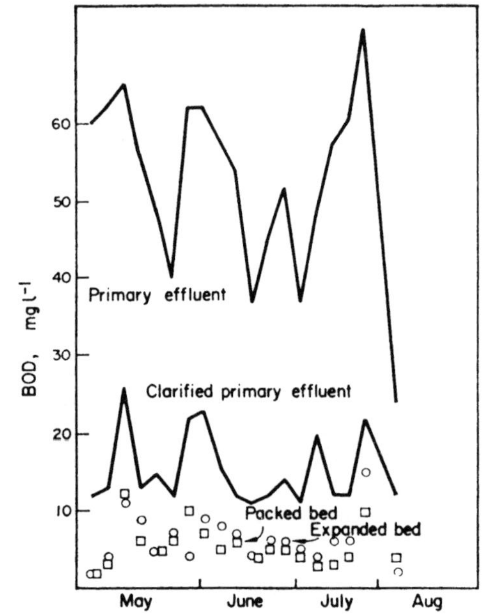

FIG. 2. Removal of BOD by chemical clarification and activated carbon. (FIG. 5 of original paper.

Blue Plains (BISHOP *et al.*, 1970)

At the Blue Plains treatment plant in Washington, D.C., a pilot physical–chemical treatment system is being operated as part of a joint effort of EPA and the District of Columbia. Lime and ferric chloride are the coagulants. The filtered effluent is further purified by ion exchange and carbon adsorption. The flocculation stage of the plant has a capacity of 100,000 gal day^{-1}. The filtration, ion exchange and downflow two-stage carbon beds have a capacity of 50,000 gal day^{-1}.

FIGURE 3 illustrates the treatment performance obtained at this plant for a variety of pollution control parameters. The average quality of the effluent was TOC 6 mg l^{-1}, BOD 5 mg l^{-1}, COD 13 mg l^{-1}, suspended solids 5 mg l^{-1}, phosphorus 0.15 mg l^{-1}, total N 4.6 mg l^{-1}.

Rocky River, Ohio (RIZZO and SCHADE, 1967)

Several studies have been made utilizing polymer addition to existing primary plants followed by small-scale pilot carbon adsorption. One such study was done at the 10 mgd primary facility at Rocky River, Ohio. An anionic polymer was added to

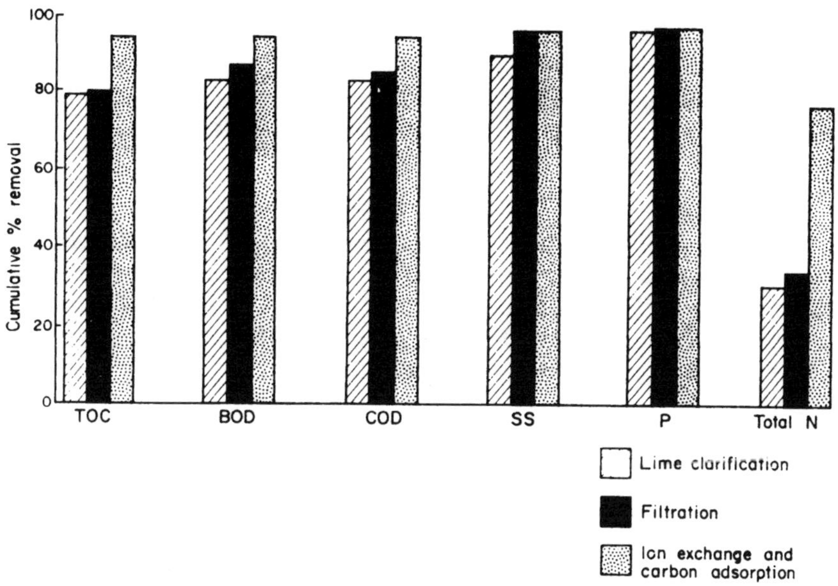

FIG. 3. Removal efficiencies physical–chemical treatment Blue Plains pilot plant.

the existing primary clarifier at a dosage of 0.3 mg l^{-1}, and a side stream of clarified effluent was applied to small carbon columns for a period of about one month. Polymer clarification reduced the suspended solids from 107 to 65 mg l^{-1} the BOD from 118 to 57 mg l^{-1} and the COD from 235 to 177 mg l^{-1}. Carbon treatment further reduced the suspended solids to 1 mg l^{-1} and BOD to 8 mg l^{-1}.

COST ESTIMATES

One of the uncertain factors concerning physical–chemical treatment is the cost of the process. Definitive data will not be available until large-scale plants have been built and are in operation for several years. SMITH (1971) has taken the available information from pilot plants and preliminary designs of several proposed large-scale plants and has made cost estimates for various size plants referenced to October 1970. Amortization was 6 per cent for 24 years. These are presented in TABLE 2. Note that ranges are given for each plant size. These ranges represent the spread of data available.

TABLE 2. PRELIMINARY COST ESTIMATE PHYSICAL–CHEMICAL TREATMENT
(total amortization + O&M, cents per 1000 gal)

Plant size, mgd	5	10	100
Chemical clarification*	9.5–9.6	7.3–9.6	4.0–5.3
Carbon adsorption	11.5–18.0	9.1–13.5	4.5–7.8
Filtration	2.9–4.5	2.1–3.3	1.0–1.4
Total	23.9–36.0	18.5–26.4	9.5–14.5

* Two-stage lime recalcination of sludge.

As a comparison Smith's estimate for primary and secondary treatment with sludge incineration is 16.5 cents per 1000 gallons at the 10 mgd level. With the addition of single stage lime for phosphorus removal the cost would rise to 23.5¢ per 1000 gal which is essentially the same as physical–chemical treatment.

FUTURE DEVELOPMENTS

As a result of the advantages of physical–chemical treatment discussed above and the favorable economic comparison, a number of full-scale treatment plants are being planned or designed. The plants at Rocky River and Painesville, Ohio, will receive Federal research funds. Others are being planned with no expectation of Federal funds other than the normal construction grant.

REFERENCES

BISHOP D. F., O'FARRELL T. P. and STAMBERG J. B. (1970) Physical–chemical treatment of municipal wastewater. Paper presented at the *43rd Annual Conference WPCF*, Boston, Mass., October.

HANNAH S. A. (1971) Chemical precipitation of phosphorus. Paper presented at the *Advanced Waste Treatment and Reuse Symposium*, Dallas, Texas, January 12–14. EPA sponsored.

MONTHLY PROGRESS REPORTS—Contract No. 14-12-585 between Eimco Corporation, Salt Lake City, Utah and EPA.

RIZZO J. L. and SCHADE R. F. (1967) Secondary treatment with granular activated carbon. *Water Sew. Wks* **116**, 307.

SMITH, R. (1971) *Internal Report, R.A. Taft Water Research Center*, EPA, February.

WEBER W. J., HOPKINS C. B. and BLOOM R. (1970) Physio–chemical treatment of wastewater. *J. Wat. Pollut. Control. Fed.* **42**, 83.

VILLIERS R. V., BERG E. L., BRUNNER C. A. and MASSE A. N. (1970) Paper presented at *ACS meeting*, Toronto, Canada, May.

Water Research Pergamon Press 1972. Vol. 6, pp. 493–494. Printed in Great Britain

REPEATED USE OF GRANULAR ACTIVATED CARBON IN WASTE WATER TREATMENT

HARTMUT J. KÄMPF

Chemviron, Maler-Becker-Str. 9, D-65 Mainz-Gonsenheim, FRG

GmbH, Dieselstr, 13–19, D-6079 Sprendlingen, FRG

THE ADVANTAGES of physical–chemical treatment of organic industrial waste waters are: Removal of high molecular weight, biologically resistant substances independent of changes in their concentration or in the throughput, removal of toxic materials, small land requirements, low investment costs and simplicity of operation. The use of granular activated carbon which has far more adsorption capacity than powdered carbon is only economical if there is a possibility to restore its activity after use. To restore the activity means to remove the adsorbed organic materials from the carbon. This can either be done by chemical (wet) means or physical (thermal) reaction. Chemical regeneration is only advisable if the adsorbed product could be sent back into production after elutriation from the carbon. This is done in some plants where phenol has to be removed from the waste water streams of phenol or phenolic production plants. In cases where the elutriate could not be recycled, additional treatment of the concentrated liquor has to be done either by incineration, catalytic oxidation etc. A further disadvantage of chemical regeneration shows up if one has more than one constituent in the waste which has to be removed. Then a different elutriating agent may be needed for each constituent adsorbed or the capacity will drop significantly after the first cycles.

Physical reactivation done by heat in a specially developed furnace has the advantage that the reactivation process is the straight reverse of the adsorption cycle. Adsorption takes place if the surface energy on the carbon, caused by unbalanced forces of molecules on the surface which are not subjected to equal forces by surrounded molecules, overcomes the kinetic energy of a molecule (impurity) in the proximity of the carbon surface. By thermal reactivation the kinetic energy of the impurity adsorbed is increased, overcomes the surface energy and is desorbed. The desorbed materials are moved out of the carbon pores by steam injected into the reactivation chamber. The impurities are oxidized by oxygen and then burned to CO_2. A high temperature (about 800–900°C) is needed to break down chlorinated hydrocarbons to CO_2, water and HCl and to reactivate the carbon.

In most cases thermal reactivation is done in a multiple hearth (Herreshoff) furnace on-site of the adsorption plant. With such a system one is independent of the transport of spent carbon back to the carbon supplier's factory and has lower carbon losses and a cleaner process, because the cycle: adsorbers—reactivation facility is a closed cycle it can be run automatically But the main advantage is: the reactivation on-site is less expensive. Reactivation on-site costs between \$0·06 kg^{-1} and \$0·11 kg^{-1} while reactivation outside ranges between \$0·30 kg^{-1} and \$0·45 kg^{-1} transportation and carbon loss included. The price for virgin carbon used in waste water treatment is in the range of \$0·75 kg^{-1}–\$1·00 kg^{-1}.

How does a thermal reactivation plant on-site work? The carbon is hydraulically transported as a slurry from the adsorbers to a holding tank, from which the wet carbon is continuously delivered by a dewatering screw to the multiple hearth furnace.

The furnace consists of a steel shell lined with refractory (either castable or in brick form) on the inside. The interior is divided by horizontal brick arches into separate hearths. The centre shaft, driven by a variable speed motor and cooled by air, rotates the rabble arms on each hearth. The rabble teeth on these arms are placed at an angle, so that the carbon is moved inwards, is dropped down to the hearth below and then it is moved outwards. The rabble arms are also cooled by air. The first two hearths are used for drying, the third and the fourth hearth for distillation of volatile material adsorbed on the carbon, and the last two hearths for reactivation and removal (oxidation and burning) of the remaining impurities on the carbon. Carbon retention time in the furnace is about 20–30 min. The regenerated carbon drops from the bottom hearth into a quench tank filled with water and is pumped from there as a slurry to a storage tank or back to the filters. The critical part of the entire re-activation process is the control of the atmosphere in the furnace, so that the re-activation is effective and efficient. Enough oxygen has to be available to oxidize the desorbed materials, but the concentration has to be low so that the carbon is not burned, which would result in high carbon losses and high operating costs. About 50 reactivation facilities on-site in various industries and applications installed run with an average carbon loss of 5–7 per cent per reactivation cycle which means that the same carbon could be loaded and reactivated at least 15 times.

The burner system consists of two pre-mix gas burners located on hearths 4 and 6, with only one air–gas mixer per hearth. The mixers are adjusted to the heating value of the fuel or gas used and to the excess air–gas ratio. The burners on hearth No. 4 run with about 20 per cent excess air while the burners on hearth No. 6 run with about 7·5 per cent excess air. Instead of introducing excess air by the burners, nozzles attached to the side of the furnace shell and discharging the air into the unfired hearths are also used in some furnaces. The burners are automatically controlled to a pre-set temperature and the furnace temperature is continuously measured by thermo-couples and recorded. Steam is added on the fired hearths at the rate of 1 kg steam per kg carbon reactivated.

In order to assure odourfree operation of the furnace an afterburner can be provided to raise the temperature of the exist gases to over 650°C.

At Lake Tahoe 30 g of activated carbon are needed to treat 1 m^3 of waste water. The operating costs for the activated carbon plant with reactivation on-site are $5·84 per 1000 m^3 One can see that the operating costs for a tertiary treatment system are about the same as for a secondary activated sludge plant. The effluent data achieved at Tahoe show about the same quality of water generally obtained from potable water treatment systems drawing from surface supplies.

Water Research Pergamon Press 1972. Vol. 6, pp. 495–497. Printed in Great Britain

DESIGN PARAMETERS FOR SLUDGE TREATMENT: INTERPRETATION OF OPERATION DATA FROM SEVEN ACTIVATED SLUDGE PLANTS FOR MUNICIPAL SEWAGE

P. WILDI

Gewässerschutzabteilung Kanton Zurich, Kantonale Baudirektion, CH-8090 Zurich, Switzerland

IN THE Canton of Zurich 72 sewage treatment plants are actually in operation (for 800 up to 400,000 inhabitants), serving about 90 per cent of the total population.

The usual basic data for sludge yield is 100 g of dry solids capita^{-1} day^{-1} (for raw primary and secondary sludge), respectively 2/3 or about 65 g capita^{-1} day^{-1} (for digested sludge). Imhoff speaks of 79–85 g capita^{-1} day^{-1} for raw sludge respectively 52–55 g capita^{-1} day^{-1} for digested sludge. In the data for the Canton of Zurich (100 g capita^{-1} day^{-1}) trade wastes in the municipal sewage is included.

The present contribution deals with the result of analysis and interpretation of the operation data and maintenance records of 7 typical activated sludge plants (ranging from 12,000 to 30,000 inhabitants), treating municipal sewage.

The data are calculated by dividing the average daily production of dry solids by the number of contributary population. The population equivalent of trade wastes has not been taken into consideration. The period of observation of these 7 plants ranges between 2 and 11 yr.

In TABLE 1 the basic data represent the average of 7 representative plants.

TABLE 1

		Average	Varying from	to
A. Raw sludge				
dry solids	g capita^{-1} day^{-1}	90·6	73	108
dry solids	%	5·9	4·5	7·7
organic solids	%	60·2	50·2	69·3
B. Digested sludge				
dry solids	g capita^{-1} day^{-1}	57·7	48	67
dry solids	%	6·3	3·7	8·8
organic solids	%	43·2	34·2	53·0
C. Gas				
production	l. capita^{-1} day^{-1}	28·6	21	36
production	l. kg^{-1} organic solids	538	401	695

The average gas production of the 7 plants is 538 l. kg^{-1} of organic solids in the raw sludge. It varies not only from plant to plant, but to a large extent from year to year.

The frequency curve of the gas production in the year 1968 for a plant for 20,000 inhabitants (FIG. 1) shows that the variation is approximately $\pm$ 50 per cent, and that the average yield is attained and exceeded on 160 days of the year. As can be seen

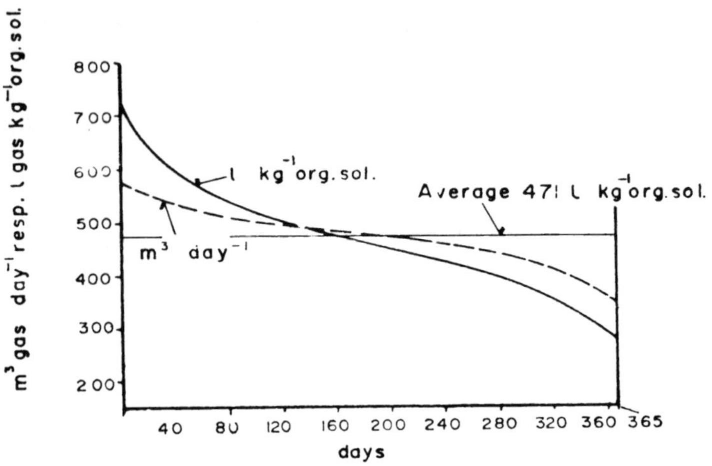

FIG. 1. Frequency curve of gas production.

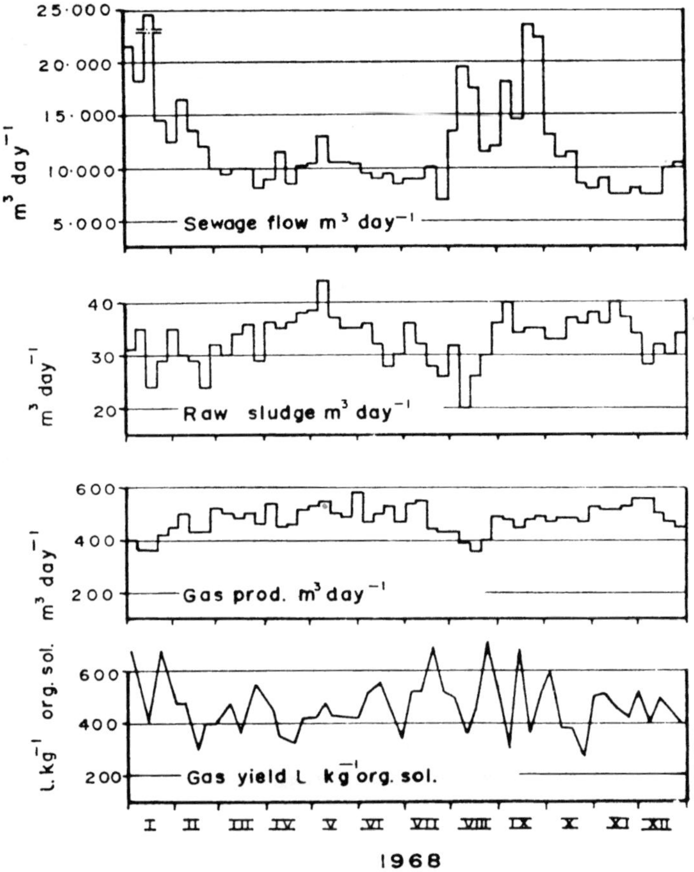

FIG. 2. Activated sludge plant for 20,000 inhabitants.

from the diagram (FIG. 2) the gas production in February, April and October was only 400 l. kg^{-1} and below for periods of more than 2 weeks.

These facts have to be considered when gas is used for heating, for drying and incineration of sludge or as fuel for gas engines in a self-sufficient plant.

Water Research Pergamon Press 1972. Vol. 6, pp. 499–502. Printed in Great Britain

STUDIES ON THE DESIGN DATA OF GRAVITY THICKENING

K.-H. KALBSKOPF
Emschergenossenschaft, D-43 Essen, FRG

PHYSICAL processes are important for thickening. With continuously operated thickeners the sludge solids pass through the zones shown in FIG. 1. Compaction of the sludge and of solid particles under their own weight and that of the sludge layers above increases until equilibrium is reached between the consolidation pressure of the sludge solids and the pore water pressure of the interstitial water. This equilibrium is obtained when the build up of pore water pressure no longer suffices to overcome the head loss of the sludge layer above.

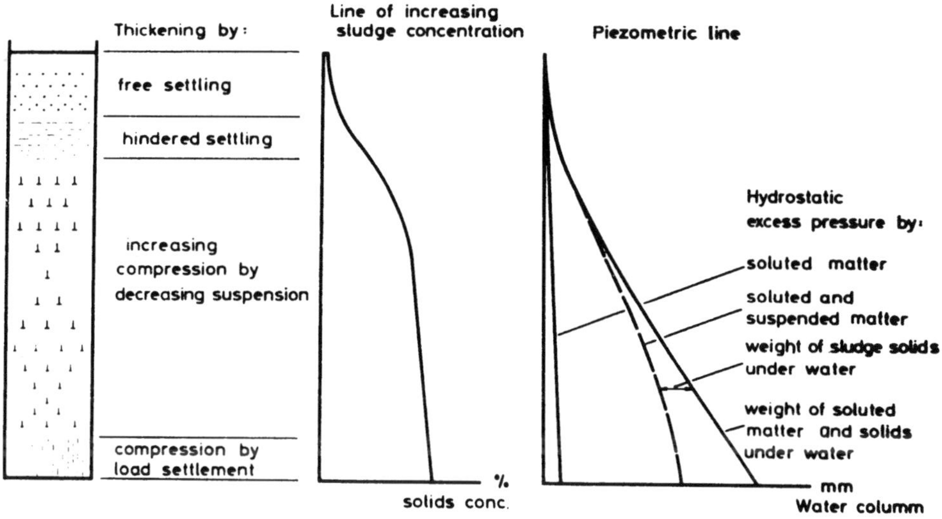

FIG. 1. Zones of gravity thickening.

The consolidation pressure existing in a certain depth of the sludge layer corresponds to the weight of the solids, subject to an uplift pressure down to the depth of the sludge layer, reduced by the corresponding hydrostatic overpressure of the solved and suspended sludge water matter. Measurements conducted by the Emschergenossenschaft showed that the effective consolidation pressures were low and amounted to only about 0·5–2 g cm^{-2}. They increase with increasing depth of the sludge layer. The attainable thickening of the sludge depends on this pressure and the time of consolidation. FIGURE 2 shows that higher solids concentrations can be reached by increased heights of the sludge layer and thus also increased consolidation pressures. As the permeability of the sludge layers quickly decreases with increasing solids concentration, as FIG. 3 shows, the equilibrium between the consolidation pressure and the pore water pressure is obtained, due to the low effective consolidation pressures, even with very low solids concentrations of 2–3 per cent, in case of light and filamentous sludges.

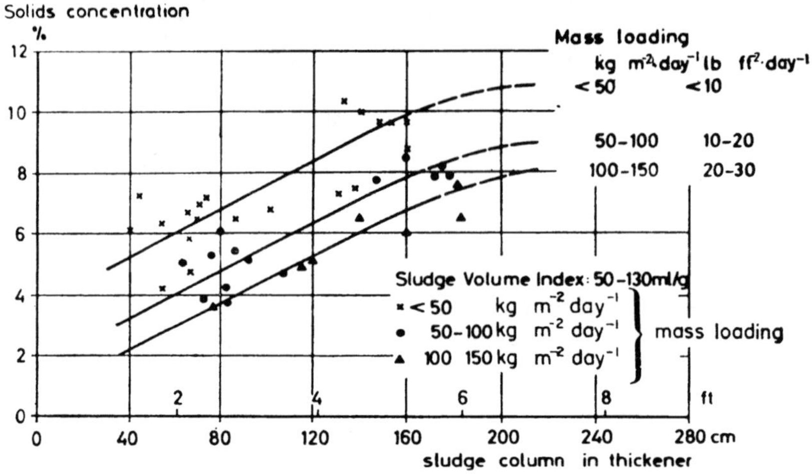

FIG. 2. Thickening tests with Emscher-activated sludge.

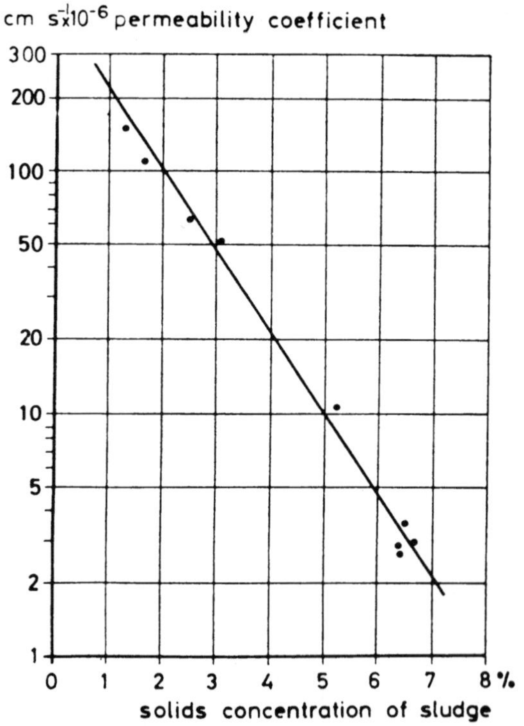

FIG. 3. Permeability coefficient of Emscher-activated sludge.

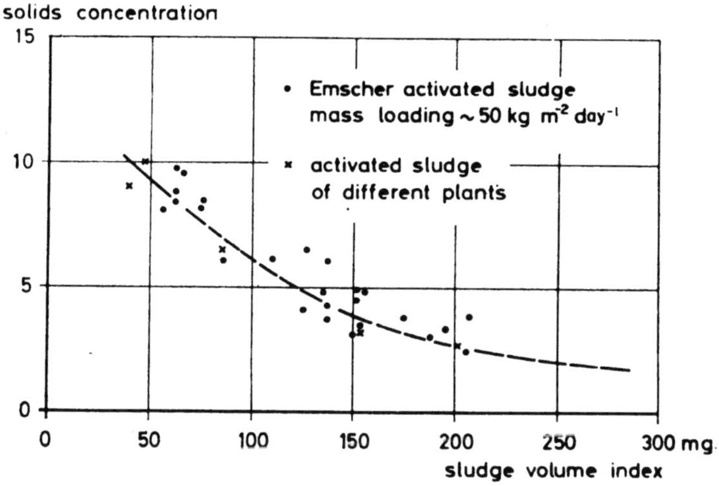

FIG. 4. Gravity thickening of activated sludge.

Sludge type	Solids concentration after sludge gravity-thickening (%)
Mixed primary and industrial sludges (coal)	10–30
Primary sludge	
volatile matter > 65%	5– 7
volatile matter < 65%	7–12
Primary and activated sludge	
SV Index > 100 ml g^{-1}	4– 6
SV Index < 100 ml g^{-1}	6–11
Sludge of extended aeration tank	3– 5
Primary and trickling filter sludge	7–11
Primary digested sludge	8–14
Activated digested sludge	6– 9
Heat treated activated sludge	10–15

FIG. 5. Solids concentration by gravity thickening.

Sludge type	: Primary and activated sludge
Primary sludge	: 200 $m^3\,day^{-1}$ (52770 gal day^{-1}) – 25% concentration pumped 2 times ÷ 100 m^3h^{-1} (26385 gal h^{-1}) 5000 kg day^{-1} 11,000 lb day^{-1}
Activated sludge	: 500 m day^{-1} (131926 gal day^{-1}) – 0·6% concentration pumped over 24 h 3000 kg day^{-1} (6600 lb day^{-1}) SV index – 80 mg l^{-1}
Solids of mixed sludge	: 8000 kg day^{-1} (17,600 lb day) – 62% volatile solids
Mass loading of solids (fixed)	: 60 kg $m^{-2}\,day^{-1}$ (12·3 lb ft^2day)
Surface area	: 8000 60 = 133 m^2 (1437 ft^2)
Diameter of thickener	: 13 m (42·64 ft)
Hydraulic overflow rate	
(a) Primary and activ. sludge 2 h-pumping time	: $\frac{100+500/24}{133}$ – 0·91 mh^{-1} (2·98 ft h^{-1}) $<$overflow rate of primary sedimentation
(b) Activated sludge 22 h-pumping time	: $\frac{500/24}{133}$ = 0·16 m h^{-1} (0 52 ft h^{-1}) $<$overflow rate of final sedimentation
Solids concentration after thickening (assumed)	: 7%
Underflow sludge volume	: $\frac{200 \times 2 \cdot 5}{7} = \frac{500 \times 0 \cdot 6}{7}$ bzw $\frac{8000 \times 100}{1000 \cdot 7}$ = 115 m^3day (30343 gal day^{-1})
Filtrate overflow volume	: 700–115 = 585 m^3day^{-1} (154352 gal day^{-1})
Average solids concentration in sludge column assumed to 75% of underflow solids concentration	: 7%0·75 = 5·25%
Sludge column volume	: $\frac{8000 \times 100}{1000 \times 5 \cdot 25}$ = 152 m^3day = 6·35 $m^3\,h^{-1}$ (1675 gal h^{-1})
Limit of solids detention time in thickener (assumed)	: 36 h
Height of sludge column	H_1 : $\frac{133(m^2) \cdot H_1(m)}{6 \cdot 35\ m^3\ h^{-1}}$ = 36h
	H_1 : 1·75 m (5·74 ft)
Height of clarified water zone (fixed)	H_2 : 1 m (3·28 ft)
Height of sludge scraping zone	H_3 : 0·30 m (0·98 ft)
Total height of thickener	: 3·05 m (10·00 ft)
Surface area of thickener	: 133 m^2 (1431 ft^2)
Volume of thickener	: 405 m^3 (14294 ft^3)

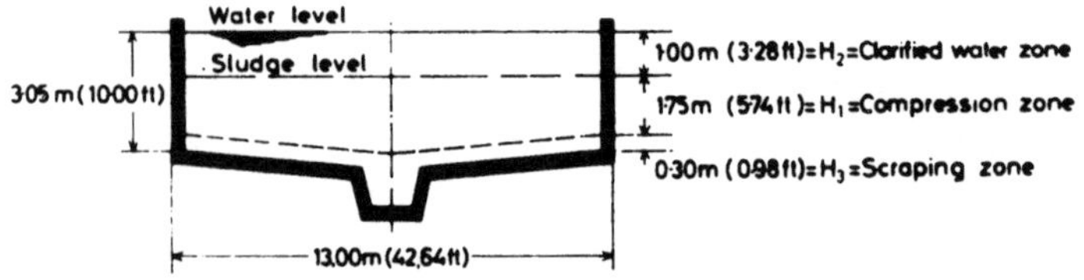

FIG. 6. Design of sludge thickener.

FIGURE 4 shows the thickening which is possible to reach with activated sludge as a function of the sludge volume index. The possible solids concentrations for other sludges can be seen in FIG. 5.

FIGURE 6 gives an example of the design of a sludge thickener using as design parameters the overflow rate, the solids loading, the detention time of the sludge solids in the compression zone and the attainable underflow concentration.

The detention time of the sludge solids in the thickener, however, must not be long enough to disturb the thickening process by biological degradation processes.

REFERENCE

KALBSKOPF K.-H. (1971) Theoretische Grundlagen, Bemessung und Verfahrensweise der Schlammeindickung. *Gewässersch., Wasser, Abwasser* **5**.

Water Research Pergamon Press 1972. Vol. 6, pp. 503–505. Printed in Great Britain

CONTROL OF THE ANAEROBIC DIGESTION PROCESS AND SUPPORTING UNIT PROCESSES

C. F. GUARINO

City of Philadelphia, Water Department, 1160 Municipal Services Building, Philadelphia, Pa. 19107, U.S.A.

IT IS our desire and intention in Philadelphia to automate anaerobic digestion and supporting unit processes. The basic processes are raw sludge pumping, sludge heating, and anaerobic digestion. We have already demonstrated at our Southwest facility, the feasibility of automating the raw sludge handling unit process. Basically, (see FIG. 1) the rectangular primary settlers are broken down into nine (9) sludge handling units; seven (7) longitudinal and two (2) cross collectors—all with the purpose of moving sludge to the front center of the tank where a submersible pump discharges through a density gage to the concentration tanks.

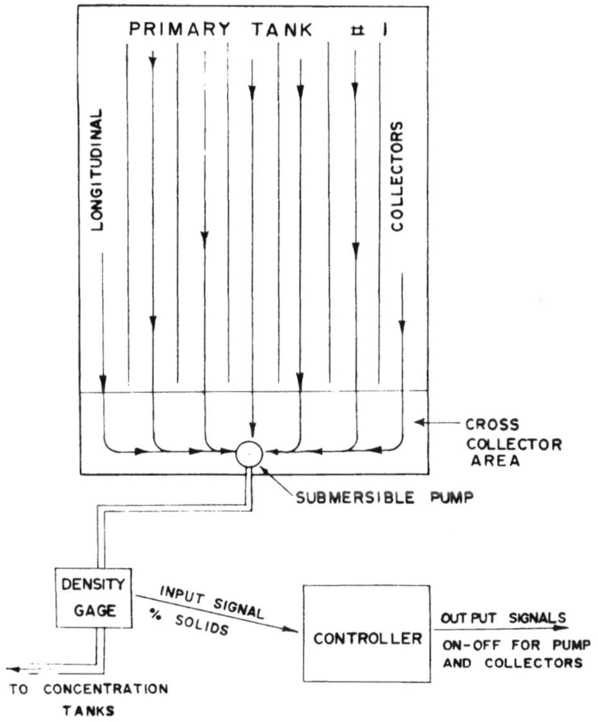

FIG. 1. Automatic sludge pumping Southwest treatment facility.

In an automated mode, the sequence of events would be:

(1) One or both cross collectors would be started simultaneously with the sludge pump so the sump could be cleared. The percentage solids of the sludge is measured by a radioactive type density gauge during this sequence to assure desired solids level—say 5 per cent.

(2) As the sludge thins out, longitudinal flights are turned on sequentially until the desired solids level is attained.
(3) At times, the sludge will be too thick. Sludge thinning is accomplished by dilution water.

If the above system were installed in all four of Southwest's primary tanks, after step 3, controllers would initiate the full procedure in an adjacent tank. The system at present is installed to control primary tank No. 1. We are assured of uniform sludge removal from the entire tank floor using the outlined system.

We feel that the second unit process, sludge heating, can be easily automated. Basically, the system would call for direct digester charging of primary sludge, temperature monitoring of all digesters and temperature adjustment via circulation of digester sludge through sludge heaters of the heat exchanger variety. Philadelphia has had considerable operating experience with temperature monitoring and sludge heating equipment. Since 1968 at our Southwest facility, two (2) 1.26 million kg-cal h^{-1} (5 million BTU h^{-1}) heat exchanger sludge heaters have been operating satisfactorily. Philadelphia also has been monitoring sludge temperatures in digesters and sludge lines since 1954. All of this experience will form the basis of plans for automated sludge heating.

Philadelphia feels that anaerobic digestion—though highly complex—can be automated. The key to digester control is the measurement of digester gas for CO_2. In Philadelphia, going back to 1951, operating records confirm that CO_2 change is sufficiently sensitive to determine the condition of a well mixed digester. For digester control, accurate on-line measurement of CO_2 is therefore a necessity.

Since 1962, Philadelphia has used a Ranarex indicator at our Southwest facility to monitor percent CO_2. This instrument is a gas analyzer which measures the specific gravity of a gas or gas mixture compared to air. Since a definite relationship exists between CO_2 and the specific gravity of a gas mixture, the scale and chart can be graduated to read in percentage CO_2. This instrument over the past nine years has required very little maintenance and compares excellently with simultaneous Orsatt measurements (see TABLE 1).

With the provision to measure CO_2 on-line, Philadelphia's digester control scheme will be structured as follows:

(1) Providing adequate space is available, raw sludge will be directly charged to the digester with the most favorable CO_2 level and gas rate.
(2) While being charged or shortly thereafter, the digester will be mixed via a gas recirculation system.
(3) If the temperature of the charged digester is below an established limit, the digester will be recirculated through sludge heaters.
(4) After temperature correction, the digester is allowed settling time before supernatant or bottom sludge is withdrawn.

To test this scheme under simulated computer controlled conditions, we have engaged the services of the General Electric Company. The scope of the joint project covers:

(1) Development of an operational flow chart.

TABLE 1. COMPARISON OF ORSATT AND RANAREX INSTRUMENTS FOR MEASUREMENT OF PERCENTAGE CO_2 SOUTHWEST TREATMENT PLANT DIGESTER GAS, JUNE 1963

Day	Orsatt %	Ranarex %
3	36.6	35.5
4	36.0	36.0
5	43.2	42.5
6	32.4	31.5
7	33.2	33.0
8	33.4	32.0
10	40.2	40.0
11	33.8	33.0
12	34.6	35.5
13	35.8	34.0
14	35.0	33.0
15	34.6	32.5
17	31.8	30.5
18	34.6	34.0
19	34.4	34.0
20	35.8	35.0
21	32.6	30.5
22	37.6	37.0
24	34.8	33.0
25	34.4	33.5
26	32.8	31.5
27	36.4	36.5
28	43.4	43.0
29	35.6	35.0
Average	35.5 per cent	34.7 per cent
Average deviation	2.25 per cent	

(2) Develop and debug a computer program which will control the operation of all digesters.
(3) Install a remote terminal at the Northeast Treatment facility for communication via telephone lines with a private process control computer.
(4) Program evaluation and modification based on operational data.

The four part control scheme will be applied simultaneously to at least eight (8) separate digester tanks, each with a capacity exceeding 7570 m^3 (two million U.S. gallons). Laboratory tests such as pH, volatile acids, and alkalinity will be manually inserted into the control program and used as a secondary source of priority establishment for digester charging. Completion of this project will give Philadelphia more insight into the problems associated with computer control of detailed unit processes, particularly anaerobic digestion. Preliminary design work is now in progress for the expansion of our three (3) treatment facilities. All three expanded plants are scheduled to be in operation by the fall of 1977. The plants will have capacities ranging from 946,000 to 454,000 m^3 day^{-1} (250–120 mgd). As the design work progresses, existing and additional information relative to plant automation will be integrated into construction specifications and drawings.

Water Research Pergamon Press 1972. Vol. 6, pp. 507–513. Printed in Great Britain

HANDLING OF AEROBIC MINERALIZED SLUDGES BY CENTRIFUGES AND BELT-PRESS FILTERS

J. ZEPER and R. PEPPING

Dwars, Heederik & Verhey, Laan 1914 Nr. 35, Amersfoort, Netherlands

1. INTRODUCTION

ANAEROBIC digestion of the sludges of wastewater treatment plants is still widely used, and one of the problems to be solved is the dewatering of the digested sludge. Two of the feasible solutions are those provided by centrifuges and belt-press filters. Mainly due to technical improvements as well as the application of new flocculation agents research in this field has made good progress. This research, however, has been limited to the handling of either primary sludges or a mixture of primary and secondary sludges or anaerobic digested sludges. Little is known about handling activated sludges alone, and no information is available on how aerobic mineralized sludges behave in centrifuges and belt-press filters. This paper will give the results of some practical tests carried out on the handling of aerobic mineralized sludges from oxidation ditches.

2. GENERAL CRITERIA

If tests results are to be compared some criteria will be needed, such as:

(a) the thickening of the sludge, expressed as the dry solids content (per cent) of the sludge volume after the dewatering process;

(b) the efficiency, being the ratio between the dewatered sludge and the sludge input, both measured as kg dry solids.

Expressed in the formula:

$$\eta = \frac{C_s\,(C_i - C_f)}{C_i\,(C_s - C_f)} \times 100 \text{ per cent in which}$$

η = the efficiency in per cent

C_i = dry solids content of the sludge input

C_s = dry solids content of the dewatered sludge

C_f = dry solids content of the filtrate/centrifugate.

(c) the quantity of input sludge that can be handled by a certain piece of equipment, expressed in $m^3\ h^{-1}$ or $kg\ h^{-1}$.

(d) the dosage of flocculation agents, expressed in grams of flocculation agent per kg sludge (dry solids) input.

(e) the cost of the process, expressed in Nfl. per ton sludge (dry solids) input (price level 1971).

3. CENTRIFUGES

3.1 *Description*

FIGURE 1 shows the decanting centrifuge and the disc centrifuge. It has already been found that although two-step centrifugation will improve the general efficiency of the process, the final thickening of the sludge is adversely affected by the second stage, i.e. the disc centrifuge. Other disadvantages of an additional disc centrifuge are:

(a) higher investment costs;

(b) higher maintenance and operation costs;

(c) frrquent cleaning operations, required to remove grease deposits from the discs;

(d) the entire installation is more complex and vulnerable.

On the basis of these experiences and considerations, the tests on the handling of aerobic mineralized sludges have intentionally been concentrated on one-step centrifugation with decanting centrifuges. This decision was taken in close consultation with Alfa–Laval Amsterdam N.V., who supplied

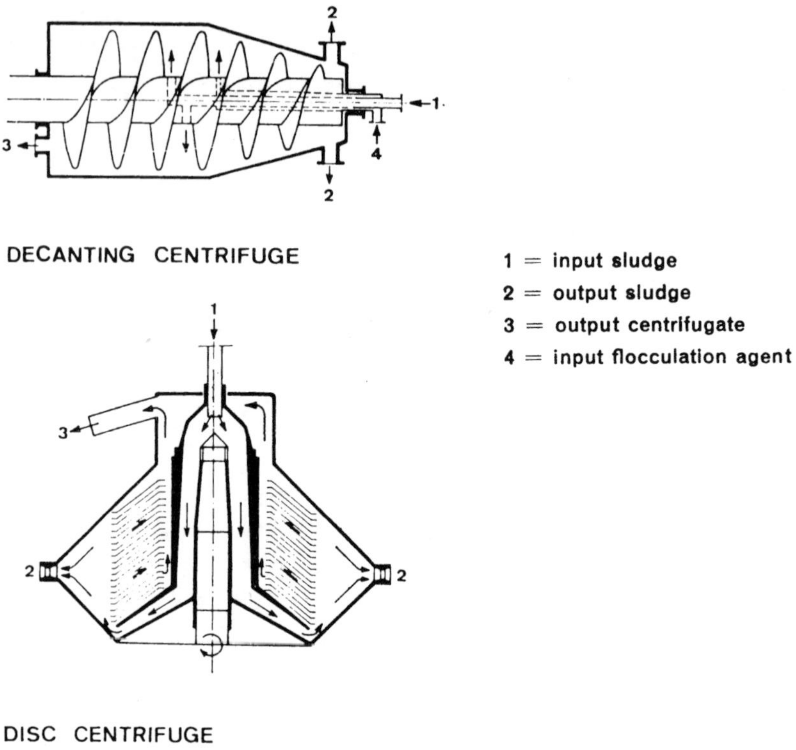

FIG. 1. Decanting centrifuge and disc centrifuge.

their decanting centrifuge type NX 214 (capacity 5 $m^3 h^{-1}$, $n = 3250$ rev min^{-1}) for the tests, which were carried out at the oxidation ditch at Woudenberg (4500 p.e.) between March and August 1969.

The experiment with the centrifuge served two purposes: Firstly, testing the operation of the centrifuge in order to find out to what extent C_s and η are related to:
(a) the dry solids load;
(b) the dosage of flocculation agents;
(c) the difference in rev min^{-1} between bowl and screw; (so called back-drive)
(d) the thickness of the liquid layer inside the centrifuge.

The speed of the centrifuge itself was not adjustable.
Secondly, testing the possible secondary effects of the filtrate with regard to:
(a) the biological process in the oxidation ditch;
(b) the quality of the effluent.
For this purpose the experiment was carried out over a period of 6 months and all excess sludge was handled.

3.2 *Results*

When the sludge (dry solids) load was increased, the efficiency decreased distinctly but the thickening increased slightly. An explanation for this is that with an increased load more coarser particles were trapped in the centrifuge, whereas more fine material was discharged with the centrifugate. When the centrifuge was fed with its design load of 5 $m^3 h^{-1}$ of sludge, having a C_i of 2·5 per cent (125 kg h^{-1} dry solids), its efficiency was only about 35 per cent and the C_s of the thickened sludge was about 15 per cent.

Therefore tests were also carried out with addition of a flocculation agent—the cation active polyelectrolyte Praestol K444 was selected for this purpose—in a 0·1 per cent concentration. This had a marked improvement on the efficiency but did not noticably effect the thickening of the sludge.

The results of these tests are shown in Fig. 2. When the dosage is related to input sludge dry solids, it appears that for an efficiency of about 70 per cent 3 g of flocculation agent have to be added per kg sludge dry solids.

The relation between C_s and the input sludge load shows a tendency of improvement of C_s with increasing loads.

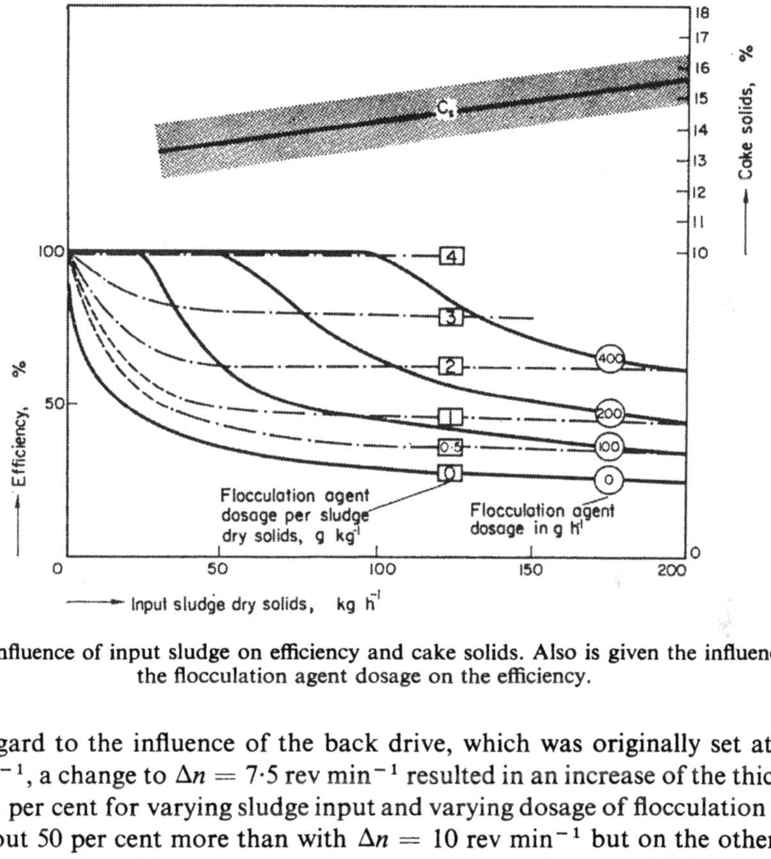

Fig. 2. Influence of input sludge on efficiency and cake solids. Also is given the influence of the flocculation agent dosage on the efficiency.

With regard to the influence of the back drive, which was originally set at Δn = 10 rev min^{-1}, a change to Δn = 7·5 rev min^{-1} resulted in an increase of the thickening by about 8 per cent for varying sludge input and varying dosage of flocculation agents. This is about 50 per cent more than with Δn = 10 rev min^{-1} but on the other hand, the efficiency decreased by 10–15 per cent. Roughly speaking an additional flocculation agent dosage of 1–1·5 g kg^{-1} sludge dry solids will be needed to overcome this disadvantage.

The effect of the thickness of the liquid layer inside the centrifuge was studied. When, without addition of polyelectrolyte, this layer was decreased from 74 to 57 mm C_s increased from 15 to 22 per cent, but η decreased from 32 to 26 per cent. Again a better thickening but at the cost of less efficiency.

The secondary effects were determined. During the 6 months experiment no change in the biological process nor in the quality of the effluent was observed. The B.O.D. load of the centrifugate was only 1 per cent of the organic load of the wastewater treatment plant. Tests using the oxidation ditch process; fish in aquaria and barley in pots, showed no ill-effects due to an overdose of polyelectrolyte.

3.3 *Conclusions*

It appears that a 5 $m^3 h^{-1}$ decanting centrifuge can be used successfully for the dewatering of the sludge from an oxidation ditch process. The capacity of the centrifuge can be more satisfactorily expressed however in sludge dry solids capacity in kg h^{-1}, and the tests have shown that the so-called 5 $m^3 h^{-1}$ centrifuge can easily handle 200 kg h^{-1} of dry solids (5 $m^3 h^{-1}$ at 4 per cent dry solids content). With a dosage of 4 g of flocculation agent per kg sludge dry solids input, an efficiency of 90 per cent can be reached and with intelligent regulation of the centrifuge a 15–20 per cent thickening of the sludge can be obtained.

The oxidation ditch at Woudenberg mainly treats domestic sewage and other tests (not discussed in this paper) on oxidation ditches with a 50/50 division between domestic and industrial sewage, show that the thickening of the sludge is limited to about 15 per cent dry solids content.

3.4 *Practical application and costs*

A 100,000 population equivalent oxidation ditch process plant (a Carrousel) will have a daily production of excess sludge = 100,000 p.e. × 40 g/p.e. day = 4000 kg day^{-1}. With $\eta = 90$ per cent the weekly quantity to be handled amounts to (dry solids) 100/90 × 4000 kg day^{-1} × 7 days $week^{-1}$ = 31,000 kg $week^{-1}$. With 2 shifts during working days and 1 shift on Saturdays and Sundays, there are 12 shifts $week^{-1}$, so in one shift the quantity is about 2600 kg dry solids.

Two of the 200 kg h^{-1} centrifuges can handle this quantity. In case of a long-lasting break-down or overhaul of one of the centrifuges, the other centrifuge is capable of doing the job, running more or less continuously. If this is not considered sufficiently safe, a third centrifuge can be added.

The cost of the handling of the sludge in this type of centrifuge including depreciation, maintenance, operation, energy and flocculation agent has been calculated at Nfl. 150 per 1000 kg sludge production. With 365 days $annum^{-1}$ × 40 g $p.e.^{-1}$ day^{-1} = 15 kg $p.e.^{-1}$ $annum^{-1}$ of sludge production the annual costs per p.e. amount to Nfl. 2·20. The thickened sludge (C_s = 15–20 per cent) can be stored for about 1 month (the storage needed is about 600–800 m^3) and is then to be removed at an average rate of 30–40 tons per working day.

4. BELT-PRESS FILTER

4.1 *Description*

The two most essential parts of the belt-press filter are the belt-filter and the belt-press. The belt-filter moves—at a little distance under the belt-press—in the same direction and with the same speed as the belt-press. When sludge is fed onto the belt-filter it is squeezed between the belt-press and the belt-filter, and the sludge is dewatered. The filtrate falls through the belt-filter. At the same time some of the sludge is pressed into the filter, but this sludge is sprayed off, as is any sludge which sticks to the belts beyond the scraping mechanism. The major part of the sludge, however, is scraped off as dewatered sludge. FIGURE 3 shows the belt-press filter.

It appears from these results that for good operation of the belt-press filter an addition of flocculation agents is indispensable. In this case Praestol K444 was used. The belt-press filter which was used for the test was a variation of the type 5/3. The "5/3" is a 0·5-m wide and 3-m long belt. In this particular test the belt-press filter was only 2·6 m long.

The speed of the belts, the sludge input and the dosage of flocculation agent can be regulated accurately. The pressure on the belt-press can be varied but cannot be measured, so the pressure is described as "relatively high" or "relatively low".

The tests were carried at the oxidation ditch plants of Scherpenzeel and Oosterwolde.

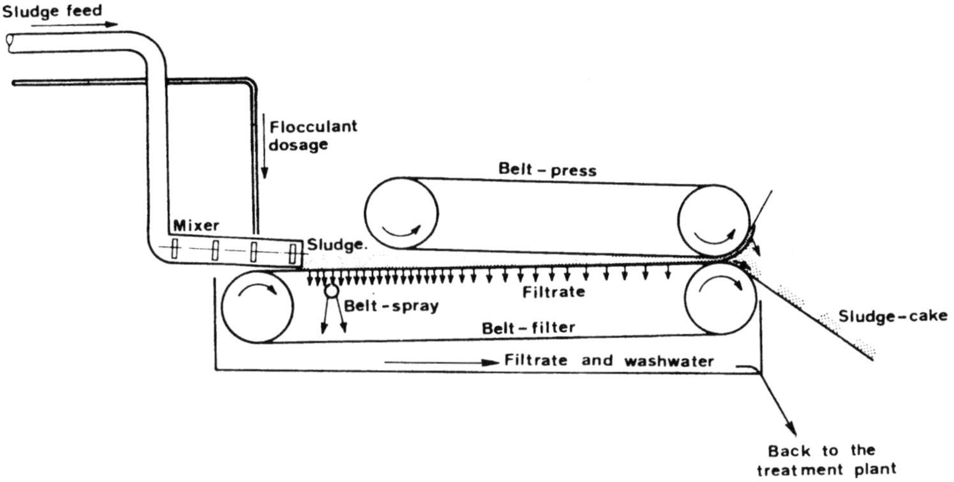

FIG. 3. Flow scheme of the belt-press filter.

4.2 *Results*

The capacity is a function of the area of filter which is available per unit of time; i.e. it is a product of the length and width of the filter and the speed of the belt. If, however, the speed of the belt increases, the period of pre-dewatering—which occurs before the sludge really arrives in the pressing zone—becomes shorter and therefore the sludge will be more fluid in the pressing zone and will more easily be squeezed off the belt, thus actually decreasing the capacity and the efficiency.

It is also obvious that this squeezing effect will be greater at a "relatively high" pressure than at a "relatively low" pressure.

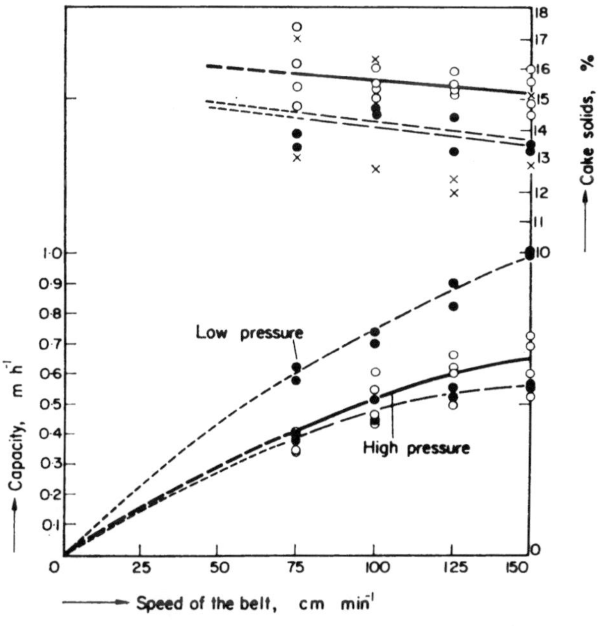

FIG. 4. Influence of the belt speed on the capacity and cake solids.

In FIG. 4 the relation between the speed of the belt and the capacity is plotted for three test runs, two at a high pressure and one at a low pressure.

No evidence could be found that the dewatering of the sludge was affected by the change from high pressure to low pressure during the tests in Oosterwolde, although the overall results were slightly less favorable than those in Scherpenzeel. This is possibly due to the fact that in the Oosterwolde plant the sludge contains about 75 per cent organic material, because of overloading of the plant, whereas the sludge at Scherpenzeel contains about 65 per cent organic material. The results are plotted in the upper section of FIG. 4.

With regard to the dosage of flocculation agent, there is a kind of threshold, below which the belt-filter press no longer functions. On the other hand an extra dosage above this minimum level seems to have no advantages.

With the tests in Scherpenzeel and Oosterwolde 3g Praestol K444 kg^{-1} sludge dry solids input was found to be the minimum value.

The efficiency of the belt-press filter was determined by the amount of sludge that was pressed into the filter, stuck to the belts beyond the scrapers and was squeezed outside the belts. When the system was adjusted properly, the efficiency was found to be 90 per cent or more.

4.3 *Conclusions*

The tests runs have shown that a belt-press filter can be used for the dewatering of aerobic mineralized sludges. With a minimum dosage of 3 g Praestol K444 kg^{-1} sludge dry solids input, the sludge can be dewatered to 15 per cent dry solids content with an efficiency of at least 90 per cent. The pressure which is applied to the belt-press plays an important role in the capacity of the belt-press filter. Based on test results the following capacities for aerobic mineralized sludge can be used for design purposes:

type 5/3 belt $0{\cdot}5 \times 3\ m^2$; cap. $1\ m^3\ h^{-1}$
type 10/3 belt $1 \times 3\ m^2$; cap. $2\ m^3\ h^{-1}$
type 15/3 belt $1{\cdot}5 \times 3\ m^2$; cap. $3\ m^3\ h^{-1}$
type 20/4 belt $2 \times 4\ m^2$; cap. $5\ m^3\ h^{-1}$.

4.4 *Practical application and costs*

The calculated amount for a population equivalent of 2600 kg of dry solids per shift can be handled by two belt-press filters of type 20/4, on the assumption that the treated product will contain 4 per cent dry solids.

The total cost of sludge treatment with the use of a belt-press filter including depreciation, maintenance, operation, energy and flocculation agent is calculated at Nfl. 140 1000 kg^{-1} sludge dry solids. At a total sludge production of 15 kg sludge dry solids per inhabitant per annum, the annual costs will be Nfl. 2·00 per population equivalent.

5. GENERAL CONCLUSIONS

The sludge from an oxidation ditch can be dewatered to 15–20 per cent dry solids with the aid of centrifuges and belt-press filters.

A flocculation agent is essential for sludge dewatering. A dosage of 4 g flocculation agent kg^{-1} sludge dry solids is required in the centrifuges and one of 3 g kg^{-1} in the belt-press filters, if the aim is a return of at least 90 per cent.

The costs of both dewatering processes is largely dependent on the cost of the flocculation agent. On the basis of the present day price of Nfl. 15 kg^{-1} flocculation agent, 40–50 per cent of the annual costs of the centrifuges are for the flocculation agent and 30–40 per cent of those of the belt-press filters.

Recently a flocculation agent costing Nfl. 3 kg^{-1} has come onto the market. Tests on a semi-technical scale with a belt-press filter have shown that the dosage with this product has to be higher but nevertheless its use can result in a reduction of approximately 25 per cent in the annual costs of the mechanical dewatering.

Tests have yet to be made to see whether this new flocculation agent produces a sufficiently stabile floc to be used in the centrifuge process.

Water Research Pergamon Press 1972. Vol. 6, pp. 515–516. Printed in Great Britain

SLUDGE DEWATERING TESTS WITH A BELT PRESS

KLAUS R. IMHOFF

Ruhrverband, Kronprinzenstrasse 37, D-43-Essen, FRG

FIVE years ago a conveyor screen press or a belt press was introduced by a German machine factory. Such devices hitherto were known only in paper industry. The application of the procedure for sludge dewatering was new.

The machine has a lower screen belt (SB) of Trevira and stainless steel with meshes of 0·1 × 0·42 mm and an upper compression belt (PB) of rubber (FIG. 1). The speed of the belts and their distance from each other can be adjusted. The wet sludge is

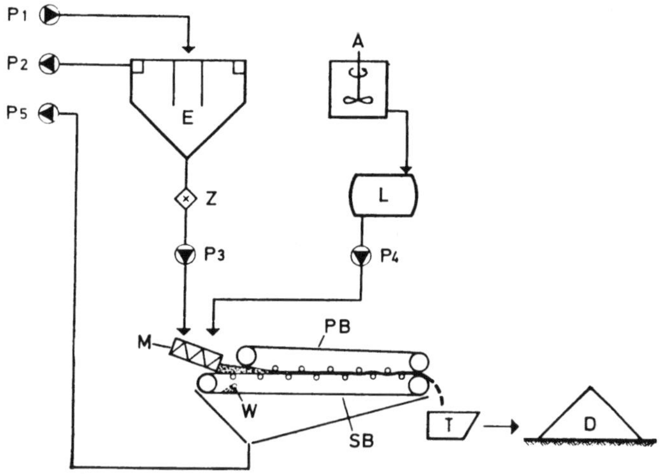

FIG. 1. Diagram of fresh sludge dewatering (conveyor screen or belt press) and a cone of the dewatered sludge.

entirely flocculated in a mixing drum with high polymer flocculants and dewatered in three zones. In the pre-dewatering zone, part of the water runs off freely. On the screen a fine filter develops. In the adjacent zone the sludge is compressed with increasing pressure. After passing the shearing zone which follows the sludge is rejected, whereas the screen belt is lead back and cleaned by a spray nozzle (W).

The development of effective flocculants was essential for the successful use of the conveyor screen press. Flocculants must be carefully added into the mixing device to achieve an effective flocculation just in the pre-dewatering zone. Mixing over a too long period turned out to be unfavourable. The flocculant delivered as powder is mixed in tank (A), stored in tank (L) and added to the sludge by the pump (P4).

The Ruhrverband carried out several full-scale tests in 1967 and 1968 with a conveyor screen press of 0·5 m width. The results are summarized in FIG. 2. The solid concentrations obtained ranged from 16 to 25 per cent. As expected aerobic stabilized sludge and excess sludge showed unfavourable results. Raw sludges consisting of primary and trickling filter sludge (No. 4 in FIG. 2) dewatered more readily than those consisting of primary and activated excess sludge (8). This statement

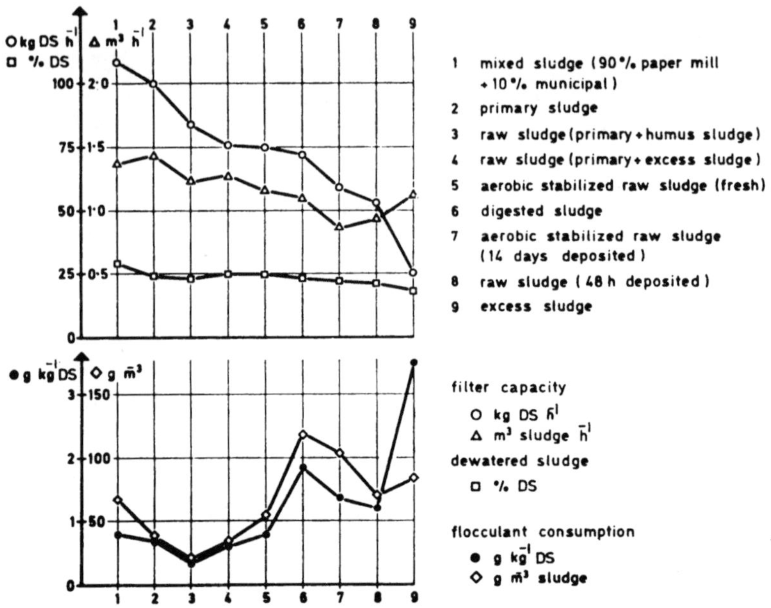

FIG. 2. Dewatering data for several types of sludge (conveyor screen or belt press of 0·5 m width).

applied also to the throughput (kg DS h^{-1}). The consumption of flocculants amounted to 40–120 g m^{-3} sludge resp. 0·4–3·4 g kg^{-1} DS.

For cleaning the screening belt 0·6–1·3 m^3 h^{-1} rinsing water was used. The solid concentration of the filter effluent averaged 1 g l^{-1}. After separation of the easily settleable solids the BOD_5 of the filter effluent amounted to 350 mg l^{-1} with raw sludge and to 1200 mg l^{-1} with digested sludge.

The dewatered sludge removed from the belt was made into a cone shaped heap and its behaviour then observed. The pile dried on the surface to 60 per cent DS, but took up water up to 25 per cent during rainfall. Neither erosion due to rain nor offensive odours developed. The three years observation of the pilot deposit showed that the upper layer of about 20 cm became more and more like soil, and digestion occurred inside the heap.

Since then the Ruhrverband has installed 2 conveyor screen or belt presses in operation for a plant of 30,000 inhabitants. In this case, due to galvanizing wastes, the wastewater is only chemically treated. Another dewatering plant for 50,000 inhabitants is nearly finished. In a third case we have ordered conveyor screen presses and centrifuges for a works dealing with the sewage from 400,000 inhabitants. Here the sludge will be burned in fluidized bed furnaces.

We regard the conveyor screen press as a useful dewatering machine, which can be mainly used in smaller and medium sized plants.

Water Research Pergamon Press 1972. Vol. 6, pp. 517–520. Printed in Great Britain

THERMAL CONDITIONING TESTS OF ACTIVATED SLUDGE AND ANAEROBIC DIGESTION TESTS OF THE FILTRATES

K. H. KALBSKOPF
Emschergenossenschaft, D-43 Essen, FRG

THE DEWATERING of excess activated sludge (about 2000 m^3 day^{-1} with 5 per cent solids concentration) obtained from the Emscher mouth treatment plant is possible only after it has been conditioned. Lime, iron chloride or ash should not be used in order to make sure the dewatered sludge maintains a higher thermal value than would be possible with chemical conditioning. This is of special importance for sludge disposal in the Emscher mouth treatment plant, as the dewatered sludge is incinerated in an electric power station.

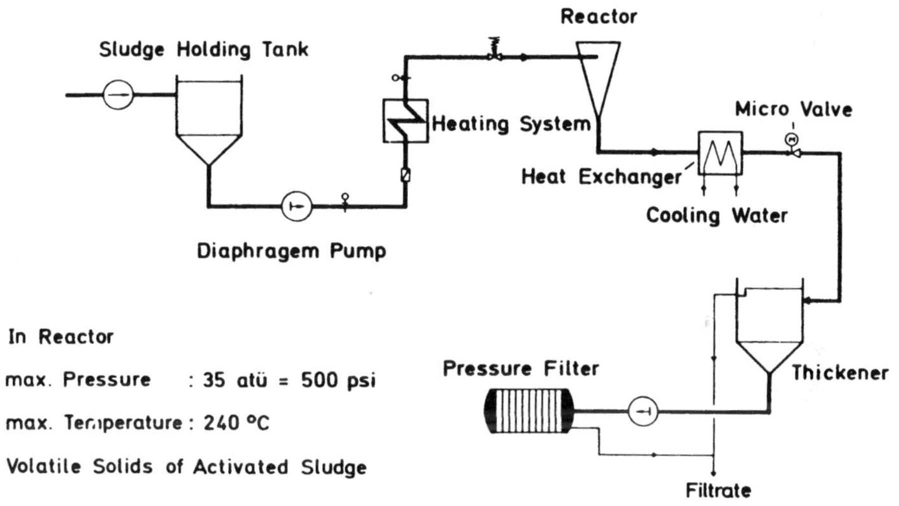

FIG. 1. Sludge heat treatment process—pilot plant.

To determine the optimal operating conditions for the heat treatment process of the activated sludge from this treatment plant, a small scale testing unit shown in FIG. 1, was built on the pilot station of the treatment plant. The sludge in the pipe system of this testing unit is indirectly heated electrically. Temperatures up to 240°C are used. The detention time of the sludge in the reactor varies between 5 and 45 min. FIGURE 2 shows the improvement obtained in the filterability of the heat-treated activated sludge as a function of temperature, and FIG. 3 as a function of the detention time in the reactor in terms of the specific filter resistance. It was found that, with a higher volatile solids content of the activated sludge, higher temperatures or longer reaction times have to be chosen in order to obtain a specific resistance below $1{\cdot}0 \times 10^{12}$ (1 cm^{-2}), when the sludge can be dewatered in pressure filters in less than 100 min to about 50–60 per cent solids content. The calorific value of such sludges amounts to about 1400–1800 kcal kg^{-1} sludge.

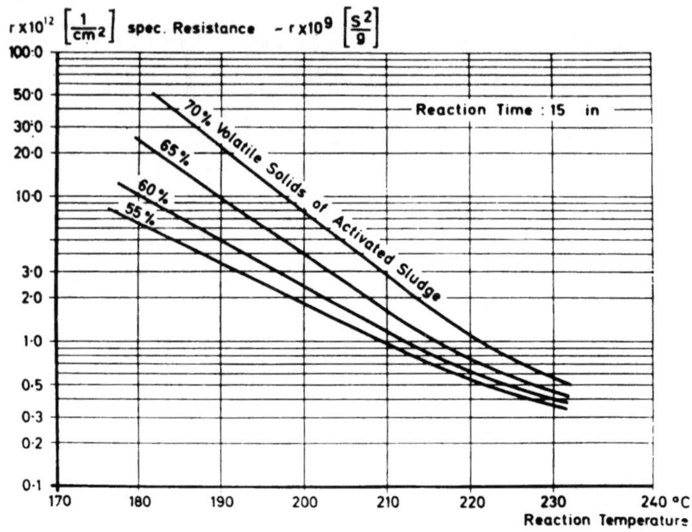

FIG. 2. Influence of reaction temperature on spec. resistance.

The heat-treated sludge filtrate is very strong, having a BOD of 15,000 mg l^{-1}, 20,000 mg $KMnO_4$ l^{-1}, 80,000 mg COD l^{-1} and 6000 mg organic C l^{-1}. Direct addition to and simultaneous treatment of the filtrates in the activated sludge tank

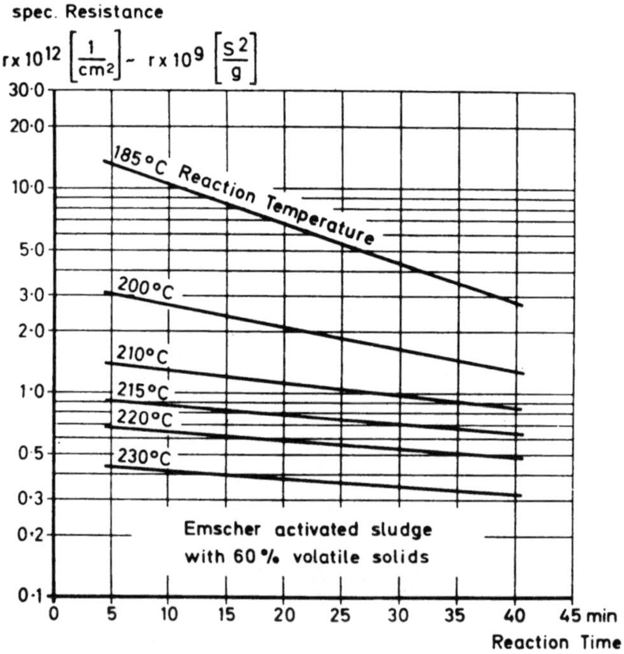

FIG. 3. Reaction time—spec. resistance.

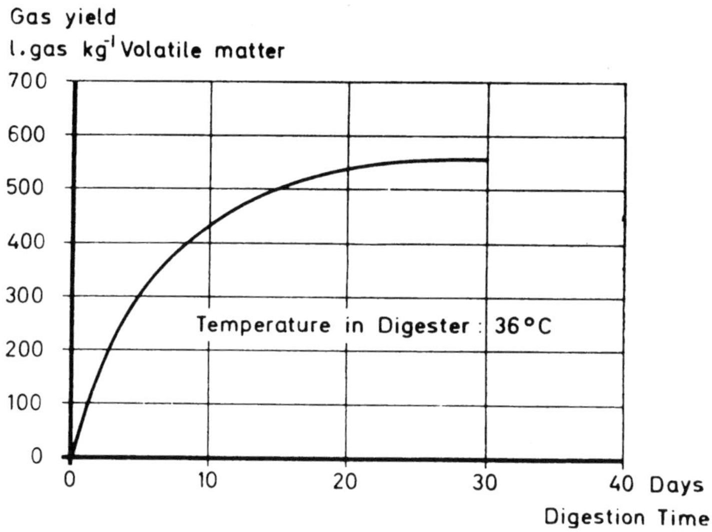

FIG. 4. Specific gas yield–digestion time of filtrates.

is possible. Tests showed, however, that pre-treatment of the filtrates by digestion is more economical. With the usual digestion process at 36°C, the removal of BOD_5 and $KMnO_4$, as a function of the detention time in the digestor, as shown in FIG. 4 could be reached. FIGURE 5 shows the gas yield obtained.

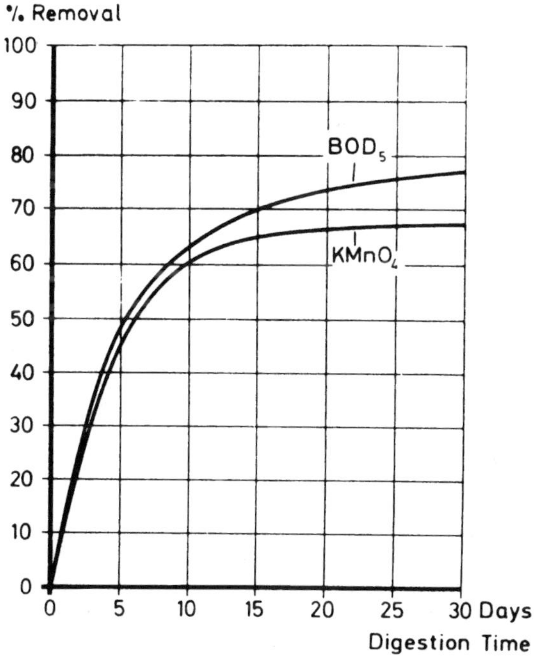

FIG. 5. BOD_5 and $KMnO_4$ removal digestion time of filtrates.

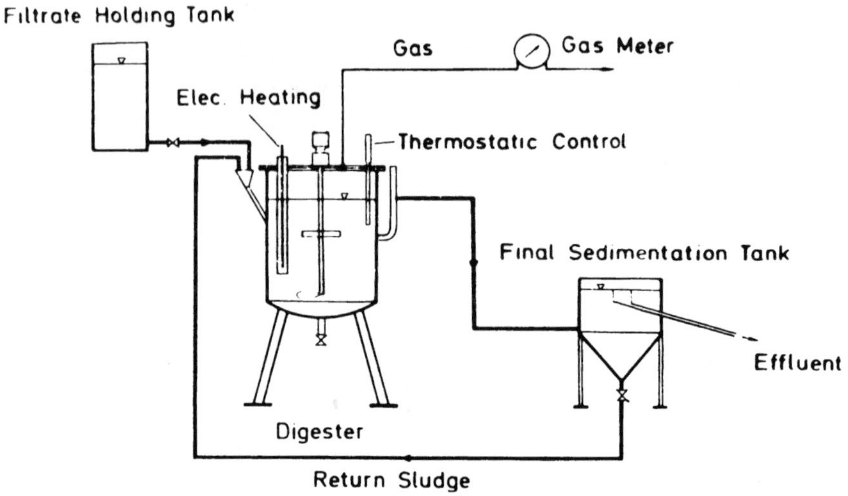

FIG. 6. Pilot plant for filtrate digestion.

The improvement of the process by recycling the anaerobic sludge according to the method of the activated sludge process, as shown in FIG. 6, resulted in a quicker decomposition of the organic matter within 2–6 days. According to the first results, further acceleration of the decomposition is possible by thermophile digestion at 55°C.

Water Research Pergamon Press 1972. Vol. 6, pp. 521–525. Printed in Great Britain

COMPOSITION OF THE "FILTRATE" FROM THERMALLY CONDITIONED SLUDGES

F. SARFERT

Berliner Entwässerungswerke, D-1 Berlin 31, Eisenzahnstrasse 32, FRG

THIS paper gives the results of an investigation on the filtrates obtained by thermally conditioning sludges from predominantly municipal plants. Analyses of such filtrates are given in TABLE 1.

Conditioning was carried out at 170–210° C, with reaction times of 30–90 min. Under these conditions no significant difference in the composition of the "filtrate" could be found. However, the filter resistance of the conditioned sludges greatly depends on the temperature and the reaction time. With an increase in temperature and an increase in reaction time, the filter resistance decreases remarkably, i.e. the sludge can be filtered far better (FIG. 1).

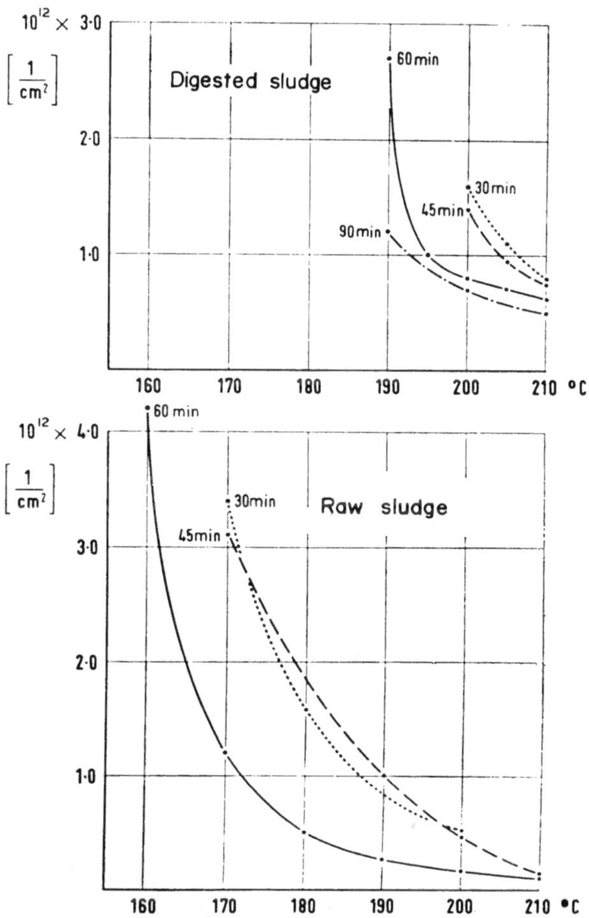

FIG. 1.

TABLE 1. CONCENTRATION OF FILTRATE (THERM. COND.)

Conditions	Author	Sludge	TS (%)	Part ÜS (%)	Filtrate constitution CSB ($mg\,l^{-1}$)	KMnO$_4$ ($mg\,l^{-1}$)	BSB$_5$ ($mg\,l^{-1}$)
Thermal (180–215°C 15–90 min)	MANN (1969)	Raw	3	20–80	13,380	—	6870
	MANN (1969)	Raw	5·6	—	15,780	—	8730
	MANN (1969)	Digested	4·3	—	16,640	—	5680
	BROOKS (1968)	Activated	1	100	5450	—	2800
	HARTMANN (1968)	Digested	7	—	—	—	8000
	EMSCHER-GENOSSENSCHAFT	Activated	5	100	13,000	20,000	15,000
	BERLINER ENTWÄSSERUNGSWERKE	Raw	5	50	12,500	13,000	10,000
	ENTWÄSSERUNGSWERKE	Digested	5	—	12,000	13,000	5000
Low oxidation (5–10% oxidation)	ERICKSON and KNOPP (1971)	Raw	?	?	11,500	—	5270
High oxidation (*ca.* 70% oxidation)	ERICKSON and KNOPP (1971)	Raw	?	?	7960	—	4435
	HURWITZ and DUNDAS (1960)	Activated	2·5–6	100	8300–11,600	—	4890–7110

TABLE 2. CONCENTRATION OF FILTRATE AND/OR DIGESTER SUPERNATANT

	Raw sludge thermally conditioned	Digestion + chemically conditioned			Digestion + chemically conditioned		
$(mg\ l^{-1})$	Filtrate	Dig. Supernatant	Filtrate	Dig. supernatant + filtrate (1:1)	Dig. Supernatant	Filtrate	Dig. Supernatant + filtrate (1:1)
BSB_5	10,000	1000	250	625	1000	5000	3000
CSB	12,500	9000	1000	5000	9000	12,000	10,500
org. C	6500	3000	300	1650	3000	5500	4250
$KMnO_4$	13,000	7500	500	4000	7500	13,000	10,250
NH_4-N	650	500	500	500	500	650	575
org.-N	1000	950	50	200	350	900	625
PO_4 (total)	85	1000	200	600	1000	150	575

The BOD_5-content of the filtrate of thermally conditioned digested sludges is 50 per cent lower than that from raw sludges. The COD-values are about the same with both types of sludges.

Mann obtained higher COD-values for digested sludge (TABLE 1). The high values are caused by the high COD-values in the filtrate of the unconditioned sludge; these values are high because of much industrial effluent in the wastewater.

We realized, as did other workers, that by conditioning activated sludge a higher amount of organic matter will go into "solution" than when conditioning primary sludge. Where primary and secondary sludges are together conditioned, the influence upon the activated sludge is such that, in our work, a smaller part of the activated sludge will go into "solution" than when it is treated alone.

With an increase in the sludge concentration of the sludge to be conditioned, the contents of organic matter in the "filtrate" also increases. We obtained, when conditioning raw sludges, e.g. the following results:

	Filtrate	
Dry matter (%)	COD (mg l^{-1})	BOD_5 (mg l^{-1})
around 3·5	9000	7500
around 5·0	12,500	10,000

The same dependence exists when conditioning digested sludges.

TABLE 2 presents data on the filtrate from a sludge with 5 per cent dry matter. This sludge contained 50 per cent primary and 50 per cent excess sludge. The sludge was treated thermally as raw sludge, another part was digested and afterwards it was divided again; one part was chemically conditioned with aluminium-chlorhydrate. The other was thermally conditioned. The data given were obtained with sludges from the treatment plant at Berlin–Ruhleben, which treats domestic sewage. The raw sludge was conditioned for 45–60 min at 190–200°C, and the digested sludge for about 60 min at 200–210°C, respectively. The data given for the digester supernatant (F_w) and the filtrate from digested sludge ($F_{f\ chk}$) are the mean of numerous results. The digester supernatant contain an average dry matter content of 1 per cent. The filtrate from digested sludge was obtained after chemical conditioning with aluminium chlorhydrate. We assumed that 50 per cent of the mixture was digester supernatant, and the other 50 per cent filtrate was from digested sludge which originated either from chemically or thermally digested sludges. The whole mixture has about the composition given in the rows 4 and 7, respectively. When conditioning raw sludge, there naturally is only the "filtrate".

The quantity of strong wastes from sludge handling normally amounts to less than 1 per cent of the total wastewater quantity. Regarding the aerobic biological treatment of "filtrates", extensive investigations have been undertaken by the "Berliner Entwässerungswerke" and results will be published soon. We can state, however, that the increase in the effluent from our treatment plant (COD, TOC were measured) was proportional to the increase in the load to our plant due to the "filtrate".

REFERENCES

BROOKS R. B. (1965) Heat treatment of activated sludge. *Wat. Pollut. Control* **67,** 592–601.

EMSCHERGENOSSENSCHAFT (Personal communication).

ERICKSON A. H. and KNOPP (1971) Biological treatment of thermally conditoned sludge liquors. In: *5th International Water Pollution Research Conference* 1970. Pergamon Press, Oxford.

HARTMAN (1968) Die Schlammbehandlung im Klärwerk Nürnberg I. *Berichte d. ATV*, H. 22 "*Nürnberger Tagung*", pp. 27–43.

HURWITZ E. and DUNDAS W. A. (1960) Wet oxidation of sewage sludge. *Wat. Pollut. Control Fed.* **32,** 918–929.

MANN T. (1969) Versuche zur thermischen Aufbereitung von Klärschlamm. Kommunalwirtsch. **9,** 374–378.

Water Research Pergamon Press 1972. Vol. 6, pp. 527–529. Printed in Great Britain

DEWATERING OF SLUDGE

W. H. MEREDITH

Howard Humphrey & Sons, Consulting Engineers, Westminster House, West Street, Epsom, Surrey, England

A DECISION on the mechanical dewatering of raw and digested sludges calls for an initial appraisal of the outline economics related to the budgetary costs of capital construction and estimate of operating costs. The latter at times can be the subject of material revision when the plant is installed and operated, and demands a firm performance undertaking from the plant contractor at tender stage.

Whilst pilot experimentation will give an indication of the feasibility of the particular sludge conditioning and dewatering arrangements, many aspects of problems associated with continuous 24 h a day operation are revealed only with full scale operation. My particular experience with thermal conditioning of sludge at a works in England demonstrates this.

The plant concerned is rated to process some 2,000,000 imperial gal (9000 m^3 say) of digested sludge per 7-day week, and the initial outline economics indicated that heat conditioning and pressing was the most favourable arrangement. The relative basic figures were:

	Heat conditioning and pressing (£)	Chemical conditioning and pressing (£)
(£100 = $240)		
Labour	5000	4000
Fuel oil	2500	—
Power costs	1200	1500
Chemicals	—	9000
Maintenance	1200	600
Annual costs	£9900	£15,100
Capital costs—M/c.	80,000	80,000
—Civil	90,000	58,000
	£170,000	£138,000

In the event, the degree of plant supervision and maintenance required has proved materially in excess for the heat conditioning plant to an extent which absorbs the originally assessed differences in operational costs. An up-to-date assessment of the comparative costings is as follows:

The heat conditioning plant as installed was geared to a duplication of capacity for which operating costs per annum are assessed on the basis of processing 18×10^6 gal yr^{-1} (820,000 m^3 say) digested sludge at 97% water content.

	Heat conditioning and pressing (£)	Chemical conditioning and pressing (£)
Labour	9000	6000
Fuel equivalent	3600	—
Power costs	2000	2000
Odour control	2000	—
Chemicals	—	9000
Maintenance	3000	1000
	£19,600	£18,000

	£	£'s per 1000 gal
Labour	9000	0·5
Power	2000	0·11
Incidental	1000	0·05
Maintenance	3000	0·16
Odour control	2000	0·11
Fuel equivalent	4000	0·22
	£21,000	£1·15

Capital costs are equivalent to £1·50 per 1000 gal (4·5 m^3 say) with handling to dump on site at £0·15. This gives an inclusive cost of some £2·80 per 1000 gal.

A similar rated plant using chemical conditioning shows an overall costing at some £3 per 1000 gal. This plant is without the attendant maintenance and odour problem of the heat conditioning plant. Arising from these costings, there is no doubt we have a much clearer understanding of the relative merits of chemical and thermal conditioning processes with the balance definitely in favour of the former.

The performance of the thermal conditioning plant is materially affected by the progressive clogging of the heat exchanger tubes particularly at the primary section of the arrangement.

The incidence of incremental clogging is materially affected by the water content of the digested sludge stock feed. Experience over some 12 months indicates that the optimum water content is at 97 per cent for the digested sludge. At this level an effective continuous operation over 21–28 days can be achieved. At 96–96·5 per cent the period is reduced to less than 4 days. This experience emphasizes the necessity for the contractual performance to be linked to a definite range of sludge water content.

The main design factor arising from the operational difficulties is the active consideration of an alternative arrangement for the primary tubes, at which the conditioned sludge gives up heat to the feed stock sludge. This primary section consists of a 2·5-in. bore inner tube set in a 4 in. bore tube. The main incremental clogging occurs at this 4 in. bore and the indications are that a sludge to water alternative arrangement at this primary section would naturally reduce the incidence of clogging, and of course, effectively double the necessary length of tubes.

The odour problem at this particular plant is mainly related to the separated decantrate waters from the heat conditioned sludge and the press filtrate which follows. A fall off in performance at the primary tubes leads to a rising temperature of the decantrate waters separated out from the conditioned sludges before pressing. If this temperature exceeds a value of 43–46°C local aerial nuisance is set up and working conditions at the press house become more difficult despite the controlled ventilation system as installed.

The overall performance of the plant is very adequate with good sludge-press cakes at 50 per cent water content at pressing periods of 3–4 h coupled to some 20 min conditioning period at not less than 188°C. A fall in this conditioning temperature calls for longer press periods and acceptance of wetter press cakes—60 per cent water content.

It is hoped that these few introductory notes which cover some salient factors of observed plant operation will give some lead to the adjustment of detailed planning and design in the field of sludge conditioning and dewatering.

Water Research Pergamon Press 1972. Vol. 6, p. 531. Printed in Great Britain

DISCUSSION

SEPARATION OF ACTIVATED SLUDGE FROM MIXED LIQUOR USING A CONTINUOUS CENTRIFUGE

A. B. WHEATLAND
Water Pollution Research Laboratory, Elder Way, Stevenage, Herts.

A FEW tests had been carried out by the author at the Water Pollution Research Laboratory to determine whether a disc-type centrifuge could be successfully used to separate activated sludge from mixed liquor for recycle to an aeration tank. The sludge used in the tests was from the Letchworth sewage treatment works.

The results of the tests are summarized in TABLE 1.

It was found necessary to screen the mixed liquor to remove particles greater than 0·38 mm dia. in order to avoid the nozzles of the centrifuge becoming clogged. Under all the conditions tested the clarified overflow liquor from the centrifuge had a higher turbidity and BOD than supernatant liquid obtained by settlement. The quality of the overflow liquor was generally satisfactory when the feed rate was 0·13 $m^3 h^{-1}$ but the condition of the separated activated sludge was very different from that obtained by settlement. It had been subjected to intense shearing forces and appeared to have been homogenized. If sludge in this condition were returned to an aeration tank it seems likely that there would be a steady deterioration in the condition of the activated sludge until the performance of the plant was completely unacceptable.

The centrifuge used was stated by the supplier to have a throughout of 1/38th of their standard machine for which a 20-h.p. motor was employed. Calculations indicated that the power required to separate 4540 m^3 (1,200,000 U.S. gal) of mixed liquor per day would be about 350 kW. The cost of this would probably be prohibitive.

TABLE 1. RESULTS OF TESTS WITH A CONTINUOUS DISC-TYPE CENTRIFUGE ON ACTIVATED SLUDGE MIXED LIQUOR WHICH HAD BEEN SCREENED TO REMOVE PARTICLES GREATER THAN 0·38 mm

Number of test	1	2	3	4	5	6	7	8
Nozzle size (mm)	0·3	0·4	0·5	0·5	0·5	0·5	0·5	0·5
Number of nozzles used	2	4	4	2	2	2	2	1
Number of nozzles clogged	2	3	1	0	0	0	0	0
Rate of flow ($m^3 h^{-1}$)		0·31	0·47	0·41	0·36	0·27	0·19	0·13
Sludge volume (per cent)	28	49	—	35	34	44	64	50
Overflow volume (per cent)	72	51	—	65	66	56	36	50
Suspended solids in feed (mg l^{-1})*	3440	3570	3570	4700	4700	4700	4700	4700
Suspended solids in overflow (mg l^{-1})	18	42	—	210	92	24	V. small	V. small
Suspended solids in sludge (mg l^{-1})	8960	7900	—	15,300	11,500	10,200	6100	8300
Volume of solids in feed settled 0·5 h (per cent)	19	19	19	25	25	25	25	25
Permanganate value (mg l^{-1})								
Feed settled 1 h	—	—	—	7	7	7	7	7
Overflow	11	—	—	40	23	11	8	9
5-day BOD (mg l^{-1})								
Feed settled 1 h	—	—	—	17	17	17	17	17
Overflow	—	—	—	100	63	36	26	24
Clarity of overflow	Reasonable	Reasonable	V. poor	V. poor	V. poor	Poor	Reasonable	Reasonable

*Rounded values.

Water Research Pergamon Press 1972. Vol. 6, pp. 533–537. Printed in Great Britain

PRACTICAL EXPERIENCES OF THERMIC SLUDGE DRYING ON THE CENTRAL TREATMENT PLANT OF NEERSEN

TH. WÜSTEN and E. ZINGLER
Niersverband, Postfach 529, D-409 Viersen, FRG

1. INTRODUCTION

THIS report should give a summary of practical experience and costs of the thermal sludge drying equipment on the sewage treatment plant of Neersen (Niersverband).

The sewage treatment plant receives an impaired-load equivalent to a population of about 600,000. The plant consists of primary settling tanks, coagulation tanks (copperas), preareation with settling tanks and the biological main stage with activated sludge treatment. The raw sludge is digested for about 21 days at 30°C (86°F). The most important characterization of the raw sludge are listed in TABLE 1.

TABLE 1. RAW SLUDGE CHARACTERISTICS

Raw sludge 5·200 m^3 week^{-1} (1,370,000 gal week^{-1})
or 1·000 m^3 day^{-1} (264,000 gal day^{-1})
Solids concentration 6 per cent—weight
Volatile solids 50–60 per cent (combined system of sewerage)
Dried solids per day 70–80 tons day^{-1}

Four-thousand and five-hundred cubic metres (1,186,000 gal week^{-1}) of the weekly volume of sludge of 5200 m^3 are pasteurized and utilized by spraying on agricultural areas. This agricultural utilization is dealt within another paper.

The remaining sludge of about 700 m^3 week^{-1} (185,000 gal week^{-1}) is dried in a selective drier and then sold to vine dressers etc.

2. TECHNICAL DESCRIPTION (FIG. 1)

2.1 *Sludge dewatering*

The digested sludge is lifted by Mohno-pumps from the storage tank to a mixing chamber. Here it is coagulated with lime (25 kg m^{-3} or 0·41 lb gal^{-1}) and a 6 per cent aluminium solution (20 kg m^{-3} or 0·33 lb gal^{-1}). At the inlet mixing is done by an impeller with high turbulence, and paddles assist in the coagulation. Another Mohno-pump lifts the sludge through a distribution system to four rotary drum filters (type Buckau–Wolf). The dewatering equipment and results are given in TABLE 2.

The capacity of sludge drying plant limits the full use of the drum filters. To this drying plant the filter cake is transported by conveyor belts. The filtrate with a BOD of about 350 mg l^{-1} lifted to the biological stage of the treatment plant.

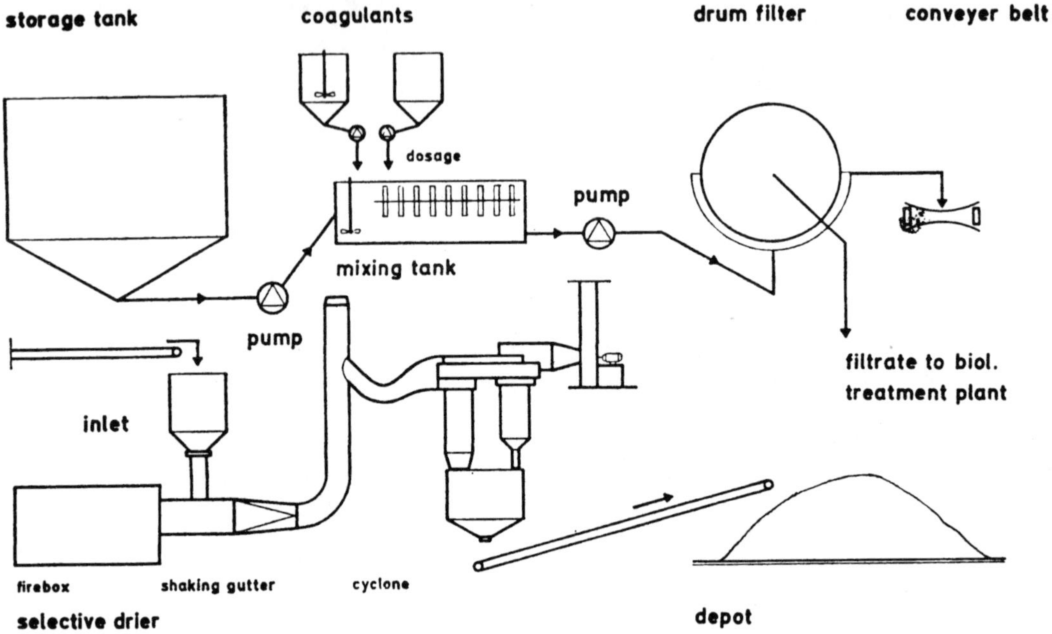

FIG. 1. Flow diagram.

TABLE 2. CHARACTERISTICS OF SLUDGE DEWATERING PLANT

Drum filters	4
Filter surface	116 m^2 (1250 ft^2)
Filter capacity	20 $m^3\ h^{-1}$ (5270 gal h^{-1})
(in dry solids)	1·2 tons h^{-1} (solids)
Specific filter capacities	170 l. $m^{-2}\ h^{-1}$ (4·2 gal ft^{-2})
(dry solids)	10·2 kg $m^{-2}\ h^{-1}$(4230 gal h^{-1})
Filter cake	4 $m^3\ h^{-1}$ (1040 gal h^{-1})
Final moisture content	70 per cent—weight

2.2 *Sludge drying plant*

The filter cake is dried in a selective drier (type Hazemag) fired with digestor gas. The performance estimated by the manufacturers has not been met. The required constancy of performance is one reason for the lower capacities, as will be demonstrated later.

Comparison of the estimated and actual performance is given in TABLE 3.

The residual product will be sold as produced after adding peat litter and artificial manure to small farmers or vine dressers by a private contractor. The costs will be given later.

TABLE 3. OPERATING RESULTS OF THE SLUDGE DRYING PLANT

		Practical	Estimated
Moisture content			
Filter cake	%	70	76
Dried material	%	45–50	40
Temperatures			
Fire box	°C	650	800
Waste gas	°C	165	180
Plant capacity	tons h^{-1}	3·0	5·3
Evaporation	tons h^{-1}	1·3	3·0
Residual product	tons h^{-1}	1·7	2·3
Digestor gas	m^3 h^{-1}	400	—

3. OPERATING EXPERIENCES

The drying process is controlled automatically. The dewatering and drying plant is operated by one qualified worker. For the filter cloths must be cleaned every 120–150 h of operation, which is costly. The durability of the filter cloths is about 6000 h of operation. Sometimes interferences are caused by the vacuum and filtrate pumps and by the steering mechanism of the drum filters.

The selective drier first started operation in 1962. Many improvements in operation have been made since 1968. The intended drying temperature of 800°C (1472°F) could not be maintained. The moving parts of the shaking gutter and the hinged supports were stressed so much by high temperatures that after a short time of operation they became defective, in spite of the use of heat-resisting materials.

Generally the following points cause interferences:

(a) shaking gutter and shock adsorbers;
(b) the charging valve of the loading station with geared engine and chain drive;
(c) outlet valve;
(d) cyclone.

The most important requirement for continuous operation is a constant moisture content of the filter cake of about 70 per cent. If this content is higher clogging occurs. Sedimentation often takes place in the cycle. The deposit must be removed manually. The causes of sedimentation are too low a velocity of the heat gases and the fact that the dried filter cake still has a high moisture content. Dust separation by the cyclone is insufficient. The design did not pay regard to the fact that dried cake is not homogeneous in grain size and moisture content. Therefore fine glowing grains are blowed out of the cyclone. Some times there were small fires on the roof of the service station.

Another disadvantage of insufficient dust separation is dust deposit on machines and gears of other equipment of the sewage treatment plant. The vent holes of motors were clogged and the machines were damaged. Therefore extensive servicing was necessary of equipment which did not belong to the drying plants.

4. OPERATING COSTS

The following operating costs are calculated for the first 4 months of this year (1971).

4.1 *Fixed costs*

Costs of:	DM yr^{-1}
amortization	
constructions 3 per cent of 700,000 DM	21,000
machines, electr. installations 7 per cent of 1,640,000 DM	114,000
interests, 5 per cent of 50 per cent of the capital invested of 2,340,000 DM	58,500
continuous maintenance 1 per cent of 2,340,000 DM	23,400
Yearly fixed costs	217,700
Fixed costs for 4 months:	72,600

4.2 *Costs of operation*

	DM
(a) Wages and costs of repairs	
wages	15,160
transportation	8500
repairs wages	18,860
material	3000
	45,520
(b) Coagulants	
224 tons lime 79·15 DM ton^{-1}	17,730
166 tons $AlCl_3$ (6 per cent) 51·20 DM ton^{-1}	8500
special salt for vacuum pumps	300
	26,530
(c) Rolling stock (4 months)	
consumption of current	7970
water	400
lubricants	270
	8640
Amount costs of operating	80,690 DM

4.3 *Summary of costs*

Fixed costs	72,600 DM
costs of operating	80,700 DM
Total costs (4 months)	153,300 DM

4.4 *Costs related to wet sludge and dried matter*

(a) Wet sludge (solids content 6 per cent).
8968 m^3 (237,000 g) in 4 months

related costs $\frac{153,300}{8968}$ = 17·00 DM m^{-3} (0·065 DM gal^{-1})

profit by selling 4·70 DM m^{-3} (0·018 DM gal^{-1})

Costs related to wet sludge 12·30 DM m^{-3} (0·047 DM gal^{-1})

(b) Dried matter (including coagulants)
653 ton in 4 months

related costs $\frac{153,000}{653}$ = 235 DM ton^{-1}

profit by selling 65 DM ton^{-1}

Costs related to dried matter 170 DM ton^{-1}

5. CONCLUSION

The costs related to wet sludge or dried matter are not transferable to other plants. It should be pointed out that more than 14 per cent of the costs are for repairs. Comparing the costs of drying with the costs of pasteurizing and spraying there is the following relation.

drying: 170·00 DM ton^{-1} or 12·30 DM m^{-3}
pasteurizing
and spraying 50·00 DM ton^{-1} or 3·00 DM m^{-3}

These figures show that sludge drying is more costly a method of disposal than spraying on agricultural areas where this is possible.

Water Research Pergamon Press 1972. Vol. 6, pp. 539–544. Printed in Great Britain

SLUDGE FILTER PRESSING AND INCINERATION AT SHEFFIELD

H. B. TENCH

Sheffield Corporation, Water Pollution Control Department, Alsing Road, Wincobank, Sheffield S9 1HF

FIVE sewage works are operated by the Sheffield Water Pollution Control Department and of these, the Blackburn Meadows works is the largest. It serves a population of 510,000, the dry weather flow being 145,000 m^3 day^{-1} (32 m. imp. gal. day^{-1}) of which 25 per cent is trade effluent, mainly from the steel industry.

The filter press plant, one of the largest in the world, was completed in 1962 with the intention that the sludge cake produced would be tipped on a site about 10 km from the works. About 250 tonnes of press cake were then produced each day and very little air drying of each day's make occurred before more cake was tipped on the top of it. As a result the tipped sludge was wet enough to behave as a viscous fluid and make tip management difficult but, more importantly, also wet enough to allow fermentation to occur. This fermentation produced a most objectionable odour and thus, as it made tipping an unacceptable means of disposal, the multiple hearth sludge incineration plant was built and commissioned in 1968.

DESCRIPTION OF SLUDGE DISPOSAL PLANT

The filter press house (HOLROYD, 1964) and incineration plant (HOLROYD, 1969; TENCH *et al.*, in press) have been fully described previously and FIG. 1 shows the sludge disposal process in outline. Only some of the more important aspects are referred to below.

As is well known, the solids content of the feed sludge has great influence on the efficiency of dewatering processes and so the primary sedimentation tanks are designed to allow the mixed activated and crude sludge to consolidate as much as possible before it is pumped to storage tanks ready for chemical conditioning. The trade effluents from the steel industry normally ensure that sufficient iron is present so lime alone is added to condition it. To do this, lime slurry and sludge are instantaneously and thoroughly mixed by a flash mixer (propeller) in the pumping main to the conditioned sludge storage tanks. From these tanks it is pumped into the filter presses by compressed air rams.

Each of the 6 air compressors installed feeds 6 filter presses by means of 2 rams, one of which draws sludge under vacuum as the other discharges to the presses at a maximum pressure of 620 kN^{-2} m^{-2} (kiloNewtons per square metre) (90 lb in^{-2}). Thirty of the presses are in use at any one time, the other 6 being out of action for cleaning. The pressing cycle takes 7 h and 420 pressings are normally done in the 120-h working week. Production amounts to about 1800 tonnes per week and this is burnt in about 104 h though an additional 24 h are worked by the plant attendants to supervise the weekend warm up and part of the cooling down procedures. Transportation to the incineration plant is by means of wagons drawn by a diesel locomotive. The wagons carry about 8·4 tonnes of cake and one wagon load is tipped

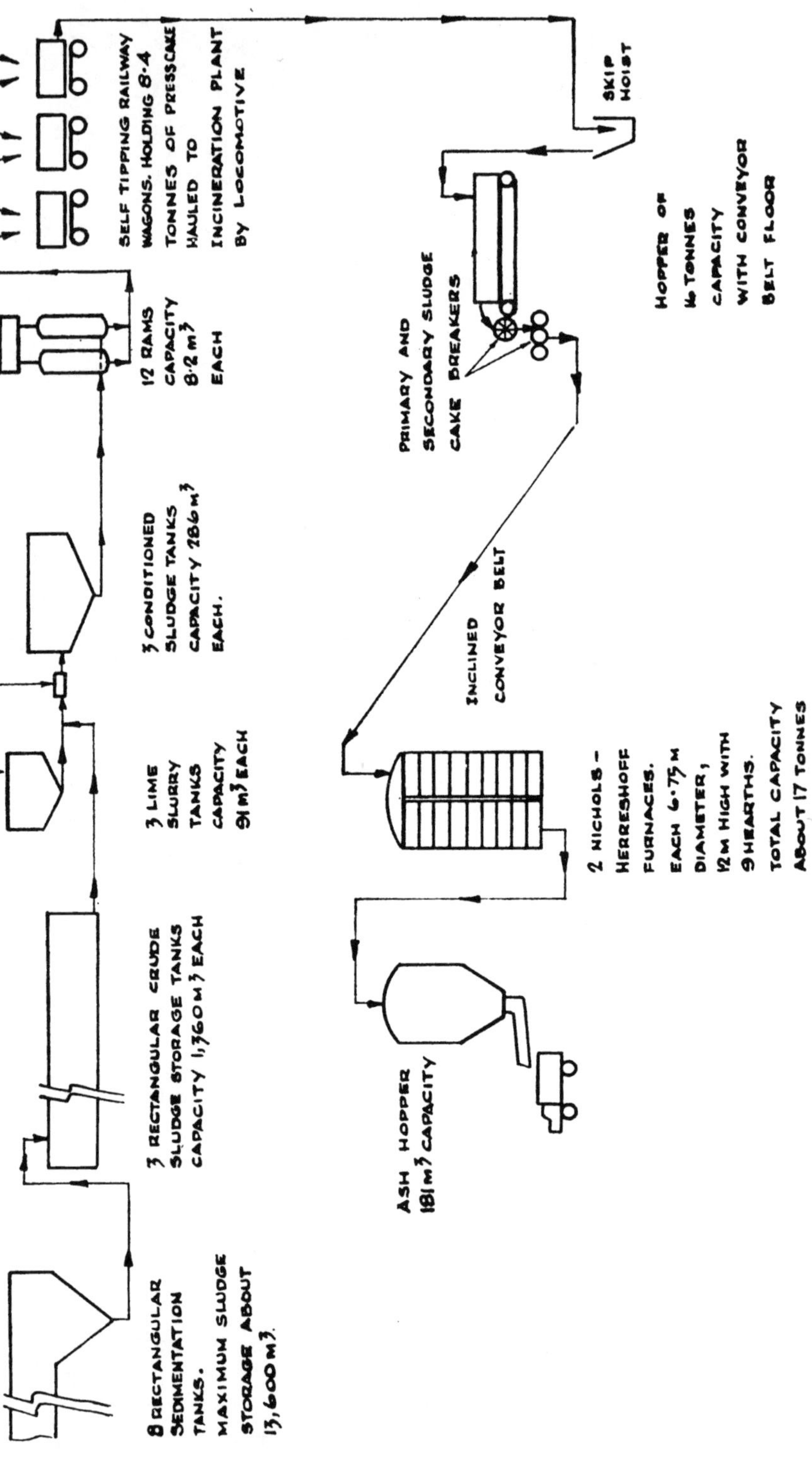
36 FILTER PRESSES 65 CAKES, 1·22 M SQUARE x 38 MM (4·2 TONNES) IN TOTAL PER PRESSING
6 AIR COMPRESSORS CAPACITY 85 L/S EACH.
SELF TIPPING RAILWAY WAGONS, HOLDING 8·4 TONNES OF PRESSCAKE HAULED TO INCINERATION PLANT BY LOCOMOTIVE
SKIP HOIST
12 RAMS CAPACITY 8·2 M³ EACH
PRIMARY AND SECONDARY SLUDGE CAKE BREAKERS
HOPPER OF 16 TONNES CAPACITY WITH CONVEYOR BELT FLOOR
HYDRATED LIME POWDER + EFFLUENT
IN-LINE PROPELLER FOR FLASH MIXING
3 CONDITIONED SLUDGE TANKS CAPACITY 286 M³ EACH.
INCLINED CONVEYOR BELT
3 LIME SLURRY TANKS CAPACITY 91 M³ EACH
2 NICHOLS-HERRESHOFF FURNACES. EACH 6·75 M DIAMETER, 12 M HIGH WITH 9 HEARTHS. TOTAL CAPACITY ABOUT 17 TONNES PER HOUR
3 RECTANGULAR CRUDE SLUDGE STORAGE TANKS CAPACITY 1,360 M³ EACH
ASH HOPPER 181 M³ CAPACITY
8 RECTANGULAR SEDIMENTATION TANKS. MAXIMUM SLUDGE STORAGE ABOUT 13,600 M³.

FIG. 1.

every half hour into a skip hoist which discharges into a hopper with a moving belt floor to provide a continuous feed to the furnaces. For efficient burning it is desirable to break the cake down into pieces no greater than about 30–40 mm dia. and to achieve this, the discharge from the hopper is passed through primary and secondary breakers and then on to an inclined conveyor which feeds the furnaces. These breakers consist of a rotating spiked drum (primary) and 3 rotating shafts with cutting blades form the secondary. The ash produced is discharged to a storage hopper by a drag link conveyor and bucket elevator, and damped by effluent in a conditioner drum to prevent dust blowing before it is transported to tip by lorries holding about 15 m^3.

OPERATING EXPERIENCE

For the first few years of the operation of the filter press plant, the major difficulties experienced were the tipping referred to above and the fact that the stock of 76 wagons was inadequate so that some sludge was pumped to lagoons, which also as a consequence produced an objectionable odour. During this period, experimental work was mainly directed towards finding a better sludge conditioner and the testing of various filter cloths. A large number of polyelectrolytes have been tested and no better conditioner than lime has been found and a "Saran" monofilament cloth has been found to be the best of those tested. These "throwover" cloths cost, at present $9.02 (£3·76) each and have to be renewed after 450–600 pressings.

Initially two pressing cycles were completed in a 12-h day but in 1967 a 2-shift system of 3 pressing cycles per 16-h day was introduced. As a result there was some increase in lime dosage, possibly due to a shortening of the pressing time or to a simultaneous increase in the bending of cloths due to an increase in the number of pressings done between press cloth and drainage surface cleaning.

The incineration plant was commissioned in late 1968 and it was soon obvious that one of the main requirements for its successful operation was consistency in the water content of the sludge cake fed to it. To achieve this, the lime consumption was increased considerably. It was realized that this increase was due to the addition of an excess of lime which was necessary only because the control of the conditioning was inadequate. In June 1969 the use of a Capillary Suction Time (C.S.T.) test on a sheared sample of the conditioned sludge was introduced instead of the Buchner funnel cracking time test previously employed as has been described by SWANWICK (1971). There was an immediate reduction in the lime consumption and a further reduction when the apparatus was equipped with automatic print out to check that the operator was in fact maintaining the C.S.T. of the conditioned sludge at the required level (which is varied from day-to-day by the plant supervisor according to the quality of the cakes produced). The advantages of the sheared C.S.T. test are that it can be done quicker, and thus more frequently, than the Buchner funnel test and that it gives a more reliable indication of the effect of filter pressing. The benefits obtained from the test are shown in FIG. 2 in which lime consumption as tonnes per pressing is plotted as a 12-month moving average (to eliminate seasonal variations). It should be noted that the cost of hydrated lime to Sheffield is currently $17.9 (£7.46) per tonne and that a difference in consumption of 0·06 tonnes per pressing is equivalent to an annual cost of $22,000 (£9,170). Whilst the financial savings resulting

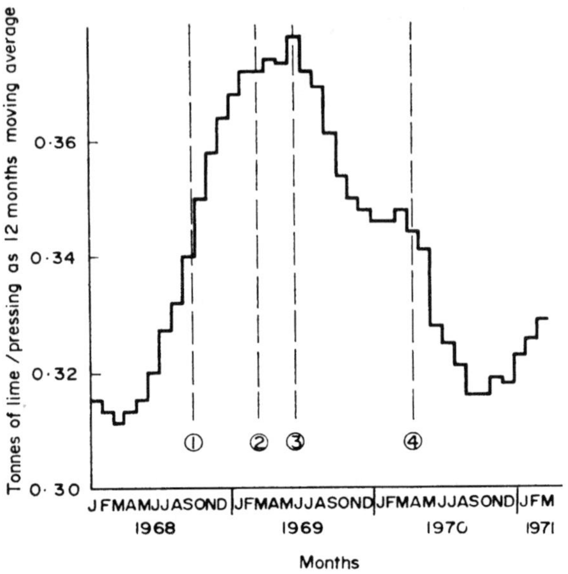

FIG. 2.

Key: (1) Commissioning of the incineration plant started.
(2) Incineration plant fully operational.
(3) Started controlling conditioning by "sheared C.S.T."
(4) Recording C.S.T. apparatus introduced.

from the tighter management control of the conditioning process have been important, perhaps of greater importance has been the greater consistency of the press cakes which has materially simplified the control of the incineration plant.

This multi-hearth plant became fully operational in March 1969 and since that date it has burnt all the sludge produced at the works. It has proved itself quite reliable in operation, though even slight variations in quality of the cake fed can lead to difficulty in maintaining the position of the fire. Initially the rate of variation in fire position could be quite alarming and even though this is no longer the case due to the measures outlined above, it can on occasions be too rapid for manual control to be effective. Experiments are therefore being conducted with a view to automatically controlling the furnaces by a combination of temperature sensing for oil burner control and oxygen in flue gas measurement to control the furnace draught by actuating the spin damper. It is expected that automatic control will reduce burner oil consumption as there is a tendency, under manual control, to use more excess air than necessary in order to maintain a clean plume from the scrubber. This of course produces a greater cooling effect than necessary.

The adequate break up of cake prior to burning has been another difficulty. Experience has shown that lumps of dry cake greater than about 50 mm dia. may not completely burn in the furnace. Apart from the resulting loss in calorific value the only other harmful effect of such lumps is that by their slow combustion, they extend the burning zone and make its control more difficult. Primary and secondary breakers are installed in series at Sheffield. The primary has been little changed since installation but experiments have been done to decide an economic design of secondary breaker which will reduce the lumps to an acceptable size.

There have been some maintenance problems. For example, damage to refractories has been caused by expansion and contraction effects resulting from the shut down practised each weekend. It is proposed to overcome this particular problem by installing automatic and fail safe devices on the burners so that they can be kept on, without attention, to maintain temperature at weekends.

Statistical details of the Sheffield sludge disposal plant are given in TABLE 1 and costs are given in TABLE 2. From these costs it can be seen the Sheffield plant provides an extremely economic means of achieving a socially acceptable disposal of an industrial sewage sludge.

TABLE 1. SLUDGE TREATMENT STATISTICS

Year ending March 31st		1970	1971
Sludge production:			
Volume (10^3 m^3)		400	390
Solid content (%)		6·0	6·2
Ash in solids (%)		42	39
Wt. of dry solids, Tonnes		24,400	24,600
Filter pressing:			
Number of pressings		22,027	20,472
Wt. of cake (tonnes)		92,310	85,800
Hydrated lime used (tonnes)		7644	6775
Ca/O/dry sludge solids (%)		23	20
Cake solids content (%)	Max	43	43
	Min	20	23
	Avge	33	34
Ash in solids (%)	Max	74	68
	Min	44	42
	Avge	57	53
Incineration:			
Electricity consumption (kWh)		1,216,280	1,130,520
Oil consumption (m^3)		1196	879
Oil consumption (l) tonne cake		13·0	10·2
Ash tipped (m^3)		Not available	29,800

Calorific value of dry volatile matter in cake (24 results 1965/71)
Maximum 27,150 kJ kg^{-1} (11,679 Btu lb^{-1})
Minimum 22,280 kJ kg^{-1} (9,620 Btu lb^{-1})
Average 25,040 kJ kg^{-1} (10,760 Btu lb^{-1})

TABLE 2. SLUDGE TREATMENT COSTS

Year ending 31st March	1970 $	1970 £	1971 $	1971 £
Filter pressing:				
Loan repayments	130,673	54,447	130,313	54,297
Operating labour	55,142	22,976	47,806	19,919
Electricity	19,781	8242	19,550	8146
Lime	106,217	44,257	101,765	42,402
Chemicals	8069	3362	5858	2441
Filter cloths	26,198	10,916	28,930	12,054
Repairs	24,490	10,204	34,646	14,436
Supplies and services etc.	5306	2211	6372	2655
Overheads	20,930	8721	22,469	9362
Total:	396,806	165,336	397,709	165,712
Incineration:				
Loan repayments	106,756	44,482	109,521	45,634
Operating labour	38,218	15,924	42,850	17,854
Transport of presscake	9682	4034	15,295	6373
Electricity	19,778	8241	19,546	8144
Gas	2770	1154	886	369
Oil	30,886	12,869	22,457	9357
Repairs	23,954	9981	29,916	12,465
Supplies and services	1728	720	3367	1403
Hire of plant	2434	1014	2074	864
Transport of ash	19,416	8090	24,206	10,086
Overheads	14,467	6028	20,138	8391
Total:	270,089	112,537	290,256	120,940
Cost per tonne of dry sludge solids:				
Filter pressing	16·25	6·77	16·18	6·74
Incineration	11·06	4·61	11·78	4·91
Total:	27·31	11·38	27·96	11·65

REFERENCES

HOLROYD A. (1964) The Sheffield Filter Pressing Plant. *J. Inst. Sew. Purif.* 159.

HOLROYD A. (1969) Recent progress in the Blackburn Meadows Reconstruction and Extension Scheme at Sheffield. *Wat. Pollut. Control* **68**, (4), 357.

TENCH H. B., PHILLIPS L. F. and SWANWICK K. H. (in press). The Sheffield Sludge Incineration Plant. *Wat. Pollut. Control.*

SWANWICK K. H. (1971) Process control at Sheffield's sludge filtration plant. *Filtration Separation* **8** (2), 137.

Water Research Pergamon Press 1972. Vol. 6, pp. 545–549. Printed in Great Britain.

"LIQUID FERTILIZER" TO RECLAIM LAND AND PRODUCE CROPS

BART T. LYNAM, BEN SOSEWITZ and THOMAS D. HINESLY*

The Metropolitan Sanitary District of Greater Chicago, 100 East Eric Street, Chicago, Ill. 60611, U.S.A.

THE Metropolitan Sanitary District of Greater Chicago collects, treats, and utilizes the waste water from Chicago and 120 suburbs. It serves an area of 2220 km^2 (858 $miles^2$) inhabited by 6 million people, plus an industrial waste load equivalent of four million people, for a total population equivalent of 10 million people. It collects and treats an average of 5.1 million kl of waste water each day (1350 mg day^{-1}) from which it removes about 958 equivalent dry metric tons of solids (1050 short tons). The District currently uses four different methods to treat and stabilize these solids, namely: (1) wet combustion (the Zimmerman process); (2) heat drying (and sale as a fertilizer); (3) Imhoff tanks (followed by air drying); and (4) high rate anaerobic digesters (followed by lagooning or farm applications).

To provide an economical solution, eliminate air pollution, and provide a beneficial use for the solids removed in its waste water treatment process, The Metropolitan Sanitary District adopted a policy in 1967 of land application for fertilization of crops and reclamation of land. Comparing the per ton cost of solids disposal methods, land applications (cost of approximately $25 per short ton) have been shown to be more economical than all other methods (costs range from $40 to $60 per short ton) presently available to the District.

Adopting a long range solution to its solids disposal problem, the District established four stringent requirements on any system which would be considered: (1) the disposal method must not pollute the air or the water; (2) it must be economical; (3) it must conserve the organic matter in sludge for a beneficial purpose; and (4) it must solve the disposal problem forever.

Of the alternative methods of solids disposal available to this District, land application is the only method that met these four criteria.

As nutrients for crops, the nitrogen, phosphorus, potassium, and zinc content in a ton of dry solids is sufficient to consider stabilized sewage solids as a "Liquid Fertilizer". The fertilizing value should be kept in mind when other means of disposal are considered.

TABLE 1 shows the chemical composition of digested and lagooned solids and the kilograms of each element applied to the soil in 56.7 metric tons per hectare (25 short tons $acre^{-1}$) application of solids. While digested solids act primarily as a nitrogen fertilizer, it is also an effective source of phosphorus for crops.

The District tested several methods of "Liquid Fertilizer" application and distribution, including ridge and furrow irrigation, spray irrigation and application by tank truck.

At the Hanover Water Reclamation Plant, the District spreads high rate digested sludge by the ridge and furrow method. An underground pipe system transports the

* Respectively, Chief of Maintenance and Operations, General Superintendent of The Metropolitan Sanitary District of Greater Chicago, and Assistant Professor of Agronomy, University of Illinois.

TABLE 1. AVERAGE ELEMENTAL ANALYSES OF SOLIDS SLURRY SAMPLED FROM LAGOON No. 29 (DRY BASIS), 1970

Element	Content (%)	Application rate to soil at 56 tons ha^{-1} (25 short tons $acre^{-1}$)	
		kg ha^{-1}	lb $acre^{-1}$
Nitrogen	1.83	1030	915
Phosphorus	3.53	1985	1765
Potassium	0.185	103.5	92
Calcium	5.11	2875	2555
Magnesium	1.25	704	625
Sulphur	0.92	518	460
Copper	0.29	163	145
Iron	4.00	2250	2000
Manganese	0.034	19.1	17
Zinc	1.02	574	510
Arsenic	0.28 (ppm)	0.0157	0.014
Cadmium	0.065	36	32
Chromium	0.50	281	250
Lead	0.23	129	115
Mercury	3.5 (ppm)	0.197	0.175

sludge from the digesters to gated irrigation pipe which lays on the soil in the field. The gates in the pipe, spaced on 1.02 m centers (40 in.), deliver the sludge to the furrows between the corn which is grown on the ridges.

Each farm plot contains three separate application areas where 0, 0.64 and 1.27 cm (0, 0.25, 0.5 in.) of slurry are applied per week. Weekly applications begin about June 15 and end about September 1. The most unusual feature of the Hanover Research Farm is that the application takes place within 30.5 m (100 ft) of residential homes, yet the District has not had a serious complaint in years.

The District and one of its contractors, SEMCO (Soil Enrichment Materials Corporation), tried various spray irrigation systems. Those systems with nozzles less than 1.9 cm (0.75 in.) in diameter failed; however, SEMCO successfully applied 288,000 wet metric tons (317,000 wet short tons) of "Liquid Fertilizer" using a traveling irrigation system with a nozzle 4.44 cm (1.75 in.) in diameter. This nozzle discharged sludge containing 10 per cent solids at a rate of about 4.3 kl min^{-1} (1200 gal min^{-1}).

The Agronomy Department of the University of Illinois conducted extensive research relative to solids on land for The Metropolitan Sanitary District of Greater Chicago at their Northeast Agronomy Research Center near Elwood, Illinois. After constructing a giant lysimeter system of 44 plots, 3 m wide by 15 m long (10 $\times$ 50 ft) they showed that applications of digested solids increased crop yields. The applications also increased nitrate levels in runoff water and soil leachate but caused no increase in fecal coliform. These applications were in some instances above the anticipated future loadings. In addition, freshly digested drawoff was used which contains a high percentage of available ammonia as compared to organic nitrogen.

The University of Illinois also conducted field research relative to applications with corn and soybeans which showed significant yield increases. TABLE 2 shows corn yields for three consecutive years for different application rates. Initially, the infiltration rate of digested solids slurry into sand is greater than in silt loam soils. However, after a few successive applications, in the absence of complete drying, the infiltration rate varied from 0.06 cm h^{-1} to 0.006 cm h^{-1} regardless of soil type. After a time, the resulting cake, not the soil type, determines the infiltration rate. The infiltration rate of a digested solids slurry into soil depends on the initial soil moisture content, the soil type, and the digested solids content of the slurry. The greater the soil moisture content and the greater the solids content, the lower the infiltration rate. However, antecedent soil moisture conditions affect the infiltration rate less on sandy soils than on loam soils or clay soils. Successive applications that do not allow time for the cake to dry decrease the soil infiltration rate. If allowed to dry between applications, the initial infiltration rate will fully recover.

TABLE 2. CORN YIELDS AND DIGESTED SOLIDS APPLICATION ON BLOUNT SILT LOAM ON N.E. ILLINOIS AGRONOMY RESEARCH CENTER, 1968, 1969 AND 1970

Digested solids application rate (cm $week^{-1}$)	Average corn yields (kg h^{-1})		
	1968	1969	1970
0	4160 (66b/A)	9000 (143b/A)	5540 (88b/A)
0.635 (0.25 in.)	6060 (96b/A)	9390 (149b/A)	7500 (119b/A)
1.27 (0.5 in.)	7180 (114b/A)	9460 (150b/A)	7640 (122b/A)
2.54 (1 in.)	7040 (112b/A)	9480 (151b/A)	8640 (137b/A)

Total liquid slurry applied: 1968—17.0 cm (6.75 in.);
1969—25.4 cm (10.0 in.); 1970—22.8 cm (9.0 in.).
Total 3 yr dry matter applied: 67.9 tons (74.9 short tons).

When the digested solids slurry temperatures are low and air humidity is high, the slurry dries at one-half the evaporation rate of a free water surface. When the slurry temperatures, however, are high and air humidity is low, the rate of drying of the slurry approaches the evaporation rate of a free water surface.

Digested solids slurry greatly improves sandy soils by increasing their organic matter content, which increases their water holding capacity and cation exchange capacity. Applications also benefit soils high in clay by increasing the organic matter content which increases the water absorption capacity and decreases the erodibility of the soil.

The danger of infection to wild-life and domestic animals from these fields depends on the persistence of pathogens on the soil surface. TABLE 3 shows the rapid disappearance of fecal coliforms on a soil surface, when dosed with freshly digested solids.

Research with laboratory scale anaerobic digesters showed that swine enterovirus could not survive in these digesters more than 4–5 days. Since the sludge detention time in the District's anaerobic digesters is 14 days, we are reasonably confident that our digested solids slurry is free of viable virus.

Our experience to date indicates that the heavy metals present will not make the soil toxic.

Relative to ammonia and nitrate concentrations in the surface drainage waters of

TABLE 3. DISAPPEARANCE OF FECAL COLIFORMS IN THE DIGESTED SOLIDS CAKE COVERING A SOIL SURFACE

Days after application of freshly digested solids	No. of fecal coliforms per g digested solids cake (dry weight)
1	3,680,000
2	655,000
3	590,000
5	45,000
7	30,000
12	700

the Elwood lysimeters, concentrations of both decreased with rainfall intensity but depended highly on the application rates and digested solids accumulations on the soil surface. The phosphorus content of the surface drainage waters was highly correlated with digested solids accumulations.

In the ground water the discharge intensity and digested solids accumulations did not affect the concentration of ammonium ions and phosphate ions, but because the nitrate ions are very mobile in the soil, their concentration in the ground water correlated significantly with discharge intensity and digested solids accumulations.

While the University of Illinois conducted sophisticated research at the Elwood lysimeter and at their Agronomy Research Farm at Urbana, the District initiated several diverse projects to demonstrate its potential usefulness. At Ottawa, Illinois, the District applied 385 equivalent dry metric tons per hectare (170 short tons acre^{-1}) to a silica sand waste pit. The applications reduced the pH of the surface 15 cm (6 in.) from 10.2 to 6.9 and allowed grass and trees to grow on an area heretofore void of vegetation. The District applied similar amounts of digested solids to strip mined soils with a pH of 3.2 and succeeded in producing grass. Success was also achieved in treating waste landfill and fertilizing trees.

Recently the District purchased 2830 ha (7000 acres) of strip mined land in Fulton County, Illinois, for this program of digested or stabilized solids on land in accord with normal farming practices. This is an initial step toward owning or leasing about 12,150 ha (30,000 acres) of land for digested solids utilization. The District's land requirements were computed on the basis of using a conservative loading rate of 45 equivalent dry metric tons (20 short tons acre^{-1}) of solids per hectare per year with allowance for runoff barriers, screening, storage reservoirs, etc. A contract for transporting digested draw-off, Wet Air Oxidized and Imhoff stabilized solids in a slurry of 6–8 per cent total solids, 290 km (180 miles) from the plants to the farm area has been awarded, together with a contract for construction of a storage reservoir. Transportation of solids slurry to the reservoir will commence during August 1971, and application to the farm field will commence in the Spring of 1972.

CONCLUSIONS

1. Utilization of stabilized sewage solids for farming is the most economical method of disposal for The Metropolitan Sanitary District of Greater Chicago.
2. Stabilized solids application on land turns disposal to utilization.
3. Properly stabilized solids can be safely applied in the vicinity of residential areas without any nuisance.

4. Stabilized sewage solids are an effective source of nitrogen, phosphorus, and micronutrients for crop growth and consequently can significantly increase corn yields.

5. Applications benefit sandy soils by increasing the water holding capacity and the cation exchange capacity. They benefit clay soils by increasing the water absorption capacity rate and decreasing the erodibility. To maintain soil porosity, each application must be dried before repeating with subsequent applications.

6. Fecal coliforms in freshly digested sewage solids die rapidly when applied to a soil surface.

7. Swine enterovirus, and probably other virus forms, do not survive more than 4 days in an efficient anaerobic digester.

8. "Liquid Fertilizer" applications provide ample nitrogen and phosphorus for crops, and they apparently provide more than enough iron and zinc for maximum crop production. To date cadmium, chromium, lead, and mercury have not been detected in dangerous amounts in the tissue of crops grown in digested sewage solids.

9. The Metropolitan Sanitary District of Greater Chicago is firmly committed to a program of land reclamation through utilization of sewage solids. Its recent purchase of 2830 ha (7000 acres) of strip mined land represents the start of a major program.

Acknowledgement—This work would not have been possible without the continued support of the Board of Trustees of The Metropolitan Sanitary District of Greater Chicago. The University of Illinois research, together with certain demonstration projects, was funded by the Public Health Service and the Water Quality Offices of the U.S. Government.

Water Research Pergamon Press 1972. Vol. 6, pp. 551–553. Printed in Great Britain

DISPOSAL OF DIGESTED SLUDGE

RICHARD WOOD and S. B. A. FERRIS

The West Hertfordshire Main Drainage Authority, Maple Lodge, Denham Way, Rickmansworth, Herts., England

THE West Hertfordshire Main Drainage Authority has successfully disposed of liquid digested sludge to agricultural land since 1952. The operation began as a relief for an overloaded plant designed to vacuum filter and heat dry both surplus activated and digested sludge.

With the development of the area served by the Authority increasing quantities of sludge had to be processed and the additional sludge disposed of by tanker. The operation proved very successful and in the last 10 yr a minimum of 118,000 m^3 (31 m. gal) (U.S. gallons are quoted in this paper) have been disposed of in this manner each year. TABLE 1 shows the development of this disposal over the past 10 yr.

Initially tankers of 4·6 m^3 (1200 gal) capacity were used but later tankers holding 6·8 m^3 (1800 gal) were added to the fleet. In order to convey large quantities of sludge to the farms owned by the Authority and to deliver liquid digested sludge to pits on private farms articulated tankers holding 20 m^3 (5280 gal) were utilized.

Distribution sites need to be carefully selected and to be, as far as possible, in the centre of a suitable farming area; that is, a good mixture of arable and grassland with the right soil conditions so that distribution can continue at all seasons. The use of a pipeline to deliver sludge to such a centre needs to be carefully evaluated economically for unless such a pipeline is essential it will probably be only economic if used in conjunction with a large treatment plant and for a relatively short distance, such as 5 km (3 miles).

When the Authority decided to close the vacuum filtration and flash drying process for economic reasons and to develop a market for liquid digested sludge as a primary means of sludge disposal it was necessary to make provision for increasing the annual output of sludge to approximately 320,000 m^3 (84 m. gal). The operation was therefore conceived essentially as that of a transport undertaking rather than a sewage treatment matter and a separate department was set up to undertake this work.

Vehicle operations by this Authority are restricted to within 6.5 km (10 miles) of the distribution point and customers (for that is what the farmer is) are developed as far as possible parallel with the main access roads.

Time is as important in making a delivery as mileage so before making deliveries to a new customer a survey is carried out to establish (a) distance (b) time required for journey (c) access to fields (d) suitability of land (e) hazards such as water courses etc. (f) will the farmer be a consistant receiver of digested sludge and (g) is he willing to help by the provision of pits, pumps and irrigation equipment to offset cost.

Time standards are then established for the new delivery point. It is possible by this method to establish in advance the quantity of digested sludge to be delivered, essential information for the manager of a treatment works.

In the area supplied by this Authority it is possible to treat by direct application with tankers for approximately eight months of the year. From December to March the ground conditions are not suitable.

Land is treated under these conditions with irrigation equipment using manure

TABLE 1. DISPOSAL OF LIQUID DIGESTED SLUDGE TO LAND

Year	Volume		Average round trip	
	10^3 m^3	mg	km m^{-3}	Miles 1000 gal^{-1}
1961–62	134·6	35·54	3·348	8·7
1962–63	122·1	32·24	3·734	9·7
1963–64	118·1	31·19	4·217	10·9
1964–65	130·0	34·32	3·654	9·5
1965–66	132·5	34·99	3·766	9·8
1966–67	139·4	36·79	3·927	10·2
1967–68	144·7	38·20	3·750	9·7
1968–69	160·0	42·25	4·378	11·3
1969–70	197·3	52·08	3·992	10·3
1970–71	245·9	64·92	2·801	7·3

guns which have a throw of approximately 21 m (70 ft) radius. The digested sludge is pumped either from a pit of minimum size of 225 m^3 (60,000 gal) or directly from the vehicles, and this can be carried up to 1000 yds from a source. A pit source can be used to dispose of approximately 225 m^3 (60,000 gal) in 8 h, while direct pumping from vehicles (which must be carefully scheduled) can dispose of 160 m^3 (42,000 gal) a day. Pits are therefore preferable in wet weather in that there is less standing time for each vehicle.

With the improvement in lighting equipment and the rising cost of vehicle operation this Authority has successfully carried out a year's trial on the treatment of land and delivery of digested sludge during the hours of darkness. By the use of shifts considerable cost savings can be made.

Three basic principles were employed in the selection of new vehicles, there were:

(a) The biggest payload that could be carried on fields under average conditions without causing excessive impact damage.

(b) That the vehicles were to be as far as possible of standard commercial design and with cross country performance.

(c) That weight be evenly distributed on all the axles when loaded.

There exists in the U.K. a range of vehicles which were originally developed for the motorway construction programme. The basic feature of these vehicles are good payload to chassis rate ratio, with cross country performance. This range of vehicles was found to be the most suitable for our type of operation.

After examining the models of this type of vehicle the Authority selected the British Leyland Guy Big J6 of 3·35 m (11 ft) wheelbase, six-wheeled concrete mixer/tipper chassis with a stripped down weight of 5·4 tonnes (6·2 short tons), making possible a tanker capacity of 12·5 m^3 (3240 gal). The unladen weight of the vehicle is 7·2 tonnes (8·2 short tons) and the laden weight 19·2 tonnes (21·8 short tons). It was therefore possible to achieve a payload increase of 91·7 per cent based upon the previous larger tanker of 6·8 m^3 (1800 gal) capacity with a decrease in capital unit cost of 33 per cent.

An additional feature of this vehicle is that it is constructed with inter-axle differential lock giving the same wheel speed on the back four driving wheels. Finally they were equipped with the largest possible super single tyres on the driving wheels. These features gave it a good cross country performance. The tanks were constructed of mild steel in maxi section which are available on a production line basis in the U.K.

Spreading is carried out through 2 × 152·5 mm (6 in.) foot valves and fish plates which again are of standard design, allowing a vehicle to dispose of its load over 4·25 m (14 ft) spread in 3·5 min.

The use of standard chassis and fittings has brought vehicle availability up to an average of 82 per cent in the first quarter of 1970, and is reflected in the reduced cost of maintenance.

The only feature of these vehicles that would not be found in a standard commercial vehicle was the use of a central pivot mounting at the front end of the tank. The tank itself is not attached to the chassis until half way down its length. This feature reduces the stresses transmitted to the tank when travelling across country. Experience had shown that tanks were liable to stress cracking caused by the flexing of the chassis frame, and this feature should eliminate this problem.

The selection of a much higher weight vehicle means that they have to be withdrawn from field operations a little earlier in the season than the smaller vehicles, but because of good weight distribution and extremely large tyres this difference is marginal and is more than offset by the extra carrying capacity.

Water Research Pergamon Press 1972. Vol. 6, pp. 555–560. Printed in Great Britain

LIQUID SLUDGE DISPOSAL

G. KUGEL

Niersverband, Postfach 529, D-406 Vcirsen, FRG

THE AGRICULTURAL disposal of liquid sludge by the Niersverband, Viersen, has been described by TRIEBEL (1968, 1970). Tank cars of 19 m^3 capacity, fast-coupling pipelines for pneumatic sludge transportation in field as well as sludgesprayer, manually operated are fairly well known. Our disposal and transportation system is briefly described below.

FIGURE 1. This shows the monthly relation for 1970 between the sludge to be delivered, and the corresponding agricultural area to be irrigated with respect to fields and pasture land. The specific rate of liquid sludge application ranges from 2 cm (at ~ 95 per cent water content) upon meadows to 4 cm (at ~ 93 per cent water content) upon fields.

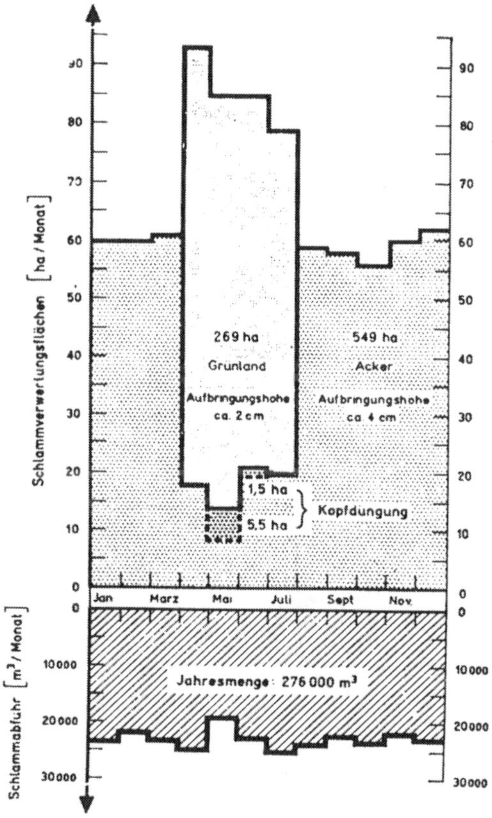

FIG. 1. Liquid sludge disposal in 1970 farming area and sludge quantities.

Grünland:	pasture land, meadow
Acker:	field
Aufbringungshöhe:	sludge irrigation rate.

The total quantity of 276,000 m^3 was sprayed over an area of about 8 km^2, 67 per cent of which were fields and the rest pasture land. The application of sludge from April to July was pasteurized before delivery by heating to 65–70°C for about 25 min.

FIGURE 2. This shows the system of sludge discharge at sewage plants, of sludge transportation and disposal. The centre is situated at our Gruppenklärwerk I,

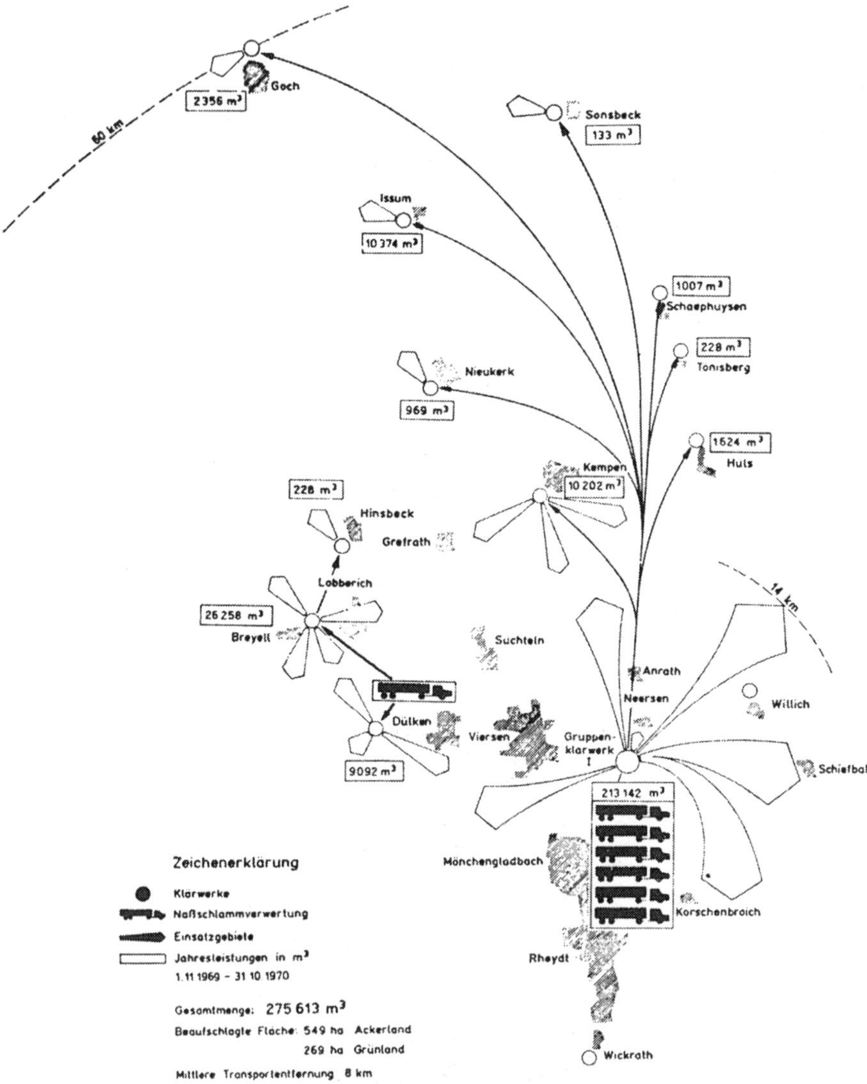

FIG. 2. Operation range of tank trucks.

Naßschlammverwertung:	liquid sludge disposal
Einsatzgebiete:	disposal area
Jahresleistung:	discharge per year
beaufschlagte Fläche:	irrigated area
mittlere Transport-entfernung:	medium transport distance.

Mönchengladbach–Neuwerk. About 75 per cent of the total sludge is produced there. The distance of transportation ranges normally between 2 and 14 km, but mostly at about 8 km. Some other smaller sewage plants are integrated in our combined disposal system up to a range of 60 km from the big centre. The greatest part of the used agricultural area available lies to the north of the town of Mönchengladbach. Some typical figures describe the situation within the 14-km range:

rain:	706 mm
soil:	loamy
agricultural area:	60 per cent fields and 9 per cent pasture land, meadows
largeness of farm:	30–50 ha farm^{-1}
population:	500–600 p km^{-2}

The area required for disposal only ranged from 2 per cent of the fields to 8 per cent of the total available pasture ground. With regard to this relationship there is no need to enter into any definite agreement with farmers. Sludge disposal is flexible and is done on a day to day basis. This has been successful for the last 10 yr.

FIGURE 3. This shows the relation between transportation capacity and the transportation distance within a range of 30 km. The upper line refers to our standard tank truck of 19 m^3 capacity. The normal "dead time" for loading and discharge—30 min each—is included. The 180 h.p. Mercedes vehicle has a speed of about 40 km h^{-1} on roads. In practice the total capacity of one unit is about 30,000–35,000 m^3 yr^{-1} at a medium distance of 8 km.

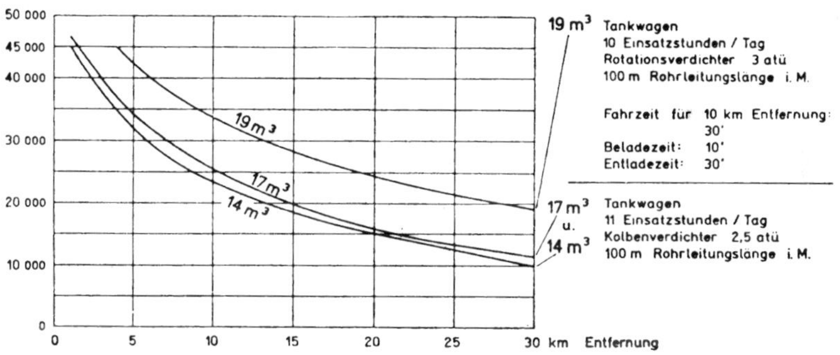

FIG. 3. Transportation capacity.

Jahresleistung:	discharge capacity per year
Schlammtankwagen:	sludge tank truck
Einsatztage:	operations time (days)
Einsatzstunden:	operation time (hours)
Entfernung:	distance
Rohrleitungslänge:	length of pipeline
Fahrzeit:	running time
Beladezeit:	loading time
Entladezeit:	unloading time
Rotationsverdichter:	air compressor (rotary)
Kolbenverdichter:	air compressor (piston).

FIGURE 4. As shown in this diagram the actual sludge disposal quantity is checked by the transportation capacity of the cars but not by the discharge capacity of the

installations in field. With regard to the decreasing amount delivered with distance, one sprayer is economically served by 2 tank trucks at a distance of more than 10 km, and by 3 tank trucks at a distance of more than 30 km.

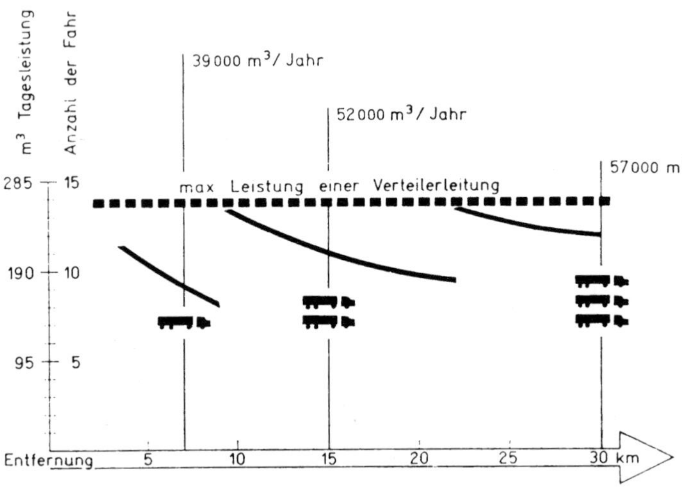

FIG. 4. Discharge and transportation capacity.

Verteilerleitung:	discharge pipeline
Leistung:	discharge capacity
Tagesleistung:	discharge capacity (per day)
Anzahl der Fahrten:	quantity of circulation of a tank truck (per day).

FIGURE 5. The costs of liquid sludge disposal—without pasteurization—are given for 1968–1970. The result in 1970 was: 2·3 DM m^{-3}. The distribution with regard to some typical elements is as following:

wages	58%
fuel	10%
maintenance, repair	10%
capital cost	16%
insurance, clothing, milk	6%

SUMMARY

This relatively inexpensive way of sludge disposal can only be realized on the base of an efficient and skillful regional organization like that of Niersverband.

The Niersverband meets the demands of sewage plant operation as well as those of agriculture in order to maintain continuous liquid sludge disposal from big sewage plants even in a region of a fairly high population density.

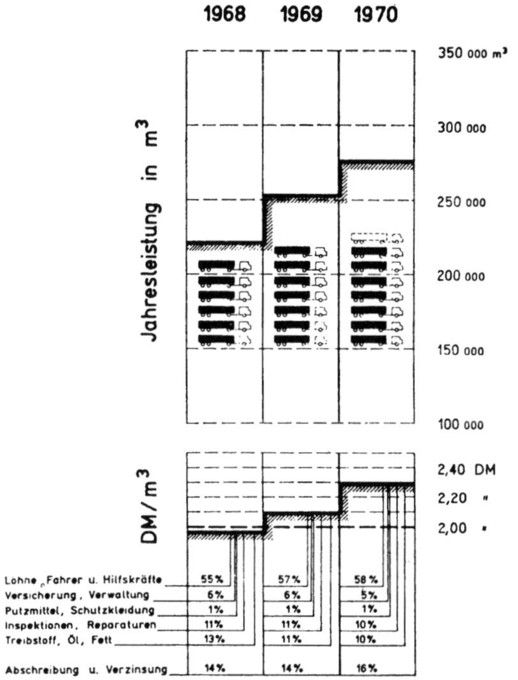

FIG. 5. Liquid sludge disposal 1968–1970 (quantities and costs).

Jahresleistung:	discharge capacity per year
Löhne etc.:	wages etc.
Versicherung etc.:	insurance etc.
Schutzkleidung, Putzmittel:	clothing, cleaning
Inspektion, Reparaturen:	inspection, repair.
Treibstoff, Oil etc.	fuel, oil etc.
Abschreibung, Verzinsung:	capital cost.

REFERENCES

KUGEL G. (1971) Technische und wirtschaftliche Erfahrungen zur landwirtschaftlichen Klärschlammverwertung. *ATV-Fortbildungskurs* II, Essen.

TRIEBEL W. (1968) Möglichkeit der landwirtschaftlichen Verwertung von Faulschlamm. *Europäisches Abwasser-Symposium*, München, 1968.

TRIEBEL W. (1970). *Lehr-und Handbuch der Abwassertechnik*, Bd. III. Verlag Ernst & Sohn, München.

Water Research Pergamon Press 1972. Vol. 6, pp. 561–563. Printed in Great Britain

PASTEURIZATION OF RAW AND DIGESTED SLUDGE

G. KUGEL
Niersverband, Postfach 529, D-406 Viersen, FRG

IF A large quantity of sludge has to be delivered day by day without storage for some months it is necessary to render it hygienic to handle before irrigating pasture land. TRIEBEL (1961, 1968, 1969) described installations for the thermal treatment of liquid sludge. Nowadays this can be done without difficulty.

FIGURE 1 shows the flow of sludge and the main installations for a discontinuous procedure; the double set allows however a semi-continuous discharge of sludge if small quantities are to be treated.

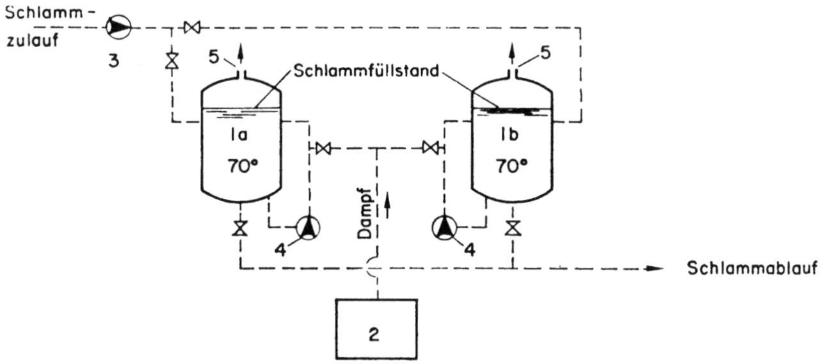

FIG. 1. Pasteurization plant, discontinuous operation.

1a, 1b	pasteurization reactor
2	steam boiler
3	sludge pump (input)
4	sludge pump (recircle)
5	air, vapours outlet.

FIGURE 2 shows the flow diagram of the pasteurization installations at our biggest plant (700,000 population equivalent) is based on a continuous process of two-step heating (reactors No. 2 and 4) and two-step recooling (reactor No. 5). For the economical use of steam efficient heat recovers recuperation combined with preheating is used. At a vacuum pressure of 0·1 atm vapours are returned from the first reactor No. 5 to the reactor No. 2. This pasteurization plant at Gruppenklärwerk I, Mönchengladbach-Neuwerk, is now charged with sludge continuously at 28 $m^3 h^{-1}$.

To use steam most economically multiple heat recovery and sludge preheating is required. There is a limit to reliability, maintenance and the cost of installation of the system. The reliability R_m of a multiple system, especially in case of series-connexion, decreases quickly:

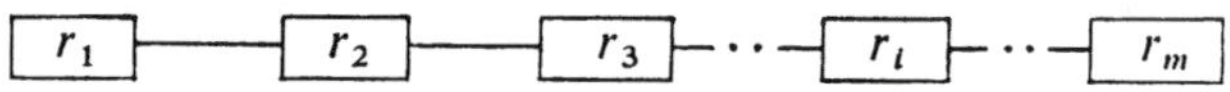

$$R_m = r_1 \,.\, r_2 \ldots\ldots r_i \ldots\ldots r_m = \prod_{i=1}^{m} r_i$$

$$0 < v_i \leqq 1.$$

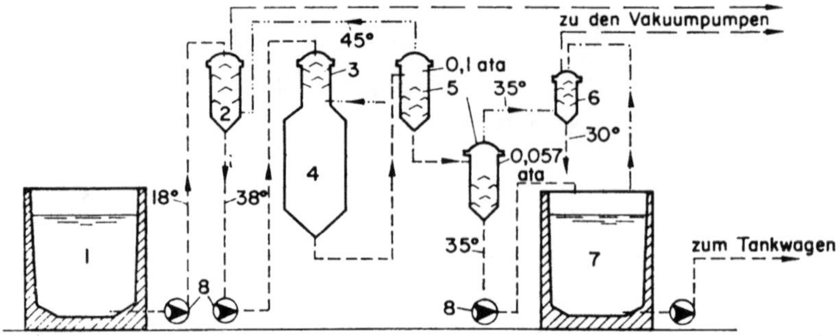

FIG. 2. Pasteurization plant continuous operation (Gruppenklärwerk I, Niersverband).

(1) digested sludge, thickened
(2) preheating reactor
(3)
(4) pasteurization reactor, detention chamber included
(5) cooling reactors
(6) condensor
(7) pasteurized sludge
(8) pumps.

Schlamm = sludge, Heißdampf = steam, Brüden = vapours, Wasser = water.

The system reliability can be improved by higher reliability of each element as well as by redundant stages, i.e. by a combination of series and parallel connexion:

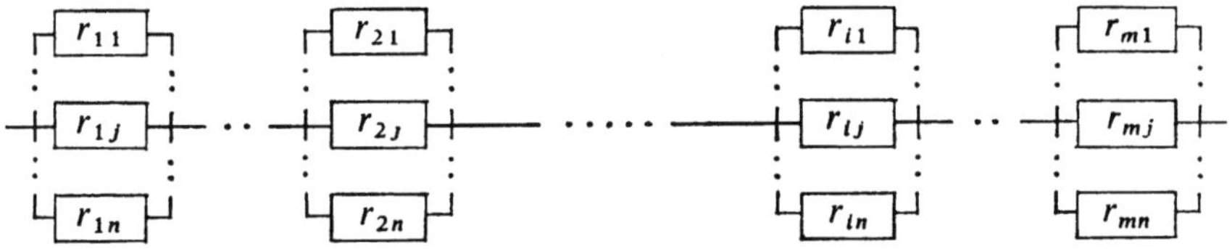

$$R = \prod_{i=1}^{m} \left[1 - \prod_{=1}^{n} (1 - r_{ij})\right] \qquad 0 < r_{ij} \leqq 1$$

Both methods, either improvement or redundancy, cost money. The same reliability aspects have to be considered also in case of other ways of sludge treatment and disposal.

The newest pasteurization installation of Niersverband is built at Gruppenklärwerk Nette and charged with raw sludge at 10 m^3 h^{-1}. The thermal process is similar to the scheme shown in FIG. 2 but without the second step of recooling. Here the warm sludge is discharged into steel storage tanks at about 40°C, where the temperature may fall to 30° or 35°C.

After 2750 operation cycles in 1970 the performance was sufficient to be described by the following typical figures:

steam	78 kg m^{-3} raw sludge
fuel	5·5 l. m^{-3}
operation cost	
fuel, electricity	0·68 DM m^{-3}
water	0·07 DM m^{-3}
chemicals (water treatment)	0·14 DM m^{-3}
maintenance, operator	0·49 DM m^{-3}
repair, inspection	0·29 DM m^{-3}
	1·67 DM m^{-3}
management (10%)	0·17 DM m^{-3}
total operation cost	1·84 DM m^{-3}.

The total cost of installation, storage tanks included, was about 550,000 DM in 1969.

REFERENCES

HOFMANN W. (1968) *Zuverlässigkeit von Meß-, Steuer-, Regel- und Sicherheitssystemen.* Verlag K. Thieme, München.

KUGEL G. (1971) Pasteurisierung von Roh- und Faulschlamm. *Abwassertagung*, ATV, Essen.

TRIEBEL W. (1961) Erfahrungen mit der Schlammpasteurisierung Technik und Wirtschaftlichkeit. *IAM-Informationsblatt* Nr. 30.

TRIEBEL W. (1968) Möglichkeit der landwirtschaftlichen Verwertung von Faulschlamm. *Europäisches Abwassersymposium*, München.

TRIEBEL W. (1969) Pasteurisierung der Klärschlämme. *IAM-Kongreß*, Basel.

Water Research Pergamon Press 1972. Vol. 6, pp. 565–567. Printed in Great Britain

RAW SLUDGE COMBUSTION OR LIQUID DIGESTED SLUDGE DISPOSAL?

W. VON DER EMDE and W. MÜLLER
Technische Hochschule Wien, Karlsplatz 13, A-1040 Wien, Austria

THE FOLLOWING sludge disposal methods were investigated for the Vienna–Simmering main treatment works:

(a) Raw sludge combustion after thermal conditioning, filter presses and fluidized bed incineration.
(b) Raw sludge combustion after chemical conditioning with organic flocculants, centrifuging and multiple hearth furnaces.
(c) Sludge digestion, pasteurization, sludge pumping (18 km), liquid sludge transportation by private enterprise (tanker transport as at Niersverband).
(d) as (c) sludge digestion and liquid sludge disposal including own electricity generation.

Local conditions in Vienna favour digested sludge utilization for agricultural purposes. The Marchfeld north of the Danube, is opposite the Simmering treatment works. Here, 400 km^2 exclusively agricultural land establishes favourable conditions for the liquid sludge utilization. Large agricultural areas are also located south of the Danube adjacent to the Schwechat airport. One company is interested in sludge transportation and many farmers are keen on fertilizing their land with sludge. Sludge fertilization experiments sponsored by the City of Vienna at the "Hochschule für Bodenkultur" gave favourable results.

For a yearly sludge production of 830,000 m $acre^{-1}$ (at 7 per cent solid content) and for an application thickness of 5 cm (50,000 m^3 km^{-2} or 35 tons ha^{-1}) 17 km^2 agricultural land is needed yearly, or 50 km^2 area for a 1-in-3 yr sludge application. The sludge would be heated to 70°C for 30 min for disinfection.

For dewatering raw sludge with centrifuges a requirement of 2·8 kg organic flocculant* additive per 1 ton solids (180 g m^{-3}) was assumed. No additional heating or extra fuel was expected to be required for the combustion of raw sludge.

Interest charges were set at 8 per cent for civil work and at 11 per cent for machinery. For personnel $5600 annum were taken for one man. Balance sheets of industry show $3200–4000 yr^{-1} per employee at present.

The benefit from electricity generation was calculated at 2 ct kWh^{-1} for 16 l. gas PE. day^{-1} and 2 kWh m^{-3} gas. This was deducted from the operational costs.

The following costs were derived for four alternatives (see TABLE 1, p. 566).

According to this table, the difference in cost is relatively small with the four alternatives. Between the least expensive, digestion and utilization as a liquid fertilizer and soil conditioner, and the most expensive solution, burning raw sludge after chemical conditioning, the difference in the estimate is 20 per cent only. When 900,000 m^3 $acre^{-1}$ of raw sludge arise compared to the 1,240,000 m^3 $acre^{-1}$ on which the estimate is based, the difference will decrease to 13 per cent ($21·4 and $24·7

* 1 kg organ. flocculants $ 2·9.

Table 1

	Raw sludge		Digested sludge	
	Therm.	Chem.	Without electricity generation ($ ton^{-1})	With electricity generation ($ ton^{-1})
Capital costs	10·6	7·8	7·0	9·6
Operational costs	8·2	13·3	13·0	7·9
Total	18·8	21·1	20·0	17·5
per 1 m^3 of sludge	1·3	1·5	1·4	1·2
Relation	1·07	1·20	1·14	1·00

Table 2. Comparison of costs
Treatment plant Vienna–Simmering, 2,5 Mio. P.E.

	Raw sludge		Digested sludge	
	Therm. ($ ton^{-1})	Chem. ($ ton^{-1})	With electricity generation ($ ton^{-1})	Without electricity generation ($ ton^{-1})
Construction cost				
Civil works	2·8	2·0	4·4	5·2
Mechanical	7·8	5·8	2·6	4·4
Total	10·6	7·8	7·0	9·6
Operating cost				
Personnel	1·8	1·5	1·2	1·8
Electricity	1·7	1·2	1·2	1·2
Chemicals etc.	0·7	8·0	0·3	1·0
Maintenance	1·6	1·7	0·7	1·2
Disposal of ash	0·9	0·9	—	—
Transportation of sludge	—	—	9·6	9·6
Utilization of electricity	—	—	—	6·9
Total	8·2	13·3	13·0	7·9

Dry weight of raw sludge	240 ton day^{-1}
Concentration of raw sludge	7%
Quantity of raw sludge	3400 m^3 day^{-1}
Dry weight of digested sludge	160 ton day^{-1}
Concentration of digested sludge	7%
Quantity of digested sludge	2300 m^3 day^{-1}

ton^{-1}, respectively). In this case, the disposal of digested sludge without electricity generation is the second least expensive solution. This is because capital expenses and charges on these capital expenses will be small. Disposal of digested sludge without electricity generation will be 6 per cent more expensive than disposal of liquid sludge with electricity generation, and disposal of raw sludge after thermal conditioning will be 8 per cent more expensive than the first solution.

Where similar difficulties in operating the thermal conditioning part as described in the paper by Meredith occur, the cost of this method will be greatly increased. At the plant Meredith reports on, the initial sludge concentration before thermal conditioning was 3 per cent, compared to 7 per cent as assumed in this estimate.

When the method of disposing of raw sludge by burning after chemical conditioning is applied, it is very difficult to estimate the cost of chemicals (eventually, there is a decrease in cost when the sludge is heated and an increase when the sludge is septic). Furthermore, it is difficult to tell to what extent additional fuel can be saved when the dewatering is done in large centrifuges (e.g. load of 30 $m^3 h^{-1}$). When filter presses are applied, higher expenses have to be considered.

The disposal of digested sludge without electricity generation can most easily, with respect to cost, be evaluated (Prerequisite: a contract with a transportation enterprise does exist). Incineration of digested sludge (e.g. after dewatering with filter presses and conditioning with ash) will be about as costly as conditioning of raw sludge (there is only 2/3 of the sludge quantity; less buffer capacity necessary by utilizing the digester; decrease in solids alone \$20 ton^{-1} by digestion). When the liquid disposal of digested sludge will be discontinued (e.g. farmers refusing to take liquid sludge or other difficulties), it will be as economical to incinerate digested sludge as it is raw sludge.

Because of simplicity in operation, avoidance of air pollution and utilization of the sludge instead of destroying decomposable organic matter, I believe that digestion, pasteurization and disposal of liquid sludge to farmers is the most suitable solution for the main treatment works Vienna–Simmering. There is no doubt this solution will require intensive preparation for the operational setup and to inform farmers.

Water Research Pergamon Press 1972. Vol. 6, pp. 569–571. Printed in Great Britain

SLUDGE DISPOSAL AT SEA

C. F. GUARINO

City of Philadelphia, Water Department, 1160 Municipal Services Building, Philadelphia, Pa. 19107, U.S.A.

THERE is international concern regarding the disposal of wastewater sludge at sea. There have been many reports that the oceans are becoming polluted and that ocean disposal at sea must be halted.

The City of Philadelphia has been disposing its digested sludge at a location 21.3 km (11.5 nautical miles) off Cape May, New Jersey, and Cape Henlopen, Delaware, since May 1961.

The decision to barge sludge to sea was made after an extensive survey performed by Philadelphia's consultants, Greeley and Hansen, in conjunction with Philadelphia Water Department personnel. Of the fourteen methods studied, ocean disposal was the most practical and economical method. The location was chosen considering favorable ocean currents for the dilution as well as an area which was relatively deep and where shell fishing was of marginal value.

Since 1961, on an annual basis, we have disposed of 265,000 m^3 (70 million U.S gal) of sludge ranging in solids concentration from 6 to 11 per cent. Our current contract allows for the disposal of 447,000 m^3 (118 million gal) of sludge per year. This takes care of the entire quantity of digested sludge generated by Philadelphia's three treatment plants. That is, the quantity of digested sludge that results from the treatment of 1,703,000 m^3 (450 million gal) of wastewater a day. The following table gives Philadelphia's cost to transport sludge to sea.

COST OF SLUDGE BARGING IN PHILADELPHIA

Year	% Solids*	Volume barged, 1000 m^3			Cost	
		Northeast	Southwest	Total	Per m^3 ($)	Per dry metric ton ($)
1961	6.2	106	—	106	1.56	25.25
1962	6.4	136	—	136	1.69	26.60
1963	5.9	162	—	162	1.24	20.95
1964	7.2	163	—	163	1.66	23.15
1965	9.5	163	—	163	0.99	10.39
1966	8.6	108	127	235	0.99	11.47
1967	9.8	183	174	357	0.99	10.09
1968	9.7	203	143	346	0.84	8.62
1969	9.2	268	136	404	0.84	9.10
1970	10.8	223	140	363	1.05	9.72

* After 1965, % solids represents the combination of both plants.

Higher costs per metric ton, in the earlier years, were due to smaller sludge volumes, shorter contract durations and inefficient techniques for sludge unloading at the disposal site. Regarding sludge unloading, the first barge contractor used a trash type pump to empty his barge at the disposal site. This technique required a minimum of 6 h if no clogging problems developed. Subsequent contractors, using bottom dump barges, were able to unload in approximately 0.5 h or one-twelfth the original time. The reduced unloading time coupled with larger sludge volumes and the initiation of 3 yr barging contracts has considerably lowered our barging cost. Our total sludge handling cost, digestion, pumping, lagooning, etc., can be estimated by adding $0.26/m^3 ($1.00 per 1000 gal) to the barging cost.

Philadelphia's sludge handling and disposal program encompasses all three of our treatment facilities; namely, Northeast (663,000 m^3 day^{-1}) Southeast (520,000 m^3 day^{-1}) and Southwest (520,000 m^3 day^{-1}). At Northeast, digested and waste activated sludge are anaerobically digested, lagooned, and pumped via a 10 in. Ellicott dredge to the barge. The barge is towed to the disposal site—an approximate distance of 204 km (110 nautical miles).

Both of Philadelphia's southern plants provide primary treatment. Our Southwest plant receives primary sludge from our Southeast facility via an 8 km (5 mile) force main. Both primary sludges, Southeast and Southwest, are anaerobically digested and pumped directly to the barge. The towing distance between the Southwest pier and the disposal site is roughly 185 km (100 nautical miles).

In the past 10 yr, sludge barging has proved to be an economical and operationally sound practice. Within the last year-and-a-half, the ecological soundness of our disposal program has been under review. Because of the great concern by citizens, various agencies and government groups, Philadelphia has contracted with the Franklin Institute Laboratories and Jefferson Medical College Laboratories—both located in Philadelphia—to take samples of the ocean waters as well as the sediment at the disposal site to determine if any harmful effects have been produced. This work has been underway since January 1971 and is scheduled to be completed December 1971. Results at the half-way mark of this investigation are summarized below:

(1) Sediment samples taken from the center of the disposal site and the immediate vicinity (including low spots) consist of clean sand, along with gravel and shell fragments.
(2) In all sediment samples taken in and around the disposal area, no black sludge or H_2S odor emitting samples were found.
(3) At the center of the sludge disposal area, starfish, sand dollars, hermit crabs, and snails were found in good health.
(4) Fish specimens collected at the disposal site included winter flounder, mackeral, stargazer, long horned sculpin, and spiny dogfish.
(5) Dissolved oxygen measurements of bottom, middle and surface waters at various points within and surrounding the site show no signs of oxygen depletion or sag.
(6) Coliform levels on all bottom mud and surface waters are zero.
(7) Animal specimens, interstitial water, and sediments from various points in and around the disposal area were analyzed for pathogens such as Salmonella and Shigella and enteroviruses; i.e. polioviruses, echoviruses, etc. All results were negative.
(8) There are no indications of significant concentrations of heavy metals in surf clams and other macrofauna collected in the area.

In summary, the results to date indicate that the disposal area does not appear any different than areas at a further distance from shore. Despite all of this information, it does appear that we will be requested to transport sludge off the continental shelf. Since barging costs are a direct function of the distance traveled, such a request by pertinent governmental agencies would increase the cost of barging and therefore waste treatment. Concluding, it is the City of Philadelphia's feeling, based on its findings at the ocean disposal site after ten years of this practice, that no harm has been done to the ocean and this should be a legitimate way of recycling waste material. It would be wasteful to terminate sea disposal and be forced to use an expensive alternate. If it is anticipated that the continual use of this site by many other municipalities would ultimately produce a problem, then the disposal of digested sludge over a suitable, wider area of the sea could be practiced. In this manner, the tremendous assimilative power of the ocean to absorb and re-use waste would be effected. This, in turn, would furnish many communities a more economical method of sludge disposal.

Water Research Pergamon Press 1972. Vol. 6, p. 573. Printed in Great Britain

DISCUSSION: DISPOSAL OF SLUDGE AT SEA

L. B. WOOD
Beckton Sewage Treatment Works, Beckton, London E6

THE Greater London Council and its predecessors have been disposing of sludge at sea since 1893, using a spoil ground some 110 km from London at the Barrow Deep in the North Sea. Like the area used by Philadelphia, this is also swept by strong currents and the sludge is rapidly dispersed.

At present about 10,000 tons of sludge with an average solids content of 3 per cent is discharged daily, about two-thirds of the sludge having been digested. A fleet of five ships is used, consisting of four motor vessels with capacity of about 2300 tons, and a stand-by steam vessel of 1500 tons capacity. The ships are so constructed that the sludge has always a hydrostatic head above the sea water level; it can therefore be discharged by gravity: the ship rises as the sludge runs out and maintains the hydrostatic head. The present cost of construction of such ships is about £1M each.

Despite the much lower solids content of this sludge, the cost of disposal is very similar to the cost given by Mr. Guarino of \$10 ton^{-1} of dried solids.

The water in the vicinity of the spoil ground has been examined regularly (at least once per fortnight) since this method of disposal was started. Analyses are made in laboratories on the ships, samples being taken at about 8-km intervals along the route of the ship's passage.

The dissolved oxygen concentration in the area has always been found to be at about 100 per cent saturation, except in the immediate area of the discharge for a few hours after it has been made.

There has been no evidence that sludge has been harmful to fish in the region, but it is my opinion it is essential to maintain a careful watch on the concentrations of pesticides, poly-chlorinated biphenyls and mercury and other metals in the sludge and in the sea water of the area.

Some guidance is also necessary on the maximum permissible concentrations in the sea water of these persistent substances, which can be tolerated without danger to fish and birds and eventually to man.

Water Research Pergamon Press 1972. Vol. 6, pp. 575–582. Printed in Great Britain

CONTROL SYSTEMS FOR WASTEWATER TREATMENT PLANTS

J. F. ANDREWS

Environmental Systems Engineering, Clemson University, Clemson, S.C., U.S.A.

INTRODUCTION

THE NEED for skilled operation of wastewater treatment plants is frequently greater than that for industrial processes because of large temporal variations in wastewater flow and composition. However, most wastewater treatment plants are in a primitive state with respect to process operation when compared with industrial processes. Gross failures, such as the bulking of activated sludge and "sour" anaerobic digesters, are all too frequent. In addition, there are significant variations in treatment plant efficiency, not only from one plant to another, but also from day to day and hour to hour in the same plant. Daily variations from 60 to 95 per cent efficiency in BOD removal are not uncommon and THOMANN (1970), by a statistical analysis of the variation in effluent quality from eight treatment plants, has shown that these variations can have a significant effect on the water quality of the receiving streams.

Maintenance of plant efficiency nearer the maximum by improved operation could result in significant decreases in the pollutional load placed on our water resources. For example, consider a plant with a maximum BOD removal efficiency of 95 per cent, a variation in efficiency of 80–95 per cent, and an average of 87.5 per cent. If, through better operation, plant efficiency could be maintained between 90 and 95 per cent with an average of 92.5 per cent, the BOD load to the receiving waters would be reduced by 40 per cent without increasing the maximum efficiency of the plant. Actual reductions may even be greater than this as demonstrated in the work reported by WEST (1969) in which the average BOD discharge from a 80,000 m^3 day^{-1} (21 mgd) activated sludge plant was reduced from 40 to 9 mg l^{-1} by improved operation. This represents a reduction in BOD load on the receiving waters of 78 per cent.

Another significant advantage of good operation can be improved productivity through an increase in the amount of waste that can be treated per unit of plant capacity. There have been several instances where operators have been able to significantly increase the capacity of their plants above the design capacity through changes in operational procedure. For example, GOULD (1953) was able to double the capacity of his activated sludge aeration basins by introducing the raw waste at intervals along the length of the basins. Another example of increased productivity is the work of TORPEY (1954) on anaerobic digestion in which he was able to triple digester loading by thickening the sludge prior to feeding it to the digester. Both Gould and Torpey's ideas have since been adopted by designers and are now standard practice in many plants.

Improved operation can be attained by increasing the quantity and quality of personnel involved and/or by using modern control systems similar to those currently used in industry. The need for an increase in the quantity and quality of personnel is well known and has been summarized for the United States in SENATE DOCUMENT NO. 49 (1967). The prospect of improving plant operation through the use of modern control systems is not as well recognized and a review of control systems which may be used is the objective of this paper.

EVOLUTION OF PROCESS CONTROL

Much of the material presented in this section has been taken from LEE, ADAMS and GAINES (1968) and is gratefully acknowledged. For a more detailed discussion of conventional control systems used in treatment plants, the reader is referred to the works of AUSTIN (1967) and BABCOCK (1968). It should be emphasized that the type of control system selected for a plant must be suited to the environment in which the plant will be operated. Use of a complex control system would be nonsensical where proper instrument maintenance is unavailable or qualified operators cannot be obtained. The more advanced control systems to be discussed, such as computer control systems, are only applicable to those plants where graduate engineers will be in charge.

Manual control

In this type of control (FIG. 1), the operator uses his senses to determine the status of the process or measure the quality of the product. Any deviations from the desired operating conditions are corrected by changing those variables which can be controlled.

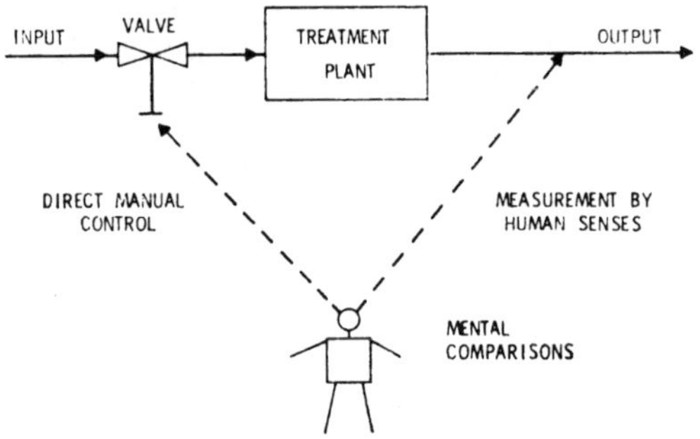

FIG. 1. Manual plant operation (after LEE *et al.*, 1968).

The operator then observes if the correction is adequate and continues to make adjustments until the process is operating properly. The operator is the feedback to control the process and good operation is therefore an art which is dependent on the skill and intuition of the operator. The control of many of our treatment plants has not advanced much beyond this stage today.

Indicating and recording instruments

The first portion of the feedback loop to be mechanized was the sensing function. Human senses are replaced with indicating instruments (FIG. 2) and automatic recording instruments added to give a historical record and indicate trends. However, these instruments serve only as an aid to the operator and it is still necessary for him to use judgment in detecting error and making the necessary corrections.

Instruments for sensing are an important segment of the feedback loop. Dependable and accurate instruments are available for many physical measurements such as pressure, temperature, and flow rate. Unfortunately, instruments for most of the bio-

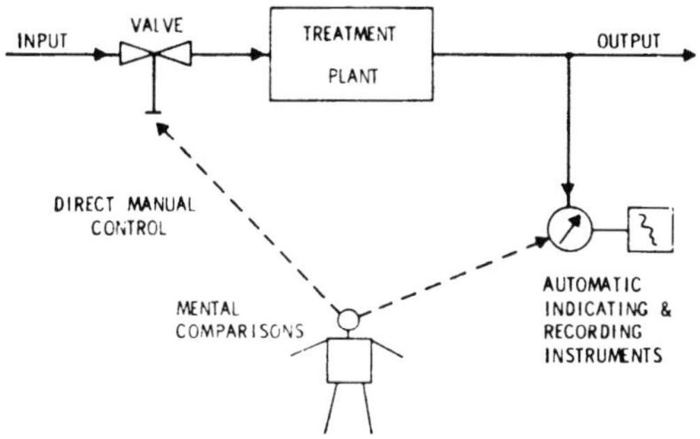

FIG. 2. Indicating and recording instruments added to plant operation (after LEE *et al.*, 1968).

logical measurements, many of the chemical measurements, and some of the physical measurements needed for plant control are still in the development stage, thus restricting the application of conventional control systems.

Local automatic control

The next step in the development of control came with the advent of the automatic controller (FIG. 3) which removed the operator from continuous participation in the feedback loop. With these controllers, it is possible to select values (set points) for the process variables and have these variables automatically maintained close to their set points. The controller performs the functions of sensing and correcting deviations from the desired operating conditions. However, the operator continues to participate in the feedback loop on an intermittent basis since he must select controller set points in the light of his judgment and experience.

Automatic controllers can be obtained with several different control "modes" or "algorithms". Included among these are: (1) on–off control in which the final control element (valve in FIG. 3) is fully open or closed whenever the measured value for the process variable deviates more than a fixed amount from the desired value (set point); (2) proportional control in which the valve opening is proportional to the difference (error signal) between the measured and desired value of the process variable; (3) integral control where the valve opening is proportional to the integral of the error signal with respect to time and therefore reflects the length of time that the error has persisted; and (4) derivative control for which the valve opening is proportional to the

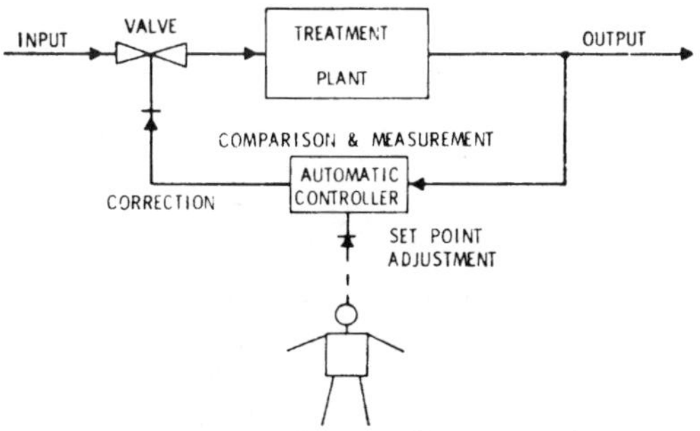

FIG. 3. Operation by local automatic controllers (after LEE *et al.*, 1968).

derivative of the error signal with respect to time thus reflecting the rate of change of the error. A combination of proportional, integral, and derivative control is known as PID control.

Problems encountered with local automatic controllers are determination of the proper set points and lack of consideration of interactions between variables, i.e. changing the recycle flow rate in the activated sludge process will influence more than just the suspended solids concentration in the aeration basin.

Centralized automatic control

Local automatic controllers are physically located near the process variables to be controlled. As the number of these increase, their maintenance and the monitoring and adjustment of set points becomes more difficult. The solution to this was the development of a centralized control room (FIG. 4) where the controllers are brought together and mounted on a control panel. Graphic panels are usually used to display

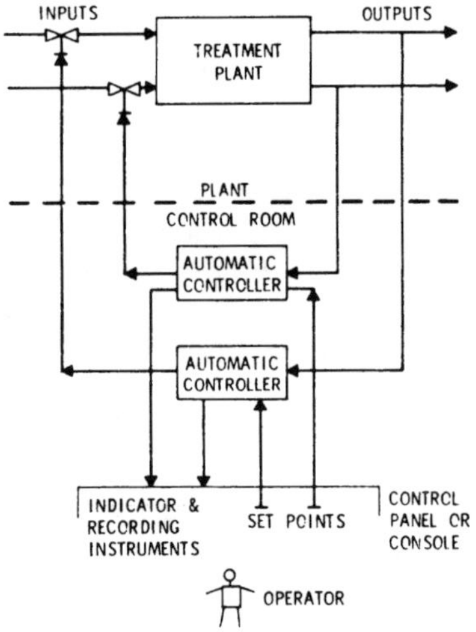

FIG. 4. Plant operation by centralized automatic controllers (after LEE *et al.*, 1968).

the key variables on a schematic flow diagram of the plant. This assists the operator in organizing and assimilating data and permits him to more quickly grasp the significance of interactions between variables so that he can take the required corrective action.

Many treatment plants currently in service have centralized control rooms with graphic display panels. However, many of the items displayed on these panels are not automatically controlled, but instead require manual action by the operator.

COMPUTER PROCESS CONTROL

By mid-1968, 1700 process control computers were installed or on order for a variety of applications in industries throughout the United States and 5900 installations are forecast by 1975 (U.S. DEPT. OF LABOR, 1970). Computers are now being actively considered for the control of wastewater treatment plants and a report on the feasibility of computer control for wastewater treatment was prepared by the AMERICAN PUBLIC

WORKS ASSOCIATION in 1970. The author is aware of at least 12 plants throughout the world which have either installed or placed process control computers on order and reports on several of these installations will be presented at this workshop. Only a brief discussion of computer control can be presented in this paper; for a detailed discussion, the reader is referred to the books of LEE *et al.* (1968) and SAVAS (1965).

The use of computers for process control can range from data processing and monitoring up to some form of optimal control and only these two types of computer control will be illustrated here. Most existing industrial computer control systems are somewhere between these two types. Installations planned for wastewater treatment plants range from data processing and monitoring up to the programming of the computer for conventional control modes such as "on–off" or "PID" control. One major advantage of using a digital computer is that the control strategy can be easily changed by reprogramming (software changes) the computer whereas conventional controllers would require replacement or rewiring (hardware changes). Still another major advantage is that the computer is better able to take into account complex interactions between variables.

Prospective users should recognize that although the potential use of computer control is great, the adaptation of a process to computer control is a difficult and time consuming task. A substantial amount of technical manpower is required to implement computer process control and it has been reported (U.S. DEPT. OF LABOR, 1970) that from 2 to 21 man-years have been required to complete computer control projects in industrial plants. BAILEY (1969) has discussed some of the problems which have been encountered in adapting industrial processes to computer control. Considerable development work will be required to take full advantage of computer control for wastewater treatment plants since these processes are complex and their dynamic behavior is poorly understood. Another problem is the lack of reliable on-line sensing instruments for many important measurements such as organic carbon, ammonia, etc. However, this problem is not as severe as might be expected since changes in treatment processes occur relatively slowly and there is frequently adequate time for chemical analysis in the laboratory with manual input of the data to the computer for subsequent control action.

Data processing and monitoring

Computer control can be installed in stages, and the first step is usually data processing and monitoring (FIG. 5). The computer collects data, processes it into a more meaningful form, and displays it to the operator. The operator then adjusts the set points of his automatic controllers. In this mode of operation, the operator is still in the feedback loop and the computer only serves to assist him in running the plant. Typical functions which may be performed by the computer are:

(1) Scan process sensing instruments, check for instrument malfunction, and convert raw data to engineering units.
(2) Process data into a more meaningful form for the operator by smoothing, curve fitting, integration, differentiation, and statistical analysis.
(3) Monitor and report on the status of process equipment. For example, the computer can monitor the on–off condition of pumps, valves, motors, and compressors. It can check for overheating of bearings and excessive vibration of motors and compressors.

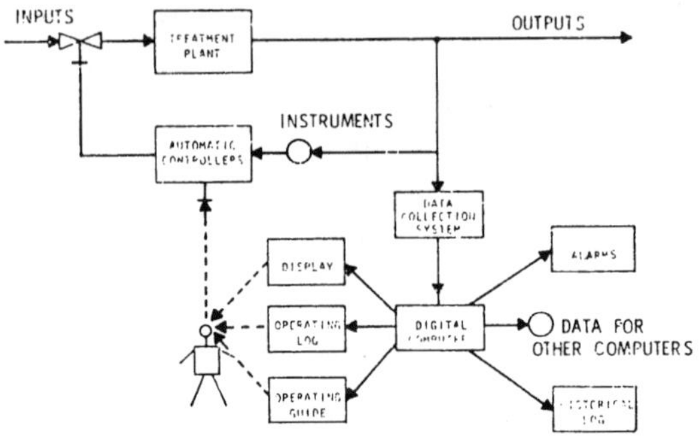

FIG. 5. Data processing and monitoring by digital computer (after LEE *et al.*, 1968).

(4) Compare process variables against high–low limits and sound alarms.
(5) Prepare operating logs and display information to the operator. This can be in tabular form, plotted trend charts of particular variables, or graphical displays on a cathode ray tube.
(6) Furnish the operator with an operating guide upon request. For example, procedures to be followed in case of a process upset.
(7) Furnish data for other computers or provide reduced operating data to higher management.

Optimal control

Although this degree of control has yet to be reached in most installations, it is a goal toward which most users are striving. In this type of control (FIG. 6) the computer takes over operation of the plant and controls it so that the overall operation is optimum. It is necessary to tell the computer what is meant by "optimum." For example, in a treatment plant, "optimum" might mean to control the plant at minimum operating cost but with the constraint that the effluent BOD always be equal to or less than $20\,mg\,l^{-1}$. The computer would take into account all significant process variables and calculate the process conditions and control changes necessary to obtain optimum performance. It would then initiate these changes.

The automatic controllers can be eliminated with the computer furnishing signals to manipulate process variables. This is called direct digital control (DDC) and, for

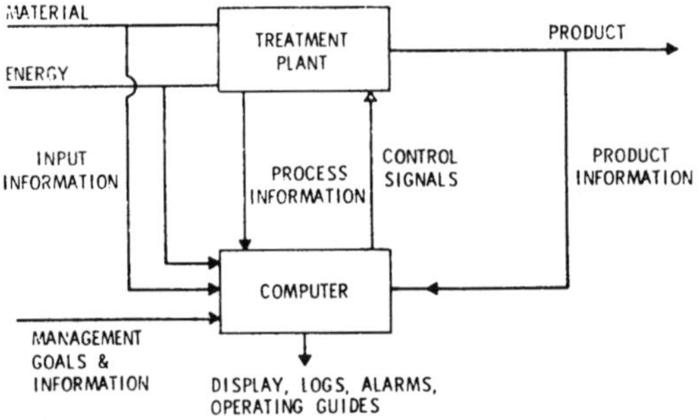

FIG. 6. A computer controlled treatment plant (after SAVAS, 1965).

processes where a large number of variables must be controlled, can result in considerable savings in capital costs for instrumentation. Feedforward control, using a dynamic mathematical model stored in the computer, can be incorporated to take control action before process upsets occur. Information for feedforward control is obtained by measuring the inputs to the plant. Some feedback control, utilizing information from the plant outputs, is usually incorporated since the mathematical models are seldom perfect. Little knowledge of the process is required for feedback control; however, considerable time may elapse after the control action is initiated before the process is restored to the proper operating condition. Although this disadvantage can be overcome by using feedforward control, considerable knowledge of the process is required to develop the necessary mathematical models. The first control programs for treatment plants will have to incorporate a considerable amount of feedback control until their dynamic behavior is more clearly defined.

ECONOMICS

The question may be raised that the improvement of plant performance by the use of modern control systems or through the employment of higher quality personnel would be uneconomical. SMITH (1971) has addressed this question by calculating the amount of money that could be spent per unit of additional BOD removed if the cost per unit removed by improved performance was the same as that currently expended for conventional treatment. His calculations indicate that improved operation is economically justifiable and these calculations should be conservative since they do not consider the additional savings which could result from decreased utility costs, operating supplies costs, and labor costs.

Another indicator of the feasibility of investing more money in control systems for treatment plants may be obtained by a comparison with investments in control systems by other industries. SILVA (1968) has indicated that in 1968 for the chemical and petroleum industry, approximately 12 per cent of the total plant cost was invested in controls. For comparison, SMITH (1971) estimated that in 1968 only 1–1.5 per cent of the total cost of wastewater treatment plants was invested in control systems. The chemical and petroleum industry apparently feels that this level of investment in control systems is profitable and it would appear that we could afford additional investments in control systems for wastewater treatment plants.

REFERENCES

AMERICAN PUBLIC WORKS ASSOCIATION (1970) *Feasibility of Computer Control of Wastewater Treatment.* American Public Works Association, Chicago.

AUSTIN J. H. (editor) (1967) *Proceedings 9th Sanitary Engineering Conference on Instrumentation, Control, and Automation for Water Supply and Wastewater Treatment Systems.* University of Illinois.

BABCOCK R. H. (1968) *Instrumentation and Control in Water Supply and Wastedisposal.* Reuben H. Donnelley Corp., New York.

BAILEY S. J. (1969) On-line computer users polled. *Control Engng* **16,** 86.

GOULD R. H. (1953) Sewage aeration practice in New York City. *Proc. Am. Soc. civ. Engrs* **79,** 307–311.

LEE T. H., ADAMS G. E. and GAINES W. M. (1968) *Computer Process Control: Modeling and Optimization.* Wiley, New York.

SAVAS E. S. (1965) *Computer Control of Industrial Processes.* McGraw-Hill, New York.

SENATE DOCUMENT No. 49 (1967) *Manpower and Training Needs in Water Pollution Control.* U.S. Government Printing Office, Washington.

SILVA R. (1968) Relating plant savings to the control hierarchy. Paper presented at the *23rd Symposium on Instrumentation for the Process Industries*, Texas A & M University, College Station, Texas.

SMITH R. (1971) Wastewater treatment plant control. Paper presented at the Joint Automatic Control Conference, Washington University, St. Louis, Missouri.

THOMANN R. V. (1970) Variability of waste treatment plant performance. *J. sanit. Engng Div. Proc. Am. Soc. civil Engrs* **96,** SA3, 819.

TORPEY W. N. (1954) High-rate digestion of concentrated primary and activated sludge. *Sew. ind. Wastes* **26,** 479.

U.S. DEPARTMENT OF LABOR (1970) *Outlook for Computer Process Control*, Bulletin 1658, U.S. Department of Labor, U.S. Government Printing Office, Washington.

WEST A. W. (1969) Case histories: improved activated sludge plant performance by operations control. *Proc. 8th Ann. Environ. Water Resourc. Engng Conf.*, Vanderbilt University.

Water Research Pergamon Press 1972. Vol. 6, pp. 583–585. Printed in Great Britain

AUTOMATIC CONTROL SYSTEMS

RICHARD WOOD

The West Hertfordshire Main Drainage Authority, Maple Lodge,
Denham Way, Rickmansworth, Herts., England

THE CONTROL on a sewage treatment plant can be grouped as follows:

1. Direct manual control
2. Remote operation
3. Automatic operation
4. Automated plant.

Each of these has its place in a modern, large, plant. The operational needs of each section of the plant need to be evaluated in terms of manpower, both operational and maintenance, effectiveness, reliability, and cost in terms of both capital and revenue. It does not follow that the criteria on which decisions are made are those of minimum manpower and minimum direct cost. In areas where unemployment is high and state financial assistance is given to the unemployed, high manpower utilization may give overall financial benefit to the community. Also the acceptable level of reliability will need to be assessed according to the type of plant being considered, the availability and cost of maintenance staff, the availability and cost of manual control staff to take over in times of malfunction, and the degree of acceptable availability of plant. There is also the need for authorities controlling large installations to be in the forefront of development of new control techniques, and the value of plant-scale experimental work in this field should be one of the factors used in system evaluation.

The following comments are based upon consideration of these factors under conditions recently prevailing in South-East England, when determining the levels of control to be installed in a new treatment plant to serve, ultimately, 250,000 people with a dry weather flow of 70,000 m^3 day^{-1} (15 mgd), a requirement to give full treatment to 4·5 dwf, and an effluent standard requirement of 7 mg l^{-1} BOD and SS in summer, relaxed to 10 mg l^{-1} BOD and S.S. in winter. Manpower requirements were to be minimal, in view of wage levels (a first-grade plant operator on a 40-h week rotating shift system could be expected to draw with bonus payments, £2000 per annum gross), and the effluent quality standard was to be maintained at all times, although in the early years of the project some by-passing of sewage flow to another plant would be available in times of emergency.

The works has the unusual feature of being situated on a hill approximately 40 m (130 ft) above the sewer level, and a requirement to pump all sludge after processing about 2·5 km (1·5 miles) to a tanker filling station for distribution on land as a liquid fertilizer.

PUMPING OF SEWAGE

This will be fully automated, a fixed proportion of the flow passing the pumping station being pumped, subject to all flows below a fixed figure being pumped, and the rate being determined at high flows according to the requirement to give full treatment to 4·5 dwf while pumping 6·0 dwf until stand-by sedimentation tanks are full and operating.

PRELIMINARY TREATMENT

In order to reduce scum formation in digesters, and pipe and valve blockages, the preliminary treatment plant will consist of comminuters followed by fine screens, with detritors for grit removal. All this plant will be automatically operated according to rate of flow of sewage, and in case of raking of the screens, by the head differential up and downstream. Sequence of operation will be set manually.

SEDIMENTATION

The rectangular sedimentation tanks will be desludged mechanically by remote operation from the Control Room. Sludge withdrawal by telescopic valves will be monitored by remote closed-circuit television. The filling (and emptying) of the stand-by sedimentation tanks will be automatic, actuated by measurement of incoming flow.

ACTIVATED SLUDGE TREATMENT

The diffused air activated sludge plant will have flow division controlled by weir settings: total air flow will be determined by the head of plant operations and set remotely, as will MLSS levels. The final settling tanks have presented interesting control problems. It was recognized that, on the basis of experience at another plant, mechanical scraping of final tanks was uneconomic: it could be effectively carried out by an operator walking round each tank twice daily drawing a weighted chain around the walls. However, the proposed installation of upward-flow sand filters dictated that no algal growths should be permitted on final tank weirs and weirs and channels as these might clog the sand filter inlets. The weirs and channels are therefore to be covered, which indicates the need for mechanical scum removal.

The removal of sludge from the tanks, and its return to the activated sludge plant will be automated: the rate of withdrawal from each tank will be in steps based upon the rate of incoming sewage flow to the plant; it will then be over-ridden by measurement of upper and lower levels of sludge blanket in the tank. This latter has been considered necessary because of the possibility of seasonable variations in sludge density index.

TERTIARY TREATMENT

Rapid upward flow sand filters will remove residual suspended matter from the effluent before it passes down a cascade (for oxygenation) in passage to the river. This process is entirely automated, the need for backwashing being determined by back-pressure on the sand filters. The filters will otherwise be washed daily in sequence, and a "memory" will determine the order of the washing. This is a standard technique for this type of plant.

SLUDGE CONSOLIDATION

Raw and activated sludges will be consolidated separately prior to digestion. The supernatant liquor will be drawn off by remote control from the Control Room, monitored by closed-circuit television. Any water separating in the secondary (unheated) digesters will be drawn off manually. This decision is based upon uncertainty of quantity of supernatant, and its location in the tanks. Provision has been made for remote operation with television monitoring should this be found economic and acceptable by experience.

SLUDGE DIGESTION

The charging of the tanks, will be carried out remotely from the Control Room, quantities being determined by levels in the consolidation tanks measured by pressure-sensing devices. The overflow from the primary digesters will flow by gravity to the secondary tanks manually selected to receive it.

POWER

It was recognized that the cost of power from public supplies was likely to remain significantly below that of power generated on site. Emergency power is therefore available from automatically-starting gas turbines. Great care has been necessary to ensure that all essential valves and units are powered by the emergency supply, and that no valves can fail in a position likely to cause operational mishaps.

CONTROL ROOM

A mimic diagram some 12·5 m (40 ft) long shows all pipelines, valves and operating units. All manually operated valves and penstocks are shown in indication only, but their position (open or shut) can be seen, and amended by self-adhesive overlay. Remotely operated valves are controlled by switches on the diagram, and automatically operating valves alter the position shown on the diagram as they operate. All valves and pipelines are numbered and colour-coded in the manner of the units on the plant and all control alterations and defects can be simply and accurately reported by reference to this. All pumps and other plant running are shown to be so doing by illuminated symbols.

It should be noted that in planning remote operation the movement of a switch should operate the unit concerned: any subsequent illumination of position or state of equipment should be actuated by the equipment itself and *not* by the controlling switch.

PERSONNEL

The operation of the plant, including maintenance of 17 hectares (42 acres) of site, will be in the hands of 5 controllers, one superintendent and 7 plant operators and attendants.

Initially, 1 fitter, 1 electrician and 1 instrument mechanic will be recruited to carry out the additional maintenance required, it being noted that all maintenance will be carried out from headquarters some 20 km (12·5 miles) away, where also will be stationed the head of plant operations and chief chemist. A small control laboratory will be built at the plant, but all major work and industrial effluent surveillance will continue to be centred at headquarters.

Acknowledgement—The Consulting Civil Engineers to the West Hertfordshire Main Drainage Authority are John Taylor & Sons of Westminster, England, and they have played the major part in control system evaluation and design for this plant, known as the Blackbirds Works.

Water Research Pergamon Press 1972. Vol. 6, pp. 587–589. Printed in Great Britain

AUTOMATIC CONTROL AT THE MILTON KEYNES WORKS

H. Roy Oakley

J. D. & D. M. Watson, Chartered Civil Engineers, 67, Tufton Street, Westminster S.W.1., England

INTRODUCTION

Milton Keynes is a New City of 250,000 people to be built 50 miles north of London. The development plan was approved earlier this year and it is expected that the population will reach over 100,000 by 1979 and that the full development will have taken place by 1989.

The Designated Area occupies 9000 ha of the upper part of the River Ouse basin and will be drained to one treatment works which will discharge effluent into the River Ouse. The estimated daily rate of flow of sewage by the year 2000 is 110×10^6 l. (4600 $m^3 h^{-1}$) which compares with 1 and 10 per cent flows in the river at the point of effluent discharge of 1900 and 3800 $m^3 h^{-1}$ respectively. All flows reaching the works will be treated. Water is abstracted from the river for potable supply at Bedford about 50 km below the point of discharge. There is further abstraction for storage and water supply below Bedford.

Construction of the treatment works will start later this year, and it is intended to commission them early in 1974.

EFFLUENT STANDARDS

The River Authority require that the effluent be fully nitrified and that the suspended solids and biochemical oxygen demand should not exceed 10 mg l^{-1}. They also wish that effluent lagoons be made out of worked up gravel pits so that up to 5 days storage can be given if required either for further treatment or to protect the river during periods of exceptionally low flow or in circumstances where the effluent quality is temporarily sub-standard.

WORKS DESCRIPTION

The treatment works will be largely conventional. Primary treatment by screens and aerated grit chambers will be followed by rectangular sedimentation tanks, diffused air activated sludge treatment and rapid gravity sand filters: space will be left for nutrient removal in case this is later required. Sludge will be treated by digestion and it is hoped that some liquid digested sludge will be disposed of to agricultural land. The remainder will be treated by centrifuge and incineration.

CONTROL OBJECTIVES

The intention of the metering and control scheme is:

(a) to provide essential facilities for operating the works;

(b) to allow for automatic control where desirable to save labour or improve efficiency;

(c) to provide information for investigation into improved methods of treatment.

COST BENEFIT

A minimum degree of information is necessary for the proper operation of the works. Beyond this the cost of additional instrumentation and control has to be justified and it is taken that capital expenditure of about £20,000 is justified for every works operative saved. To offset this the cost of maintaining instrumentation is assessed on the basis that every £60,000 of equipment requires an instrument mechanic or equivalent cost in maintenance contracts.

CONTROL METHODS

Three characteristics are identified: physical, chemical and biochemical.

The control of works and equipment by physical parameters such as level, rates of flow etc. is common place and requires no elaboration in this note.

Control by chemical parameters is more difficult, largely because of the limited range of reliable and practicable sensing instruments; at the present time these are restricted to pH, dissolved oxygen, COD, ammonia and one or two specific ions. For convenience, temperature and suspended solid concentration are included in this section, although strictly physical in nature.

Control by biochemical parameters is not yet generally feasible, although consideration is being given to the possibility of a continuous toxicity test on the effluent which would give early warning of unsatisfactory conditions for discharge to the river.

Methods of control can be identified as local, hand or automatic and remote, hand or automatic. Remote control requires expensive transmission to a control centre and is only justified when saving in man power is realized or improved efficiency of operation results.

A third subject for discussion under method is the display and storage of data. Visual display is convenient, but storage of paper charts is not the best way of handling a quantity of data.

OUTLINE SYSTEM

The principal features of the system which is intended are:

Control centres

There will be a main control centre in the administration block which will accommodate a display board and data logging units. There will be sub-centres at the sand filters and in the sludge treatment building.

Inlet works

The numbers of screens and grit channels in use will be automatically selected according to the rate of flow through the works, with remote indication only at the control centre. Continuous monitoring of pH and suspended solids will take place at this point with indication and recording in the main control room; there will also be an automatic sampling station.

Sedimentation tanks

The numbers of units and the operation of machinery will be controlled locally, but there will be remote control facilities for decanting tanks. Sludge withdrawal will be on a time cycle with overriding control from solids concentration meters, either nucleonic or ultrasonic.

Aeration units

The numbers of units will be selected manually, but air flow will be controlled by the dissolved oxygen concentration at selected points in the units. The rate of return of sludge from the final tanks will be governed by the rate of flow through the works.

Sand filters

These will operate on a batch principle and will be fully controlled automatically according to the rate of flow and the head loss through each filter. There will be automatic sampling of the effluent and continuous monitoring to the control centre of the final effluent pH and suspended solids and ammonia concentration.

Sludge treatment

This is still being designed and details of the control scheme have not yet been determined.

Electrical and mechanical

There will be normal metering and monitoring of electrical and mechanical devices with alarm conditions relayed to the central control stations.

Data recording, storage and retrieval

Analogue signals will be transmitted to a computer control package which will log in digital form up to 20 selected items including flow rates, liquid levels and chemical characteristics. Alarm conditions of data logged items will be printed out on teletype and linked to an alarm panel with audio and visual signals. Either punched or magnetic tape will be used for the storage of logged items. It is intended that the computer be designed to enable feed back of both physical and chemical characteristics to the control of works units when appropriate relationships have been established, and the reliability of the instrumentation confirmed.

A visual display showing the works operation will be given on a mimic panel and where appropriate remote operation of units will be operated from this panel.

Chart recorders will be installed, but linked to the central computer system so that chosen parameters can be recorded for visual display on a selected basis as required.

Water Research Pergamon Press 1972. Vol. 6, pp. 591–595. Printed in Great Britain

COMPUTER CONTROL—ARAKAWA TREATMENT PLANT

SATORU TOHYAMA

Association of Ara-river Basin-wide Sewage Works, 48, 1-Chome, Nishihara, Shibuya-ku, Tokyo, Japan

1. THE PURPOSE OF COMPUTER CONTROL SYSTEM

(1) Saving of manpower
(2) Decreasing of operating cost
(3) Rationalization and high-grading of wastewater quality and quantity control
(4) Management system—electric department, instrument department, water quality department
(5) Prediction of the incoming shore water quality and quantity change, and optimizing planning of treatment equipment.

2. THE CONSTITUTION OF MANAGEMENT SYSTEM

The system aims to supervisory control of wastewater treatment, sludge treatment, power equipment, and electric equipment. For this purpose, we have adopted the centralized supervisory control system and this system consists of following equipment.

(1) Central supervisory control board and control desk
(2) Instruments and its panel board
(3) Computer control system
(4) Telemeter, telecontrol system
(5) Relay control system
(6) ITV equipment
(7) Broadcasting equipment
(8) Local operating board.

This consists of:

(1) *Central supervisory control board and control desk*

Mosaic, graphic panel, 2350 mm height × 15,000 mm width. Supervisory and control of electric equipment, power equipment, wastewater treatment equipment, sludge treatment equipment.
Each piece of equipment is distinguished and the operating state shown.
Operation desks are provided for the divisions of electricity, wastewater treatment, and sludge treatment. Every division is supervised and telecontrolled.

(2) *Instruments and its panel board*

Instruments are classified as follows:

wastewater treatment	flow	33 loops
	level	17 loops
	water quality	31 loops

sludge treatment	flow	10 loops
	level	15 loops
	water quality	5 loops
	temperature	40 loops
	pressure	11 loops

The instrument data are data processed at the central process computer.
The panel has the function of a backup for the computer. As few controllers and recorders as possible are installed.

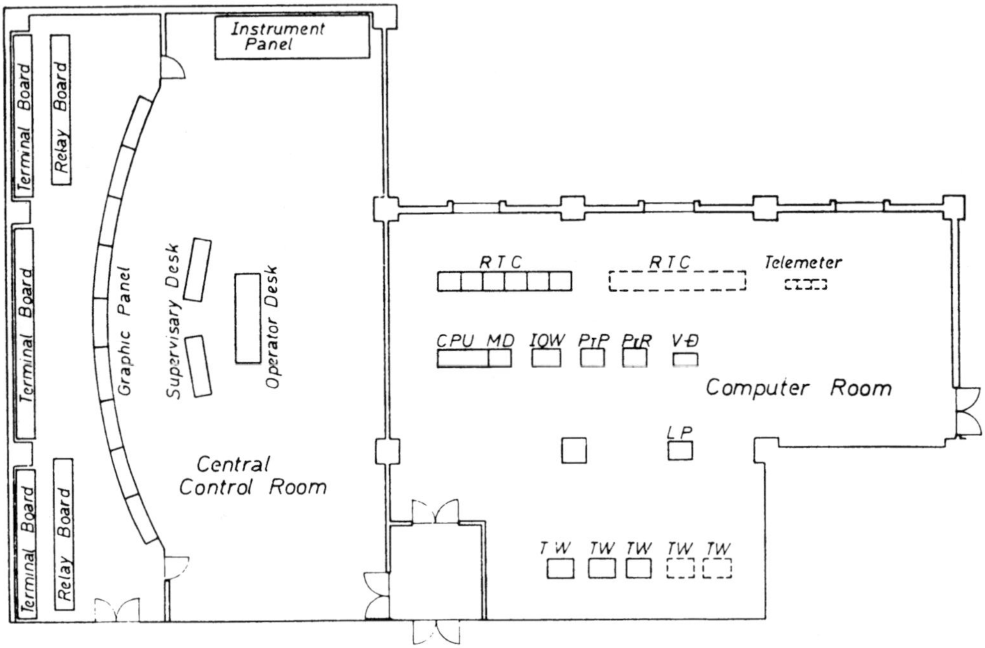

FIG. 1. Central supervisory control room.

(3) *Computer control system*

This system is composed as follows:

1 × central processing unit (32 kW, 1 W = 16 bits), 1 × magnetic drum (512 kW), 6 × typewriters, 2 × vidio displays, 1 × paper tape puncher, 1 × paper tape reader, 1 × line printer, 1 × real time controller, 1 × operator desk.

On-line supervisory, control, operation are treated at CRT display, operator desk and

Off-line jobs such as daily logging, program editing, scientific calculation are treated as hard copies at typewriters and lineprinters.

(4) *Telemeter, telecontrol system*

The situation of the sewerage system, and the data for sludge treatment works and relay pump station and operation are performed by the telemetering and telecontrol.

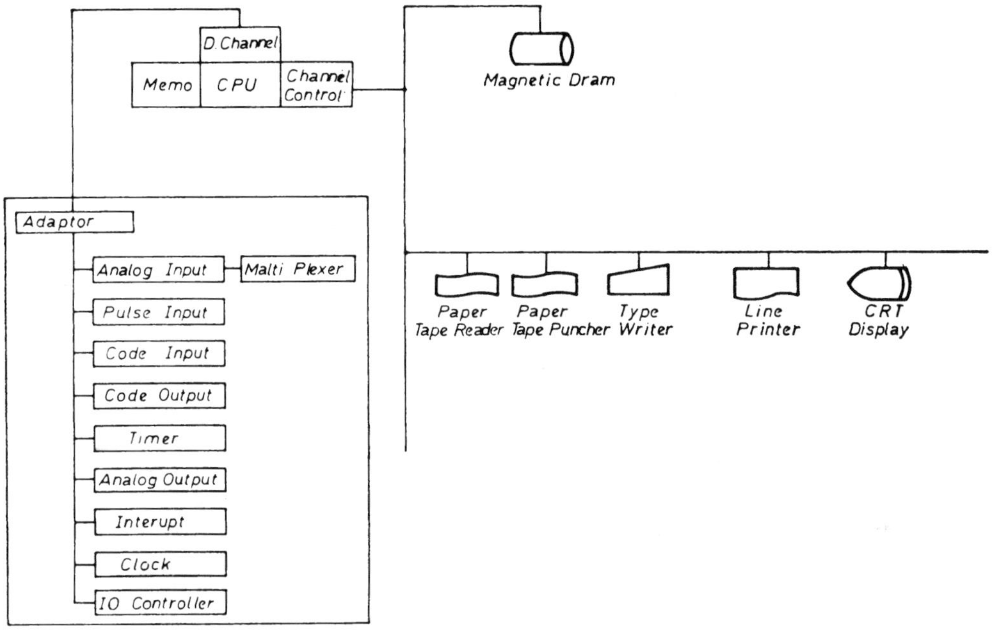

FIG. 2. Computer system diagram.

The sewerage system data are transmitted by TM = 50 ch, SV = 100 ch, TC = 60 ch.
Sludge treatment data are composed by TM = 35 ch, SV = 60 ch, TC = 40 ch.
Pump station data are by TM = 150 ch, SV = 200 ch, TC = 150 ch.
Private communication is provided for telemeter and telecontrol, and they use a digital system.

(5) *Relay control system*

Sludge treatment consists of decanting, heat treatment, sludge dewatering and incineration.

The combination of this equipment and treatment may change according to the influent. Therefore, the equipment is flexible.

Static universal sequence equipment is used as a protection against corrosion.

(6) *ITV equipment*

In order to supervise the operating state of the machines by eye, we use the ITV equipment.

The works are as follows:

Electric division	4 points
Wastewater treatment	15 points
Sludge treatment	6 points

(7) *Broadcasting equipment*

These are used for communication between the central control room and every location.

(8) *Local operating board*

This is equipped for the purpose of backup to the computer and emergency operation. They consist of an electric division, wastewater treatment division, sludge treatment. division.

3. COMPUTER FUNCTIONS

These are described as jobs of *normal-level, optimizing-level, management-level,* i.e.

(1) *Normal-level*

Data-acquisition, Supervisory of process status, Abnormal state display, Abnormal state previous announcement, Operation Guide, Sequential Control, Conventional Control.

Process control items.

Incoming flow-control, grit chamber operating control, pump well operating control, gate control, flow control of wastewater treatment, storm water storage tank control, confluence tank control, fan control, chlorine control, sludge treatment control, relay pump station control.

(2) *Optimizing-level*

Supervisory control—setpoint instruction to local—controller; (a) Relay pump station control by predicting incoming sewage, (b) Air control, (c) Return sludge control.

(3) *Management-level*

Decision of operation of wastewater treatment.

Analysis of correlation and prediction of control; (a) Trend changing index, Seasonal changing index, irregular changing index of the quality and quantity of incoming wastewater, Equipment operating planning, Equipment expanding planning, (b) Correlation between irregular changing index and rainfall, (c) Analysis of the effect of the water quality on operation of the activated sludge process, (d) Calculation of the optimizing value of the BOD load, (e) Analysis of the correlation between the change of the sludge boundary level in the final settling tank and SVI, (f) Inference of the treatment ability, and operating scheduling, (g) Instrumentation of incoming sewage quality and quantity, and prediction of the control of the works.

4. FURTHER POINTS

(1) *Reliability*

(a) The computer system adopted is the sinplex system based on our process computer experience. It is backed up by analog controllers, unit sequencers, and manual operation at the centre or locally. Analog controllers are used for backup use and their use minimized. Scanning control is used, controlling all processes by switching the controller to every process.

As unit sequencer, we use small-sized universal sequences throughout. It is possible to operate these separately.

A big-sized central sequencer controls the correlation between each sequencers. As a result, every sequencer continues to work in case the central sequencer fails.

(b) Because of the bad atmosphere, we use 1C circuit and solid state elements.

(2) *Operation guide*

To facilitate operation, we pay close attention to man–machine communication. For this purpose we use mainly CRT display with a keyboard. We also use a mosaic graphic panel, ITV, broadcasting apparatus, operator console, and a typewriter.

The CRT display function is as follows:
Data display (many items at one time); Alarm display; Alarm warning; Delay display; Operation guide, operation announcement; Trend display; Plant state graphic display; Program read-in or modification.

(3) *Saving manpower*

As the wastewater treatment works is well divided into units enlargement is simple by the combination of these unit processes once we establish the unit process sequence. This will save manpower. An increased amount of information on enlargement of the plant is obtained with the computer.

On completion of the initial investment we aim at decreasing operating costs. Automatic operation is especially effective at the sludge treatment works and relay pump stations.

Water Research Pergamon Press 1972. Vol. 6, pp. 597–600. Printed in Great Britain

THE USE OF COMPUTERS IN PHILADELPHIA'S WATER POLLUTION CONTROL ACTIVITIES

C. F. GUARINO

City of Philadelphia, Water Department, 1160 Municipal Services Building, Philadelphia, Pa. 19107, U.S.A.

SINCE 1966, Philadelphia has been using computers to maintain proper operational records, keep abreast of the condition of our mechanical equipment, furnish information for the design of future plants, and to apply mathematical disciplines to analytical data which would otherwise be impractical.

Since we can now look back on 5 years' experience with computers, we are in a position to review and evaluate their usefulness. Even though we have itemized the many uses that we have made of the computer, we really had two main objectives in mind; one, to reduce the paper work required in record keeping of large treatment facilities; and, two, pave the way for the eventual instrumentation and automation of the treatment facilities.

Our first project which we termed "Datamation" was to prepare a program which would enable the daily tabulation of all of the data collected from the various operator and laboratory log sheets. We found that each day, 384 pieces of information were recorded at our Northeast Water Pollution Control Plant. By means of suitable programming, a monthly report was generated which tabulated all the above data and performed all the necessary calculations including desired statistical analyses. This program is called NELOG (GUARINO and RADZIUL, 1968). The program included provisions for data screening and subsequent rejection. From this monthly report, the computer generates a monthly summary. The twelve summaries are also combined by the computer into an annual report. These three programs reduce the technical and clerical record keeping time, minimize human errors, and give us better and more readily accessible data.

An outgrowth of the operationally oriented NELOG was a maintenance based data logging report termed MEDICO (GUARINO and CARPENTER, 1970). This program, established in 1969, lists the condition of every large piece of equipment at the Northeast Treatment facility. MEDICO serves as a management tool to evaluate, control and direct maintenance in waste treatment. An excerpt from MEDICO is presented in TABLE 1.

The City of Philadelphia has embarked on a $200-million program to improve its treatment plant facilities. In the Delaware Valley drainage area, all municipalities and industries has been allocated a fixed quantity of carbonaceous biochemical oxygen demand as well as suspended solids which they would be permitted to discharge into the receiving stream. To enable the proper design of the proposed treatment facilities to meet these effluent requirements, data available in the memory of the computer has been statistically analyzed and furnished to the designer. This, of course, is not a new technique but simply one of the many advantages that the computer makes available to us. We expect this information to be extremely useful in the design of the new treatment facilities.

The first of such reports "Variations of Input Parameters at the Northeast Water

TABLE 1. NORTHEAST WATER POLLUTION CONTROL PLANT—MAINTENANCE MANAGEMENT REPORT. MEDICO. Special report May 1969–January 1970

Equipment blower building	Avail (h)	Oper (h)	Down (h)	Type of maintenance M	E	D	I	C	O	% Time avail.
Return sludge pump 18	6191	4961	193	192	1	0	0	0	0	97
Blower No. 6	4896	4798	1488	1488	0	0	0	0	0	77
Air conditioner unit	6058	5400	326	206	0	120	0	0	0	95
Electric air filter (east)	456	312	5928	5448	408	72	0	0	0	7
Motor generator set 1	6000	3933	384	144	240	0	0	0	0	94

M—Mechanical. E—Electrical. D—Diagnostic. I—Instrumentation. C—Contractural. O—Operator.

Pollution Control Plant" (NELSON, 1970a) studied fifteen individual input parameters such as flow, influent suspended solids, organic loading, etc. for 730 consecutive days. The distribution of each parameter about the mean was determined by cumulative frequency analysis.

Cumulative frequency analysis is used as a refinement of the design process. For example, in the design of a wastewater treatment plant, it can be used to generate several alternatives, based on different probabilities of loading. Instead of using average values as a basis for design, the plant can be designed to treat 95 per cent of the loading conditions to which it would be subjected. The sizing of the plant and construction costs for this probability level can be calculated and compared to the costs corresponding to probability levels of 85 or 90 per cent. Cost benefit ratios can then be derived for each alternative. This will provide a means of selecting the best trade-off between construction costs and plant efficiency.

A second design report "Secondary Treatment Summary" (NELSON, 1970b) highlighted 2 yr of sanitary engineering parameters used to describe the Northeast secondary facilities. The report extracted on a daily basis, 54 out of 490 parameters such as sludge age, sludge density, organic loading, etc. and organized these same parameters into spans of time that represented various treatment modes for our Northeast facility. Compilation of such a design report without assistance from the computer would be a major undertaking. The computer within an 8-h period produced 138 pages of tabulated averaged data on *stencils*.

Preparation of detailed reports for design purposes assures the public works manager that design decisions will be based on all available information. Without computer assistance, preparation of such reports may become too tedious and result in time and labor saving generalizations which may mislead the designer.

Philadelphia is using its computer facilities to help develop and evaluate sensors vital to wastewater process control. One of this city's greatest stumbling blocks in its plant automation program is the lack of proper sensors, for instance, an on-line biochemical oxygen demand (BOD) monitor. A great deal of work throughout the field has been aimed at correlating BOD with one easily measured parameter such as, chemical oxygen demand, total carbon (TC), total organic carbon (TOC), etc. The success of such work depends primarily on the consistency of the particular wastewater analyzed.

Philadelphia feels that most likely no one instrument will predict the BOD and thus is researching the feasibility of predicting BOD by a combination of COD and total carbon. The computer is used to run multiple regression analysis on data sets containing BOD, TC and COD. Results for industrialized primary effluent are shown in TABLE 2. Equations are based on 45 data sets. It will be seen from TABLE 2, though

TABLE 2. PREDICTING BOD OF INDUSTRIALIZED PRIMARY EFFLUENT

(1) BOD = 0.34 COD + 48

Deviation breakdown

% Deviation	% of time
0–10	46.6
0–21	77.7
0–30	84.4

(2) BOD = 0.97 TC + 54

Deviation breakdown

% Deviation	% of time
0–10	37.7
0–21	64.4
0–30	88.8

(3) BOD = 0.19 COD + 0.58 TC + 27

Deviation breakdown

% Deviation	% of time
0–10	48.8
0–21	80.0
0–30	86.6

Above equations based on 45 data sets

COD and TC are fairly good predictors independently, by combining same, the predictability is enhanced. Equation (3) can predict the BOD to within 21 per cent over 80 per cent of the time. This, in itself, may not seem significant until applied to a real life situation. Recent work at our Northeast Plant indicates that the organic loading to the secondary system fluctuates 150 per cent from the daily average. Thus, by the use of equation 3 or one of a similar form, the fluctuations high and low in respect to time can be ascertained and adjustments made in the aerators to compensate for such fluctuations.

In our continual field evaluations of various sensors such as dissolved oxygen probes and turbidimeters, large volumes of comparative data are generated. The computer with its various correlation programs can quickly determine the value of a particular sensor and point out relationships that manually would be difficult to establish. Thus the data interpretation phase of sensor evaluation is eased.

Concluding, in the 5 yr of computer use in Philadelphia, operational reports for large treatment works have been maintained with minimum effort. We have used the final product of our data logging system as a foundation for design reports and to better understand the inner workings of some of our unit processes.

The computer has aided us in sensor development and has brought us one step closer to automation of our treatment facilities.

Acknowledgements—I would like to acknowledge the help of MICHAEL D. NELSON of the Water Pollution Control Division engineering staff in the preparation of this paper, especially in the statistical and correlation analyses sections.

REFERENCES

GUARINO and RADZIUL (1968) Data Processing in Philadelphia. *J. Water Pollut. Control Fed.*

GUARINO and CARPENTER (1970) Philadelphia's plans toward instrumentation and automation of the wastewater treatment processes. *5th International Water Pollution Research Conference.*

NELSON M. D. (1970a) Variations of input parameters at the northeast water pollution control plant. *Philadelphia Water Department internal report.*

NELSON M. D. (1970b) Secondary treatment summary. *Philadelphia Water Department internal report.*

Water Research Pergamon Press 1972. Vol. 6, pp. 601–604. Printed in Great Britain

EDUCATIONAL SYSTEMS FOR PLANT OPERATORS

J. H. Austin
Environmental Systems Engineering, Clemson University, Clemson, S.C. 29631, U.S.A.

CONTENT AND SCOPE

Construction of new facilities necessitates staffing by technicians with advanced training over and above the fundamentals required in older, relatively simpler plants. A wide variety of training programs are in existence.

PRESENT TRAINING EFFORTS

In 1968, Sacramento State College in Sacramento, California, was awarded an $83,000 grant by the Federal Water Pollution Control Administration for a 3-yr program to develop a correspondence training program for operators of wastewater treatment plants, in a joint effort with the California Water Pollution Control Association.

In 1966, the University of Michigan was awarded a $70,000 grant from the Federal Water Pollution Control Administration for a 3-yr program to develop a series of programmed learning lessons dealing with the chemistry of analysis and treatment of water and wastewater. This effort was described by Pursglove (1969).

The state of South Carolina has instituted a series of correspondence courses McLellon, Austin and Andrews, 1969; Austin, 1970a, 1970b; Austin, McLellon and Dyer, 1970).

BEHAVIORAL OBJECTIVE APPROACH TO TRAINING

The behavioral objective or performance objective approach to solving training problems is one of the major advances in educational technology of the past decade. The approach does just what its name implies; it states in clear and concise terms what the behavior or performance of the trainee is expected to be. It has a major advantage over normal teaching methods in that both the instructor and the student know the nature of the desired behavior.

The behavioral objectives approach satisfies three needs. They are: (1) the basic continuum of performance of operators will have been defined so as to facilitate the preparation and maintenance of job descriptions; (2) the basic content will have been identified for course curriculum development and for instructional media selection and design; and (3) the conditions and criteria for evaluating the posttraining performance (licensing examinations) of the operator will have been determined.

IMPLEMENTATION OF THE BEHAVIORAL OBJECTIVE APPROACH

Clemson University is presently engaged in the development of a 2-yr post high school training program for operators of wastewater treatment plants (Clemson–EPA, 1971a, b). This work is being carried out with the assistance of highly qualified wastewater treatment plant operators as well as three technical education centers in the United States. Initially thirty-seven broad behavioral objectives were developed which

TABLE 1. BROAD BEHAVIORAL CATEGORIES

Normal operation procedures
Abnormal operation procedures
Preventive maintenance procedures
Corrective maintenance procedures
Laboratory control procedures
Systems interaction procedures
Management/supervisory procedures

included the majority of the knowledge and skills required of an operator. Next a composite model plant (CMP) was developed. This plant is composed of process units representative of those found in practice. Representative process units were chosen, operation of which was designed to afford the trainee the most comprehensive mastery

TABLE 2. EXAMPLE OF BEHAVIORAL OBJECTIVES

S	101 →	R
When given the task of performing the *normal* operation procedures, and confronted with an actual process unit, or a common name, functional or physical description, or other standard representation thereof . . .		Trainee, from recall, will describe the routine he will follow and its frequency, the conditions he will look for, the actions he will take, and for certain actions the "whys" that are most relevant; from recall, trainee will actually perform the procedure.
S	401 →	R
When given the task of performing the *corrective* maintenance procedures, and confronted with actual, mechanical indications of a malfunction of a process unit or component within the unit, or a verbal description or other standard representation thereof . . .		Trainee will recognize the mechanical malfunction as such and will recall the name(s) of relevant tools (e.g. guides) available for use, and using the reference tools will describe the immediate and/or long term operational adjustments that may be necessary, will name the most probable component(s) needing repair or replacement, if more than one probable cause of malfunction, will list in order of probability from highest to lowest, if component(s) need repair, will briefly describe the nature of the repair; using the reference tools trainee will actually perform the indicated procedure.
S	501 →	R
When given the task of performing the laboratory control procedures, and confronted with the need to conduct any common laboratory analysis, identified by name, functional description, or other standard representation thereof . . .		Trainee will recall the name(s) of relevant reference tools (e.g. guides) available for use in the conduct of the tests, and using the reference tools will describe the conduct of sampling and analysis and the facilities, equipment, and supplies involved; using the reference tools, trainee will actually perform the sampling and analysis in accordance with the most current recommendations as reported by the relevant professional organization.

TABLE 2 *continued*

S	601 ↔	R
When given the task of planning the overall operation of a specific plant for a certain period of time, and confronted with the actual condition(s) of the wastewater entering or of the wastestream within the plant, or a verbal description or other standard representation thereof . . .		Trainee, from recall, will identify by common name, physical or functional description, or other standard representation thereof, the process unit(s) directly involved in dealing with the condition(s) and will describe how the unit(s) deal with the specified condition(s).

S	709 →	R
When given the task of accomplishing employee dismissal, and confronted with an actual justifiable cause, by name, verbal description, or other standard representation thereof . . .		Trainee will recognize the cause as such, and, from recall, will describe the reason(s) why the cause justifies employee dismissal and will describe action(s) relating directly to the actual dismissal procedure.

S	715 →	R
When given the task of responding to criticism of plant operation, and confronted with an actual complaint from private citizens and/or public officials regarding conditions both relevant and irrelevant to the operation of a wastewater plant, or a verbal description or other standard representation thereof . . .		Trainee will recognize the complaint as relevant or irrelevant, and, from recall, will describe the type of data appropriate to the complaint in question and will describe *how* and *why* the complaint is relevant or irrelevant; using the data noted trainee will demonstrate an actual response to the complaint.

of all process units of this type. In other words the probability of a trainee's transferring his knowledge and skills from the CMP Process Unit to a similar process unit has been maximized. Once the behaviors had been set and the process units determined it was possible to develop specific behavioral objectives for each process unit and its components. Specific behaviors were developed to indicate to the instructor just what behavior is expected of a trainee with each component of each process unit.

With this information available a curriculum was devised and at the present time lesson plans are being developed for the training sequences.

In TABLE 1 are the seven broad categories considered a part of plant operations. TABLE 2 shows examples of the behavioral objectives.

TRAINING GUIDE

Behaviorally oriented training guides explicitly state what needs to be done on which piece of equipment to what level of proficiency. Guides of this type serve as a training guide for the new trainee before and during his hands-on training. They serve as a refresher for the experienced man and are often useful in determining errors in an operational procedure.

SUMMARY

Technicians engaged in programs concerned with environmental control need to continually upgrade themselves to keep abreast of the latest advances in technology.

It is the duty of the professional educator to continually update the training materials both in technical content and in method of presentation.

The behavioral objective approach appears to have the greatest potential for the development of systematic training systems and guidance materials for wastewater treatment plant operators.

REFERENCES

AUSTIN J. H. (editor) (1969) *Correspondence Course Manual for Wastewater Plant Operators—Class D.* 424 pp. South Carolina Water and Pollution Control Association, Columbia, S.C.

AUSTIN J. H. (1970a) Current university activities in wastewater treatment plant operator training. In: *Proceedings, Education Systems for Operators of Water Pollution Control Facilities*, pp. 133–149. Clemson University.

AUSTIN J. H. (1970b) Training—how to do it. *J. Am. Wat. Wks Ass.* **62**, 431–432.

AUSTIN J. H., MCLELLON W. M. and DYER J. C. (1970) Training the environmental technician. *Am. J. Publ. Hlth.* **60**, 2314–2320.

BROADWELL M. M. (1969) Questions trainers ask. *Training in Business and Industry* **6**, 43, 62–63, 66.

CLEMSON UNIVERSITY AND THE ENVIRONMENTAL PROTECTION AGENCY (1971a) *Criteria for the Establishment and Maintenance of Two Year Post High School Wastewater Technology Training Programs.* Vol. I. *Program Criteria.* 45 pp.

CLEMSON UNIVERSITY AND THE ENVIRONMENTAL PROTECTION AGENCY (1971b). *Criteria for the Establishment and Maintenance of Two Year Post High School Wastewater Technology Training Programs.* Vol. II. *Curriculum Guidelines.* 637 pp.

FITZPATRICK J. W. (editor) (1965) *Sewerage Seminar on Plant Maintenance.* Rocky Mountain Water and Sewage Operators School, Boulder, Colo.

MCLELLON W. M., AUSTIN J. H. and ANDREWS J. F. (1969) School and State Cooperate for Effective Operator Training. *Water Sew. Wks* **116**, 46–49.

MAHLIE W. S. (editor) (1964) *Manual for Sewage Plant Operators.* Texas State Department of Health, Austin, Texas. 782 pp.

MARKLE S. M. and TIEMANN P. W. (1966) *Programming Is a Process*, 15 pp. Office of Instructional Resources, University of Illinois at Chicago.

NEW YORK STATE DEPARTMENT OF HEALTH. *Manual of Instruction for Sewage Treatment Plant Operators*, 247 pp. Health Education Service, Albany, N.Y.

PURSGLOVE L. (1969) Programmed manuals. In: *Proceedings, Educational Systems for Operators of Water Pollution Control Facilities*, pp. 315–328. Clemson University.

REDEKOPP A. B. and AUSTIN J. H. (1971) Systems approach to training and licensing of water works personnel in Ontario. In: *Technical Papers of the* 1971 *Annual Conference*, pp. C-1 to C-17. Ontario Section, AWWA.

TIEMANN P. W. (1969) Task analysis requirements for upgrading occupational instruction. In: *Proceedings, Educational Systems for Operators of Water Pollution Control Facilities*, pp. 275–281. Clemson University.

Water Research Pergamon Press 1972. Vol. 6, pp. 605–609. Printed in Great Britain

AN AUTOMATIC SYSTEM FOR THE CONTINUOUS DETERMINATION OF ORGANICS IN WATER AND WASTEWATER

H. BLEIER
Technische Mochschule Wien, KaMplatz 13, A-1040 Wien, Austria

THERE are different methods used to measure the concentration of organic matter in water and waste-water.

(a) physical methods (adsorption, refraction index, turbidity)
(b) biochemical methods (BOD_5)
(c) chemical methods (COD, permanganate index, TOC).

Because of the low specifity of the physical methods and of the slow reaction rate of the biochemical methods only the chemical methods can be used as a basis for an automatic system to determine the amount of organics.

Different instruments for COD measurements have been developed so far. More and more the determination of total organic carbon (TOC) is used in water and waste-water analysis. The first instruments built have two disadvantages, (1) they are neither automatic nor continuous; (2) they use high-temperature combustion. For complete oxidation only small sample-volumes can be injected into the combustion tube. Suspended solids and fibres in the sample cause difficulties in the analysis with these instruments. Only the organic carbon is of interest and therefore the inorganic carbon has to be expelled from the sample or has to be measured separately and subtracted from the total amount of carbon.

The new system has regard to the special problems of waste-water analysis. The principle of this system is shown in FIG. 1.

The samples are mixed and homogenized (unit A). Sulfuric acid is added and the inorganic carbon is expelled by sparging the acidified sample with purified air (unit B). The sample then is mixed with the reagents (potassium dichromate/silver sulfate/ sulfuric acid) and the mixture goes through a reaction unit, which consists of a glass

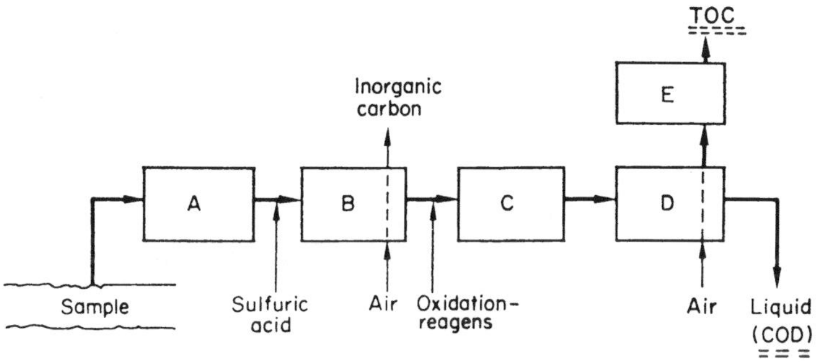

FIG. 1. Principle of the system.

helix heated to 170°C (unit C). The organic carbon is oxidized to carbon dioxide, which is expelled in the separator (unit D). While the COD is determined in the liquid by continuous potentiometric titration, the gas goes to a relative conductometric carbondioxide-analyzer (unit E). The results are recorded on a strip chart recorder. By using standard solutions the chart is calibrated in mg organic carbon per litre. Samples and chemicals are transported through the system by use of a multi-channel peristaltic pump.

The instrument has been used so far both for automatic continuous measurement of different effluents and for the determination of the TOC of individual samples. In this case a series of samples is analyzed by using an automatic sampler, which turns to the next sample every 6 min.

More than 80 different pure organic substances have been analyzed so far, including amino acids, carbohydrates, aliphatic and aromatic aldehydes, ketones and acids, aromatic compounds and heterocyclic substances. As the conditions of oxidation are much stronger than those of the standard method, in all cases more than 94 per cent

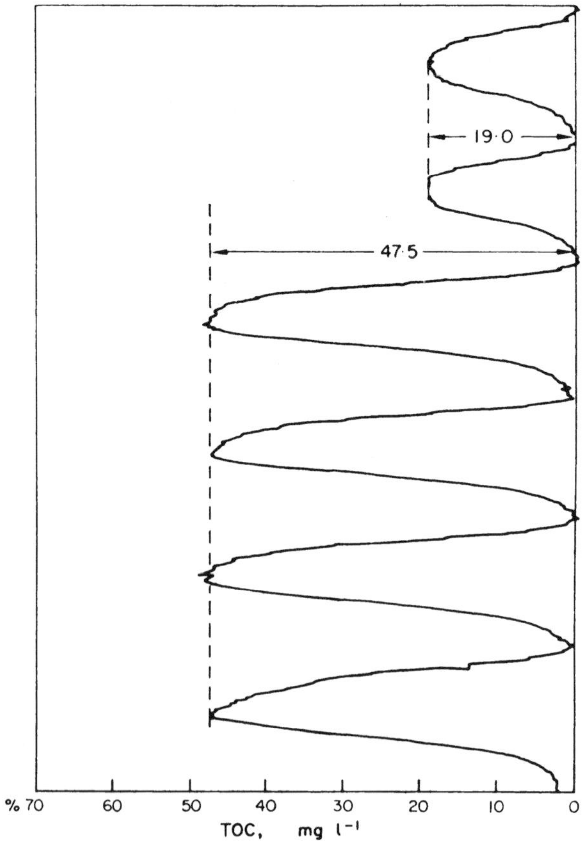

FIG. 2. Reliability of the TOC-determination.

of the theoretical value was obtained. Besides that, the disintegration time could be cut to 15 min, so the results of the analysis can be seen 20 min after input. The reliability of the system is better than $\pm 0{\cdot}5$ per cent, as it is shown in FIG. 2.

TABLE 1. BOD_5 AND TOC AT VIENNA–BLUMENTAL ON 2 SEPTEMBER 1971

	BOD_5	TOC		BOD_5	TOC
0–2	15	15	12–14	13	14
2–4	13	15	14–16	14	15
4–6	12	15	16–18	11	13
6–8	13	12	18–20	13	12
8–10	12	12	20–22	15	14
10–12	9	12	22–24	16	15

Analysis of secondary effluents at different treatment plants with both domestic and industrial wastewater, showed a very good correlation between BOD_5 and TOC. In TABLE 1, the results of a 24 h-test at the effluent of the treatment plant in Vienna–Blumental are summarized.

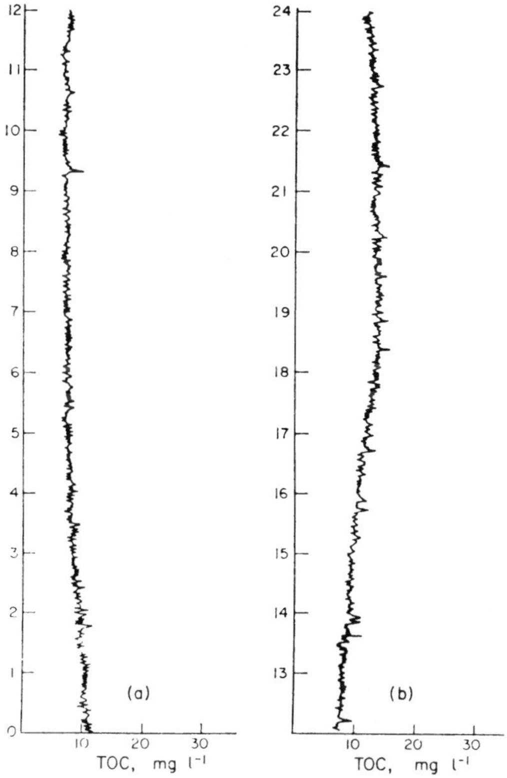

FIG. 3. Continuous TOC-monitoring at Vienne–Blumental, secondary effluent 27 October 1971. (a) 0.00–12.00 (b) 12.00–24.00.

The results in TABLE 1 show the good correlation between BOD and TOC. Daily controls at the effluent of Blumental treatment-plant for several months showed the ratio BOD_5:TOC = 0·8 ± 0·1. Because of this good correlation the continuous TOC-determination of the effluent makes it possible to give the actual BOD after 20 min. The accuracy is about 16 per cent. In FIG. 3, the recorder strip is shown, where the TOC is registered automatically over a period of 24 h.

Tests were also made to replace the determination of MLVSS by TOC-measurement of the activated sludge. Seventy-two samples of activated sludge from Vienna–Blumental were analyzed. The ratio TOC:MLVSS in per cent was determined. All results are drawn in the statistical net in FIG. 4. The average of this ratio is seen to be 46·75 per cent TOC in 100 per cent MLVSS. The small standard deviation in this ratio is of great importance, because it shows that the TOC-measurement can be used with good accuracy instead of a MLVSS-determination. It shows that the system can be used as a sensor in automatic control of activated sludge plants (see EMDE: Control of activated sludge plants).

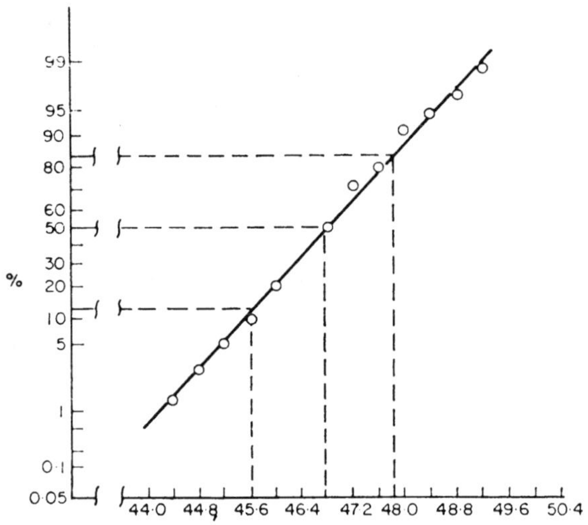

FIG. 4. The ratio TOC:MLUSS of 72 samples from Vienne–Blumental in a statistical net.

For continuous control of organics in the raw wastewater of a paper and pulp mill in Austria, we used the combined instrument for simultanous TOC and COD determination. FIGURE 5 shows a part of the recorder strip. Besides a very good correlation between COD and TOC a high alteration in the concentrations can be seen. The peak at 18·22 was shown to be caused by washwater of the pulp after disintegration of the wood. The big advantage of continuous determination is well seen in this figure.

The new system will be built by H. Wösthoff OHG., Bochum, Hagenstraße 30, D-463 Bochum, called "Hydromat-TOC" and "Hydromat-COD". A combined system for simultanous determination of COD and TOC will also be available.

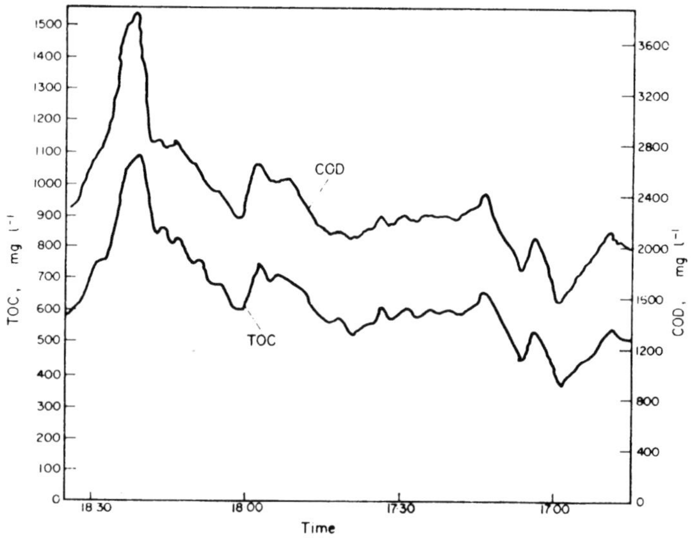

FIG. 5. Effluent of a paper and pulp industry, COD and TOC monitorship.

Water Research Pergamon Press 1972. Vol. 6, pp. 611–615. Printed in Great Britain

SEWER CONTROL AND PLANT AUTOMATION

J. J. ANDERSON
Watermation Inc., 2304 University Avenue, St. Paul, Minn. 55114, U.S.A.

1. INTRODUCTION

EXCEPT where storage basins have been constructed at the inlet to treatment plants, plant design has been based on anticipated wide ranges of flow and load dictated by sewer system function.

Recent experience has shown that large combined sewer systems can effectively be controlled during storm runoff events. The same techniques may be useful for operating the sewer system during dry weather to improve plant operation by dynamically modifying the flow rate and instantaneous strength of the plant influent. By properly scheduling (dispatching) the load through the sewer system, flow rates and hourly mass flux of pollutional material can be equalized over daily periods. In the absence of in-sewer storage, normally found in combined sewer systems, strategically located storage basins will be required. Experiments with two controlled combined sewer systems would indicate that external storage may not be required for effective dry weather flow modification.

2. SEWER CONTROL SYSTEMS

Several sewer systems have been or are currently being placed under central computer control. The first operating system at Minneapolis Saint Paul was placed in operation in the spring of 1969 (CALLERY, 1971). A system at Detroit was subsequently placed in operation, and in Seattle, a system is to be placed in operation in 1971. The City of Cleveland is in the early stages of installation of a comprehensive sewer control system.

Although each of the above control systems utilize quite different approaches in detail, they are all similar in general concept. Thus, the Minneapolis Saint Paul system can be used as an example for discussion.

The Minneapolis Saint Paul sewer system serves about 1.5 million people and receives an average annual flow of approximately 230 U.S. mgd (870,000 m^3 day^{-1}). The combined sewer area is approximately 30 $miles^2$ (80 km^2). The total length of interceptor sewers is approximately 30 miles (48 km).

The control system uses inflatable rubber dams and cylinder operated gates in major sewers at 15 locations. These devices are operated by a central computer over leased telephone lines. The leased line system also connects to sewer level measurements, rain gauges, river quality monitors and gate position indicators. 140 measurements can be taken and 40 gates controlled. The computer is a small computer, equipped with 24k of 18 bit word core memory, a 1.5×10^6 word disc, two small magnetic tapes, a line printer, and other input output devices. The computer contains a mathematical model of the hydrology and hydraulics of the sewer system.

During installation of the control system, hourly samples at the treatment plant influent and at key locations in the sewer system were collected in discrete automatic samplers. Each discrete sample, a total of about 35,000 samples, was automatically analyzed for several chemical parameters (ANDERSON *et al.*, 1967).

3. INFLUENT TIME VARIATIONS

FIGURE 1 shows the average daily flow, strength, and daily flux of chemical oxygen demand, to which this discussion will be limited, entering the interceptor system at key

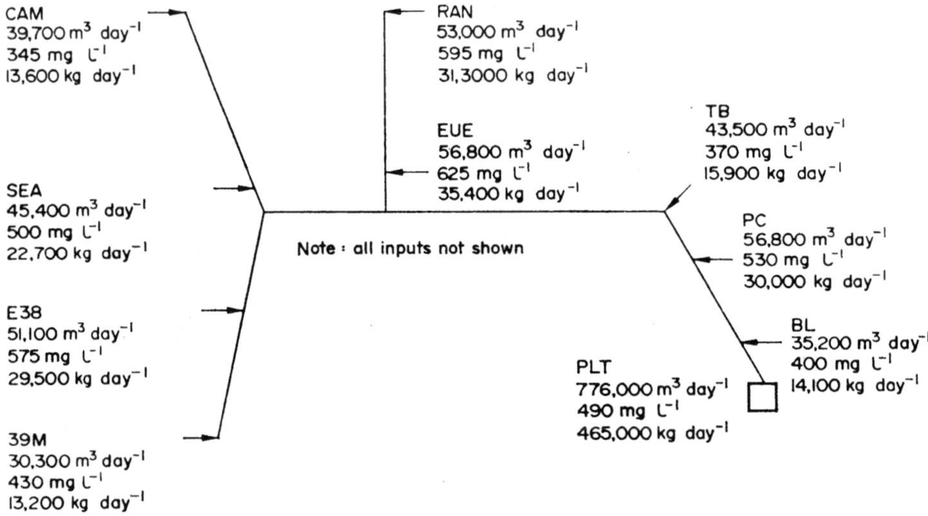

FIG. 1. Schematic diagram of interceptor sewer system showing average flow and COD inputs.

points. All inputs are not shown, but those selected are representative of the key locations which are controlled. While the strength of the average values shown vary nearly 2 to 1, the hourly values at a given location varied more than 16 to 1. FIGURE 2 shows the hourly flow rate variations and FIG. 3 shows the hourly COD load variations for a

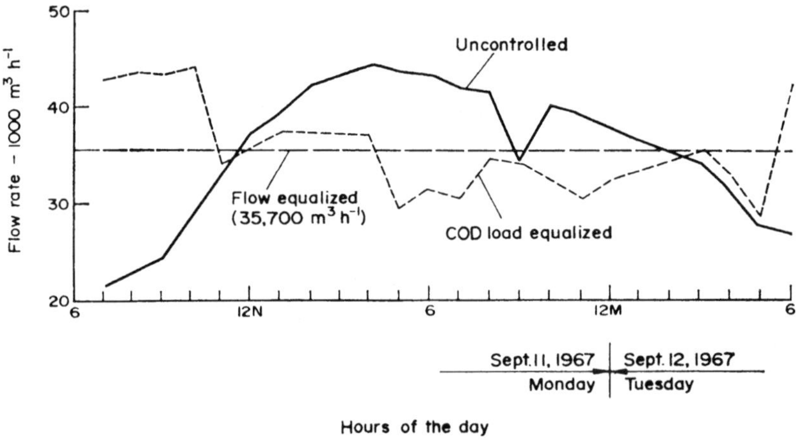

FIG. 2. Hourly flow rate variations.

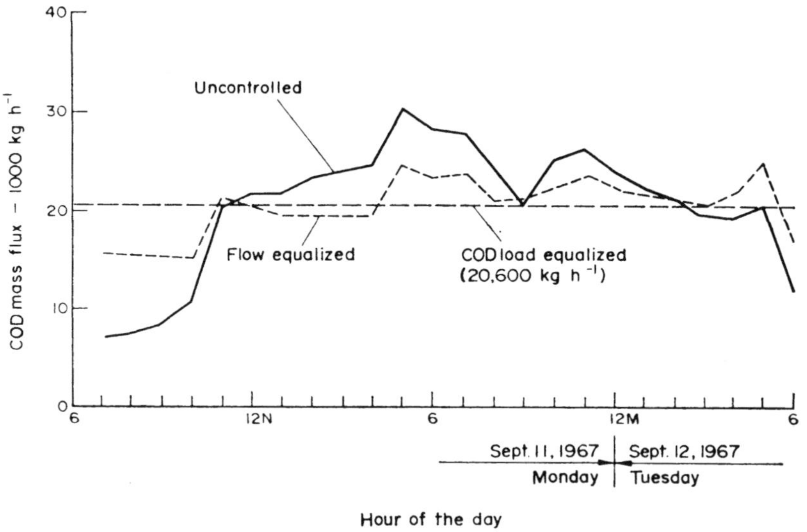

FIG. 3. Hourly COD load variations.

selected day at the treatment plant. The ratio of maximum to minimum flow is typically about 1.8 to 1, and COD flux ratio is about 3. These variations are less than was found at the control locations.

Control can be exercised at any location, but subsequent discussion will be limited to the plant influent.

4. STORAGE REQUIREMENTS

Calculations, using simplified assumptions, were made to determine the volume of storage and the effect on plant influent at the treatment plant inlet to (1) equalize hourly flow rate, and (2) equalize hourly COD mass flux. Since the flow and flux cannot both be equalized simultaneously by inlet storage, calculations of the altered flows or strengths were also made. These are shown on FIGS. 2 and 3 previously presented.

About 8 per cent of the daily flow volume is required for hourly flow equalization. While this will improve volumetric loading on plant unit processes, the resulting daily ratio of peak to minimum mass flux will be about 1.4.

About 15 per cent of the daily flow volume is required for hourly mass flux equalization, double the volume for flow equalization. The resulting daily ratio of peak to minimum flow is 1.6.

5. EFFECT ON PLANT OPERATION

In either case an improvement in plant performance may result, but daily variations of this extent will tax the capabilities of many automation schemes. Further, the effect of change of loads must be considered on a rate basis. Not only must the control system have a sufficient response capability, but the effect on unit process operational performance must be determined. In the case of COD mass flux equalization, high flow rates at low strengths are required. This will have a detrimental effect on volume sensitive unit processes. Intuitively, it is possible that any given series of commonly used unit processes will operate "best" at a compromise between flow and load equalization.

6. TREATMENT PROCESS TRANSFORMS

One of the ways in which the effect of revised input waveforms on a process can be determined is through the use of transforms. A known input waveform and a known output waveform can be subjected to Fourier analysis to determine the system transfer function (ROTH, 1970, 1971; THOMANN, 1967). A trial modified input waveform can then be analyzed to predict a new output waveform based on the system transfer function.

Using data collected during the Minneapolis Saint Paul project, a preliminary Fourier analysis was made of the activated sludge plant to see what effects flow equalization might have on effluent discharge flux. The modified input COD mass flux was used as input to the system transfer function to determine an effluent mass flux.

The preliminary results are not conclusive, but indicate that some process instability might result and that the total daily discharge of COD would be considerably reduced.

Several problems arise when using this technique. The transfer function (and also a mathematical model) of a real treatment plant constantly changes because of operating adjustments. The use of a transfer function implies linear response at a given frequency, a condition not found in practice because of wide load variations and the nature of treatment processes. Both of these problems can be solved in an iterative procedure if a set of optimum operating rules can be determined over a limited range of load variation. The analysis presented used data from an uncontrolled experiment, and considered the entire treatment process as a single unit operation. Undoubtedly considerably more definitive results could be obtained with the influent load controlled over a narrow range and by considering each unit operation separately.

7. IMPACT OF SEWER CONTROL

Field experiments in Minneapolis Saint Paul have shown that 10–20 per cent of daily dry weather flow can be stored in the combined sewer system. Detroit (SUHRE, private commun.) has found parts of the combined sewer system where up to eight hours of dry weather flow can be stored.

In the writer's judgment, it would appear that a number of benefits to plant operation and automation can be obtained through sewer operating control. By temporary manipulation of influent load, experimental data can be obtained useful for determining optimum operational settings and for determining unit process transfer functions over known operating ranges. In normal operation, partially or fully equalized flow or mass flux will improve standard operation and minimize the operating range and the rate of load change capabilities required of a plant automation and control scheme.

Acknowledgements—The combined sewer control project was installed by the Minneapolis Saint Paul Sanitary District (now Metropolitan Sewer Board) under a matching grant from the Federal Water Pollution Control Administration (now Environmental Protection Agency, Water Quality Office).

REFERENCES

ANDERSON J. J. *et al.* (1967) Application of the AutoAnalyzer to combined sanitary and storm sewer problems. *Proceedings of the* 1967 *Technicon Symposium*, New York.

BOWERS C. E., HARRIS G. S. and PABST A. F. (1968) The real time computation of runoff and storm flow in the Minneapolis St. Paul interceptor sewer. *St. Anthony Falls Hydraulic Laboratory Memorandum* No. 118, University of Minnesota.

CALLERY R. L. (1971) Dispatching system for control of combined sewer losses—final report. *Water Pollution Control Research Series*, 11020 FAQ 03/71, US Govt. Printing Office.

HARRIS G. S. (1968) Real time estimation of runoff in the Minneapolis St. Paul areas. *SAFHL Memo.* 119, University of Minnesota.

ROTH P. R. (1970) Digital Fourier Analysis. *Hewlett-Packard J.*

ROTH P. R. (1971) Effective measurement using digital signal analysis. *IEEE Spectrum* 62.

SUHRE D. G. Private communication.

THOMANN R. V. (1967) Time series analyses of water quality data. *J. SED, ASCE* **SA1**, 1.

WALLACE A. T. and ZOLLMAN D. M. (1971) Characterization of time-varying organic loads. *J. SED, ASCE* **SA3**, 257.

Water Research Pergamon Press 1972. Vol. 6, pp. 617–618. Printed in Great Britain

SYSTEMS ENGINEERING APPLIED TO A NEW LOS ANGELES PLANT

DONALD C. TILLMAN
City of Los Angeles, 200 Spring Street, Room 800, Los Angeles, Cal. 90012, U.S.A.

DURING the past 4 yr the City of Los Angeles has embarked on a program of design for three major treatment plants to complement the existing Hyperion plant near the Los Angeles International Airport. The largest of these new plants is the Sepulveda Water Reclamation Plant, which is to be constructed upstream in the San Fernando Valley to reclaim the wastewater. The present status of this plant is midway in design, and construction is planned for the latter part of 1972. It will be the first of five plant modules on 35.6 ha (88 acres), have a flow capacity of 151,400 m^3 day^{-1} (40 million U.S. gal day^{-1}), and is estimated to cost about $24 million U.S.

The site is within a flood control basin belonging to the Federal government, with a lease to the City's Recreation and Parks Department for recreation purposes. The ecology protection required, the committees, commissions, and agencies involved in approvals, the variety of design alternatives all added up to the need for a systematic control of management decisions and earned approvals.

This was the City of Los Angeles' first attempt at systems engineering applied to a major project. There were three basic steps in the process for Los Angeles:

(a) Management must decide to use systems approaches and then enforce their use;
(b) Design personnel must be trained in the technique;
(c) Continuous monitoring of the system and management control of major decisions are mandatory for success.

An aerospace consultant skilled in teaching systems engineering was used to conduct a course for about forty of our designers in six 2-h sessions. An average of twenty-five City engineering personnel have been working on the plant design. A close alliance of the project with computer operations required activity numbering, fragmentation networks, and computer abbreviation definitions. The specifics of the engineering operation are included in the following comments:

The project was developed in parts prior to beginning design. Each part was separated geographically or by discipline from the others and was identified one of 19 Design Elements.

Since design had to be delayed for some time in order to acquire the land site, it was expeditious to construct the plant in several portions so that construction and design could take place in an overlapping sequence. The most favorable breakdown of the project was established as six Construction Elements. These were selected because they could be constructed without undue interference from the other Construction Elements. In this case, the elements were named such as Grading, Process Tanks, etc.

To facilitate design of the project and to maintain management control, a PERT/CPM network was established for each design element. Milestones from these networks were incorporated into a management control network and were time-phased. This time-phased network was supplied to the private engineering firm selected as co-designers on the project and used by them to establish interfacing schedules and networks of their own. The City of Los Angeles engineers are performing the civil or sanitary engineering design and the consultants are designing the architectural, mechanical engineering, and instrumentation features of the plant.

The design element networks were used by the designers to control the project design. Each of the over 700 activities in the networks was assigned a manpower requirement in man-days which is used on the designer level to prevent over-expenditure of manpower.

A management report form was programmed into the computer identifying each design element and other manpower use functions. Every 2 weeks this form is circulated among the designers and submitted to management displaying the manpower assigned to each design element, the manpower used in the last two-week period, and the amount of manpower remaining to complete each element.

As an easy means of reporting to management, a chart summarizing the information on the computer print-out is occasionally circulated, as are graphs showing the total manpower expenditure by the City and total funds paid to the private engineering firm.

In conclusion, it is my judgment that the multiple interfaces of today's society and government with any large project demand the use of an organized system for project control. When properly utilized, systems engineering will help management with a project breakdown into tasks, utilization of personnel, adherence to time schedules, trade-off studies of alternatives, data-based management decisions, and a control not realized in trial-and-error methods. However, there are hazards in the application of systems engineering if management abdicates its responsibility of project control knowledge.

The primary risk is that the maintenance of the system approach can become a major task in itself. If all personnel involved work as a team, with systems engineering as a tool, the goal of the treatment plant in operation will be achieved on schedule, with effectiveness, and the best of all possible designs.

Water Research Pergamon Press 1972. Vol. 6, pp. 619–626. Printed in Great Britain

A PILOT STUDY OF PHYSICAL–CHEMICAL TREATMENT OF THE RAW WASTEWATER AT THE WESTERLY PLANT IN CLEVELAND, OHIO

A. J. SHUCKROW,* W. F. BONNER,* N. L. PRESECAN† and E. J. KAZMIERCZAK‡

INTRODUCTION

THE CITY of Cleveland authorized a program to develop the preliminary engineering and design of a new advanced wastewater treatment plant to be located on the eight acres occupied by the present Westerly plant. This new plant was to have a capacity of 50 mgd and be expandable to 100 mgd. Treatment objectives were 80 per cent phosphorus removal, 90 per cent BOD removal, and 90 per cent suspended solids removal.

Following a careful evaluation of all of the physical–chemical systems available for the treatment of sewage a pilot plant with a maximum flow of 50 gpm was constructed at the Westerly site. The current report describes pilot plant studies of physical–chemical treatment of raw sewage at the Westerly Wastewater Treatment Plant conducted during the period September, 1970–March, 1971.

Pilot plant operations demonstrated that a high quality effluent can be produced from raw sewage at Westerly by a physical–chemical system employing lime clarification, recarbonation, filtration, granular carbon sorption, and disinfection.

PILOT PLANT DESCRIPTION

The pilot system consisted of a clarification stage followed by two-stage recarbonation, mixed media filtration, downflow granular carbon adsorption, and disinfection. FIGURE 1 is a schematic diagram of the pilot plant system. Influent to the pilot plant was obtained from the main Westerly flow after it had been screened and degritted. Lime slurry was injected upstream of the rapid mix tank; the slurry feed rate being controlled on the basis of the desired clarification pH. A 0.2-per cent polymer solution was injected in the rapid mix effluent pipe prior to flocculation. Sedimentation could be accomplished in a 6-ft diameter reactor-clarifier or a 12-ft dia. flocculator-clarifier, which were operated alternately. A 1.5-ft^2 tube settler could be operated in parallel with the flocculator-clarifier on a small slip stream from the flocculator.

Carbon dioxide feed to the recarbonation basins was controlled on the basis of pH in each of the recarbonation stages.

Following recarbonation to pH 7, the stream was filtered through a 2.5-ft dia. dual-media or a tri-media pressure filter operating alternately or in parallel. Filter effluent

* Water and Waste Management Section, Battelle–Northwest, P.O. Box 999, Richland, Washington 99352, U.S.A.

† Zurn Environmental Engineers, Centerline Building, 19200 Villaview Road, Cleveland, Ohio 44119, U.S.A.

‡ Engineering-Science Inc., Suite 888, 600 New Hampshire Avenue, N.W., Washington D.C. 20037. U.S.A.

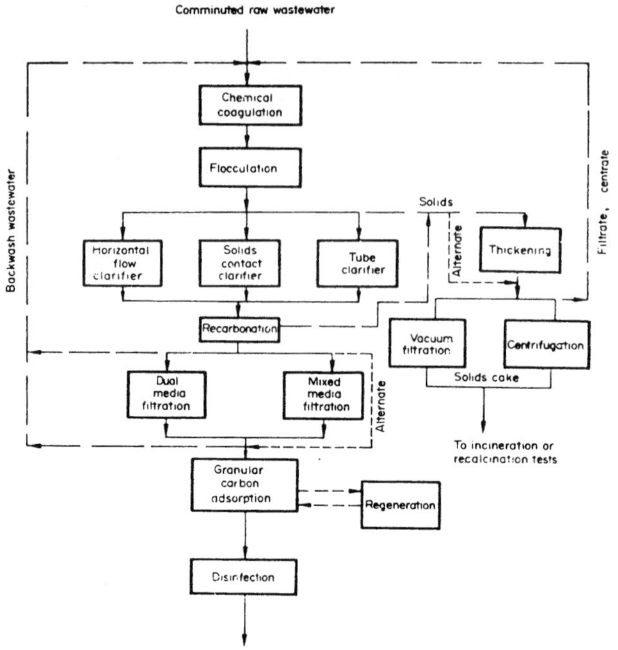

FIG. 1. Unit processes for pilot evaluation.

was then passed through two 4-ft. dia. columns packed with 8 × 30 mesh granular activated carbon. The main carbon effluent stream was disinfected by HOCl generated electrolytically from brine. In addition to the main system, a 6-in. dia. mixed media filter and four 5-in. dia. minicarbon columns were available for small scale pilot studies.

Clarifier sludge was stored in a holding tank until enough accumulated for dewatering studies. A 6-in. cylindrical–conical scroll centrifuge and a vacuum filter were both available for dewatering studies. Prior to dewatering and frequently during the course of any dewatering run, the sludge in the holding tank was mixed by air sparging. A 0.1-per cent polymer solution was fed to both dewatering units by a gear pump, which was chosen to eliminate the possibility of polymer degradation from a centrifugal pump. Polymer was injected in the sludge feed line approximately 1.5 ft upstream of both the centrifuge and the vacuum filter.

Backwash water and centrate were collected in a holding tank and fed back into the rapid mix tank at a controlled rate.

Samples for analytical purposes were manually composited over a 24-h period from grab samples taken at various points in the system at 2-h intervals. During compositing, samples were refrigerated to minimize biological activity. Analytical determinations were made by the City of Cleveland's Bureau of Industrial Wastes.

Operation of the pilot plant began in mid-September, 1970 following a short shakedown period. Ferric chloride was employed for coagulation in the early phases of the study. Lime clarification studies commenced on October 9, 1970, and were continued through March, 1971.

TABLE 1. SUMMARY OF WASTEWATER CHARACTERISTICS AT WESTERLY

Monthly average	Alkalinity at $CaCO_3$ (mg l^{-1})	Chloride (mg l^{-1})	Iron (mg l^{-1})	COD (mg l^{-1})	BOD (mg l^{-1})	TS (mg l^{-1})	TVS (mg l^{-1})	SS (mg l^{-1})	Total P (mg l^{-1})	NH_3 (mg l^{-1})	Sulfate (mg l^{-1})	Total N (mg l^{-1})
September, 1970	160	90	6.4	501	214	746	280	206	6.2	9.1	98	23
October, 1970	158	86	3.1	408	222	687	271	178	5.2	9.1	82	23
November, 1970	177	105	4.1	655	264	811	313	224	5.4	10.5	102	26
December, 1970	166	144	4.5	510	235	831	296	190	5.0	9.2	137	22
January, 1971	173	166	3.4	566	243	904	321	206	5.4	9.3	133	25
February, 1971	131	269	3.7	481	210	1026	275	233	4.4	5.7	83	18
March, 1971	166	228	4.3	542	251	1071	322	215	5.3	8.6	110	24
Average pilot plant operation	162	155	4.2	523	235	867	297	207	5.4	8.8	106	23
Minimum day, 1970	100	60	1.0	71	30	438	76	22	0.6	4.2	34	9.0
Average for 1970	167	134	5.3	564	206	826	—	211	6.4	—	—	—
Maximum day, 1970	232	800	26.4	1984	680	1884	604	578	17.4	17.9	192	45

WASTEWATER CHARACTERISTICS

The physical and chemical characteristics of the raw wastewater at Westerly were observed to be highly variable. Average monthly characteristics of the raw wastewater during the pilot studies as well as the maximum, minimum and average values for 1970 are presented on TABLE 1.

CLARIFICATION—PHOSPHATE REMOVAL

Ferric chloride at a dose of 15–20 mg Fe l^{-1} was employed as a primary coagulant in the early phases of the pilot study. Based on prior laboratory results, Purifloc A-23 at a dose of 0.5–1.0 mg l^{-1} was employed as a coagulant aid. Pilot operations substantiated the laboratory conclusions on the effectiveness of $FeCl_3$ with regard to phosphorus removal. However, several problems with ferric chloride coagulation became evident during the course of the pilot plant operations. Significant quantities of iron (up to 4 mg l^{-1}) were consistently present in the plant effluent. In addition to the undesirability of discharging iron in the plant effluent, this created the problem of FeS, a black precipitate, formation in the plant effluent when H_2S was present. Precipitation of hydrous iron oxides at various points in the system after clarification contributed to early filter plugging and higher than anticipated effluent suspended solids residuals. A prefilter polymer injection of 0.5 mg l^{-1} was required in order to effect good suspended solids removal.

Lime clarification was found to be generally more effective than ferric chloride coagulation in achieving better suspended solids removal at higher clarifier overflow rates. No prefilter polymer injection was required in the case of lime clarification. An effluent consistently low in turbidity and low in phosphorus was produced. Effective clarification with lime was accomplished in the pH range 11.5–12.2 with no polymer addition. Since BOD and COD removal in the adsorbers was found to be independent of clarification pH selection of the most desirable operating pH must be based upon other considerations. Pilot plant experience indicated a linear relationship between clarification pH and effluent phosphorus concentration as shown in FIG. 2. However, even at the lowest pH investigated, pH 10, phosphorus removal was consistently greater than 80 per cent. Therefore, selection of the operating pH should be based on economic considerations relating to lime and polyelectrolyte dose requirements and sludge handling characteristics. The economics of sludge handling and disposal may well dictate the mode of operation of the liquid handling system. Chemical costs associated with operation at pH 10.5 with polymer addition and at pH 11.5 with no polymer addition are comparable. However, operation at the higher pH results in the production of approximately twice as much sludge as is produced at pH 10.5. This high pH sludge was dewatered satisfactorily by vacuum filtration without a conditioning polymer but did require conditioning when centrifugation was employed for dewatering. Sludge conditioning was required for both dewatering methods in the case of the lower pH sludge. There is no apparent advantage to operating the lime clarification process at a pH > 10.5 escept for the greater phosphorus removal achieved.

During periods of high runoff resulting from storms or snow melt, the clarification efficiency deteriorated at pH 10.5. At these times it was necessary to increase the pH to 11–11.5 to maintain clarification efficiency at any given overflow rate.

Based on the above considerations, it was concluded that lime clarification should be

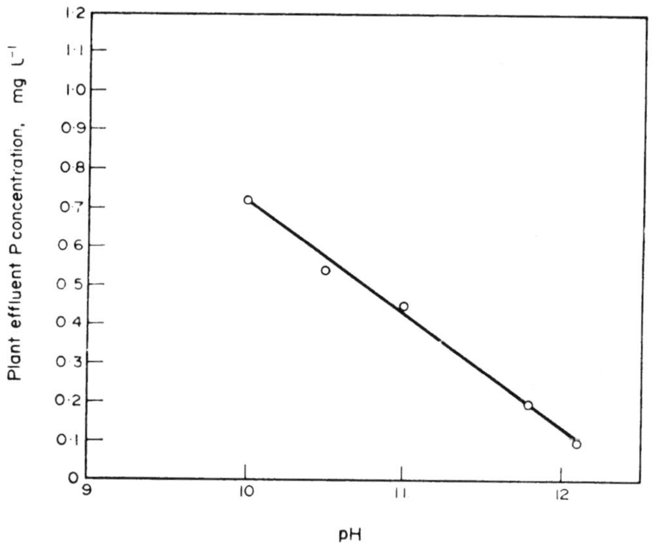

FIG. 2. Relationship between lime clarification pH and residual phosphorus.

adopted for the proposed Westerly plant. Normal operation should be carried out at pH 10.5.

CLARIFIER PERFORMANCE

The flocculator–clarifier, which was operated as a conventional settling basin, performed adequately at maximum hydraulic loading rates of 650 gpd ft^{-2} and 1000 gpd ft^{-2} for iron and lime clarification, respectively. Iron sludge, which represented about 1 per cent of the plant flow, ranged from 2–3.6 per cent solids and averaged about 3 per cent solids, 60 per cent of which were volatile solids. Lime sludge withdrawal rates ranged from 0.3 to 0.5 per cent of the plant flow. The solids concentration in the sludge ranged from 8 to 14 per cent solids at a clarifier pH of 10.5.

Hydraulic loading rates of 1200 gpd ft^{-2} and 1000 gpd ft^{-2} could be achieved with the reactor clarifier at operating pH values of 11.8 and 10.5, respectively. Frequent upsets of the sludge blanket were experienced during operations with this unit. The only measure which was found to be effective in reducing the level of the sludge blanket to its proper position after an upset was a decrease of 15–20 per cent in the hydraulic loading rate. Sludge withdrawn from the reactor–clarifier had a dry solids content of less than 1 per cent. No accurate measure of sludge withdrawal rates could be obtained due to frequent upsets during the operation with this unit.

In view of the fact that, even with constant attention, it was impossible to prevent frequent sludge blanket upsets in the reactor–clarifier, a solids contactor type of clarifier was not recommended for the full-scale design at Westerly.

RECARBONATION

Two-stage recarbonation following lime clarification was employed. The first stage was operated at a pH of 9.3 for $CaCO_3$ precipitation, while the pH was reduced to

7.0 in the second stage. A CO_2 controller maintained the pH within ± 0.2 pH units of the desired value.

FILTRATION

Filtration following both iron coagulation and lime clarification provided a significant fraction of the overall treatment efficiency of the pilot plant.

Parallel operation of two types of filters demonstrated that the tri-media filter outperformed the dual-media filter in all respects. Observations at a filter loading rate of 4 gpm ft^{-2}, indicate that the tri-media filter has considerably longer run times and throughputs then the dual-media filter.

Average effluent quality data for the parallel runs as given in Table 2 indicate that the tri-media filter is more effective in removing suspended matter, P, BOD, COD, and turbidity, than the dual-media filter.

TABLE 2. COMPARISON OF DUAL-MEDIA AND TRI-MEDIA FILTER EFFLUENTS*

	P	BOD	COD	SS	Turbidity
Dual	1.3	94	178	36	24
Tri	1.0	90	168	28	17
$\frac{\text{Dual}}{\text{Tri}}$	1.3	1.05	1.06	1.28	1.41

* All values are in mg l^{-1} except turbidity, which is in JTU.

CARBON SORPTION

Throughout the course of the study, the carbon columns were operated downflow in series at a flow rate of 20 gpm. This corresponds to a 30-min superficial detention time or a throughput rate of two column volumes per hour. Carbon contact time requirements were investigated using the 5-in. dia. mini-carbon columns as well as the main pilot columns. Sorption is essentially complete after a contact time of 15–20 min. The effect of clarification pH on carbon sorption efficiency was studied in the main carbon columns and in the 5-in. mini-carbon columns. In the case of the large columns, the data indicated that BOD and COD removal were independent of the clarification pH. Subsequent to these findings, virgin carbon was tested in the mini columns. The results show that there is no substantial difference in carbon column effluent BOD and COD at clarification pH values of 10.5 and 11.8. BOD removal averaged slightly less than 90 per cent.

Hydrogen sulfide production in the columns was a major problem in the carbon sorption process. Experience with a laboratory carbon column and small pilot columns indicated that it should be possible to control H_2S production by means of daily backwashing. This result was not observed with the large pilot columns.

Chlorination of the carbon column backwash water was studied as a method to control H_2S formation. Initially, daily backwashing of the carbon for 30 min with water containing NaOCl at a concentration of 27 mg Cl_2 l^{-1} reduced the average H_2S concentration to 2 mg l^{-1}. After 3 weeks of daily backwash with chlorinated water, the

H_2S concentration was observed to increase to 5–10 mg l^{-1} toward the end of the 24-h loading cycle. As time progressed, this increase in H_2S concentration occurred about 6 h after backwash.

CARBON REGENERATION

Approximately 0.45 lb COD lb^{-1} carbon and 0.17 lb BOD lb^{-1} carbon were loaded before the carbon column effluent quality data indicated that the carbon was loaded to the point that carbon regeneration was required. Biological growth in the pilot carbon columns may have contributed toward increasing the capacity of the carbon in the system and such growth may have to be eliminated or reduced in the full-scale plant to control the H_2S problem.

DISINFECTION

Electrolytic generation of chlorine from an NaCl brine was carried out successfully on the pilot plant scale.

SLUDGE DEWATERING

Only a limited amount of data related to sludge handling aspects was obtained for the hydrous ferric oxide sludge. It was determined that iron sludge volumes were 2–3 times greater than lime sludge volumes. Iron sludge withdrawn from the clarifier averaged about 3 per cent dry solids. Direct centrifugation of this sludge without conditioning polymers was not feasible.

Lime sludge volumes represented 0.3–0.5 per cent of the plant flow. This sludge ranged from 6 to 14 per cent solids and 41 per cent volatile solids at a clarification pH of 10.5. For design purposes, a 1 per cent sludge withdrawal rate was recommended. No thickening prior to dewatering by centrifugation or vacuum filtration was necessary in the pilot studies but probably would be required in the case of vacuum filtration in the full-scale plant. Conditioning polymers were found to be necessary for both dewatering devices in order to effect good capture.

Vacuum filter performance was dependent on influent solids content, polymer dose and media speed. Cake solids ranged from 19 to 36 per cent dry solids; performance being mainly dependent on influent solids content. Polymer addition somewhat reduced the cake solids content, the effect being less at high septum speeds. Wetter cake was produced at higher septum speeds due to the shorter drying time. At 50 rph the cake solids averaged about 75 per cent as much as the dry solids content at 10 rph.

Filter loading rates varied from 3 to 19 lb h^{-1} ft^{-2}. Sludge conditioning with polymer prior to filtration greatly improved the loading rate, often as much as 3-fold. Polymer dosages of less than 2 lb ton^{-1} dry solids were investigated. Higher dosages resulted in overtreatment of the sludge. Maximum loading rates were obtained at the fastest septum speeds. At higher speeds, however, the higher moisture content of the cake caused it to adhere to the media. The septum speed at which the cake would no longer drop off the media varied with each batch of sludge. The cake solids content at this point varied from 25 to 31 per cent averaging 27 per cent.

Centrifugation was not as sensitive to influent solids concentration as vacuum filtration. Polymer addition was required to attain good centrate quality. The polymer dose necessary to obtain 95 per cent solids capture varied from about

1.2 to 1.7 lb ton^{-1} dry solids at sludge flow rates of 2 and 4 gpm, respectively, and at a rotation speed of 3500 rev min^{-1} The required polymer dose increased rapidly at sludge flow rates greater than 4 gpm. At a rotation speed of 3500 rev min^{-1} the cake solids ranged from 20 to 28 per cent. Centrifugation was recommended for the final design.

SLUDGE DISPOSAL

A limited study of recalcination of lime sludge was conducted during the program. Several hundred pounds of lime sludge was collected at two clarifier pH values, pH 10.5 and pH 11.5. BSP Corporation was contracted to conduct recalcination studies on each of these samples. The BSP report indicated that recalcination is feasible and 100 per cent CaO recovery is possible. However, analysis of the data suggests that recalcination may not be economical when compared to incineration and ash hauling unless a highly efficient method of separating the calcium oxide from the large quantities of inerts found in the sludge is available.

Water Research Pergamon Press 1972. Vol. 6, pp. 627–636. Printed in Great Britain

NOTES ON THE DISCUSSIONS RECORDED BY ALEXANDER DUFF

County Engineer and Master of Works, County Council Offices, Dumbarton, Scotland

FIRST SESSION

RALPH C. PALANGE: Lack of communication between Research Worker and Operator causes concern and it is imperative that results of research must be used for the benefit of all. Vigorous action is being taken in the United States. By 1974 an expenditure of $12 million will have been made. The intention is a policy of national water control of stormwater run-off, mixed systems, domestic, industrial and agricultural wastes providing beneficial effects on environment and population with the minimum damage to the environment. The effectiveness of Schemes is investigated by cost analysis methods before work is commenced and this has resulted in linking certain communities together for a combined programme and for the provision of separate schemes for others. A recurring problem is the reluctance of designers to accept new methods and techniques and to co-ordinate and to raise standards. In September 1970 a manual setting down guide lines for design and management was published. The speaker stressed the necessity for the dissemination of information by process design manuals and the involvement of the general public in the campaign to improve the environment.

PROFESSOR ANDREWS: Regretted that while money was being put into plant design it was not being used for the training of plant operators. The requirements for an efficient plant depend upon: (1) Operating Personnel; (2) Process Stability; (3) Availability of Space; (4) Reliability; (5) Finance; and (6) Flexibility.

Large plants require flexibility and it has to be remembered that a plant using chemicals in the treatment process requires too much labour.

PROFESSOR VON DER EMDE: The findings on this paper were not wholly applicable to the United Kingdom as electricity rates had not been steady over the last decade and standards of effluents required to be reached are higher than those laid down on the Continent of Europe and in the U.S.A.

There was a difference of opinion between ANDREWS and VON DER EMDE on the question of plant size and the effect on the standard of effluent in the event of one tank unit going out of commission. On the one side the advocacy of large tanks, on the other small tanks, percolating filters preferred to activated sludge process. Agreement that B.O.D. would rise if the loading increased on the tanks still operating, if one tank went out of commission—but VON DER EMDE was of the opinion that as a rise in B.O.D. would be for temporary period he did not consider it to be too serious a problem. ANDREWS held that percolating filters were better for small plants but models were being tested to compare types.

RADEMACHER: Advocated spreading the discharge of effluent into a watercourse over a period of time to help environmental control. VON DER EMDE considered that characteristics of the river should not be used when considering standard of effluent to be discharged—he advocated a plant with ample oxygen available and of high efficiency.

GUIVER: Spoke of the importance of the final settlement of percolating filter sludge.

ZEPER: In his paper on Design Period considered the present policy of providing plants to be in existence for a long period of time (40 yr) was unrealistic, and wants cheap designs which can be written off as equipment can go out of date quickly and that a borrowing period of 7 yr would be more realistic.

KAYSER: Spoke on the evaluation of design data and quoted a plant which had been designed on observed detail but within 2 yr of completion was grossly overloaded, the moral being that continuous checks should have been made in the actual design period when the change in the characteristics of the sewage would have been observed.

WHEATLAND: Expressed the opinion that adequate allowance must be made for industrial loading (Note in Scotland 75 persons per acre at 60 gal per head), that there must be flexibility in the aeration tanks, short retention period provided that there is concentration of micro-organisms and sedimentation.

TILLMAN: In his paper on Systems Engineering, as applied to the new Los Angeles Plant, spoke of the method adopted. The plant which would cover an area of 88 acres has a capacity of 40 million gal day^{-1} and will cost \$24 million. Sludge + 5 per cent water sent into Pacific. The works sits in a flood control basin—a recreation area surrounded by residences. The effluent from the works will pass down the usually dry Los Angeles River. U.S. Army Engineers control flood basin.

Three basic requirements in management: (1) Management must decide to use systems and enforce it on the designers; (2) Designers must be trained; (3) Continuous control of systems and of major decisions must be maintained.

Having decided to adopt this technique 40 men were given six 2-hr sessions. One-half of the design team are Los Angeles Engineers working on the works design, the other half is made up of private practice Architects, Landscape Architects and Mechanical Engineers. After protracted consultations it was agreed that a Japanese Garden layout should be evolved to screen the new plant.

This Critical Path method required Fragmentation Network daily reports, cost analysis, monitoring of the process by management, the consideration of aesthetics because of public concern, and re-use of water.

It has to be appreciated that the finished works will not look like a sewage treatment plant.

The Speaker pointed out that System Engineering principles were not being used on the design of small plant and made it clear that the computer can never replace "Team Work".

The RYDER and MCGOVERN paper on a problem in Hawaii brought out the point that in this case the reclamation of water for irrigation use required less treatment than discharge to the sea and that the local labour was unskilled and difficult to train.

GUIVER spoke of the parameters concerning: (1) effect of boron on sensitive crops but could be used if irrigation not deeper than 6 in.; (2) effect of chloride. In Hawaii re-claimed water used on Citrus Crops. In California there is concern regarding boron but ion-exchange is being considered. Because of chloride, water supply for the Islands is taken from the surface layer of fresh water moving out to sea. The chloride/salt ratio is expected to be critical because of percolation into sub-strata.

In answer to MALAN it was pointed out that power costs are high in Hawaii.

OAKLEY: Was concerned regarding the operation and maintenance of the plant as it was of a highly sophisticated nature. In reply it was stated that there would be four or five plants in the area and that a spare would be kept available as the modules are removable. It was not thought to be over-mechanized and the stocking of spares would be simplified as the comminutors were the same and the sedimentation tanks the same. One superintendent would be in charge over all and the individual operators would spend portions of their time at each plant.

RADEMACHER: In his very critical paper on the efficiency of waste treatment plants in the United States made strong criticism of the policy makers who reduce the amount of money available with the result that the ultimate efficiency of the plant suffers. There was also criticism of the standard of operator skill which falls short of required standards for complex plants. It was established that the requirements were well designed plants maintained and operated by well trained personnel.

It was also pointed out that when costs are being compared the cost of providing interceptor sewers and site preparation had to be considered as all the cost is not going into plant.

It was stated that similar problems were being experienced in Poland.

Mr. Tench in Chair

VON DER EMDE: On the question of providing primary tanks pointed out that there were small activated sludge plants operating successfully without such tanks. The Vienna-Blumental Works do not have primary tanks but there is twice as much sludge produced (if chemical treatment is used much more sludge is produced). There is a marked increase in the power consumption but it was claimed that this was off-set by construction costs. A further criticism was the settling out of fine sandy material in channels and tanks with damage to scrapers.

VON DER EMDE: Claimed that if in an area with a large number of works, sludge handling and thickening can be concentrated, no primary tanks are required. IMHOFF was concerned that, if no primary tanks exist more nitrates and phosphates will pass into the effluent. von der Emde claimed that de-nitrification can be controlled but there was difficulty with the phosphates.

TENCH: Pointed out that Ardern and Lockett designed the activated sludge plant so that it could be fed with crude sewage—but it never worked.

VON DER EMDE: Continued with a paper on the basic principles of aeration tank design bringing out the design criteria of the Activated Sludge Process' Committee of the German Sewage Works Association emphasizing that plants should be built with spare capacity.

ZEPER: Introduced a paper on the "Carrousel" method based on the Pasveer oxidation ditch but showing that the previous necessity of a large area had been reduced by using deeper tanks.

IMHOFF: Considered the treatment of three times Dry Weather Flow as practised in the United Kingdom was a good idea and considered that Europe lags 10–15 yr behind the United Kingdom. He also criticized the energy input for aeration purposes being less than 0·1 W m^{-3} and considered that there would be a danger of settlement.

WHEATLAND: Spoke of effluent standards and the necessity to think of effluent quality. In the United Kingdom in the Year 1900 the Royal Commission required nine out of twelve samples had to conform to the 30/20 standard but to-day the Prevention of Pollution Acts requires the standard be met at all times. Over a period of 1 yr a study of plants showed no significant difference between percolating filters and activated sludge.

MOORE: Introduced his paper on the deep aeration tanks adopted for the New York City and which are to be 28 ft in depth with adjustable diffusers. KALBSKOPF expressed concern at the problems which would arise from the presence of detergents and queried the decrease in flows on the floor of the tanks. Jet acceleration is used to increase the velocity and VON DER EMDE stated that baffles before the rotors improves velocity near floor, but it was pointed out that this required extra power, 1–3 kW in a 30–50 kW plant. The flat bottomed aeration tank was compared unfavourably with the United Kingdom practice of a sloping floor.

WHEATLAND: Stated that a 7 in. cone was being tried in a 27 ft deep tank and it had been shown that diffused air was better than cones in deep tanks. The Water Research Laboratory had shown there to be economy in deep tanks with diffused air, vertical baffles being used to improve flow but with increased power costs.

Herr Kayser in Chair

KALBSKOPF: Paper on Flow Velocities of Aerator Tanks with Mechanical Aerators again brought the need for adequate velocity along the floor of tanks with particular consideration being given to the effect of the concentration of detergent present. It was pointed out that errors can arise in the Settleable Solids Tests as the sample is taken at the tank but the test is carried out in the laboratory which may be some distance away. KLEFFNER raised the question of density currents—where high density flow drops to bottom of tank with a stream effect along the bottom and up the side of the tank. In reply it was stated that the main problem of feed design is at its worst in rectangular tanks, the outlet weirs are situated along side of tank to overcome difficulties—similar procedure used in U.S.A. In answer to a request for an overflow rate it was stated that this depends on location and length of weir—in a length of 126 ft one third of length used as outlet weir; the effect of wind on a sharp crested weir has to be considered whereas there is no problem with a V-notch.

Mr. Lynam in Chair

ISAACS: Paper on Tertiary Treatment brought out a good discussion on the relative merits of micro-straining and sand filtration. With the increase of trade effluent from industry it was becoming more essential to provide tertiary treatment to protect the environment and fishing interests in watercourses. The degrees of treatment depends upon the standard required but as high flows tend to occur during wet weather when the watercourse is more able to cope with the load it was suggested that flows in excess of twice Dry Weather Flow should by-pass the tertiary treatment unit. Comparisons of the mechanisms of removal and flow rate and efficiency are shown:

Mechanics of removal		Flow rate and efficiency	
Rapid sand filter	Micro strainer	Rapid sand filter	Micro strainer
1. Straining (undesirable) 2. Interception 3. Sedimentation 4. Flocculation 5. Biological (slight)	1. Straining only	Decrease of efficiency with increased flow	Efficiency less affected by rate of flow

Comparison of costs of tertiary treatment

Rapid sand filter	Micro strainer
Capital costs–£m^{-2} of filter area	
400–500 (without pumping)	950–1100
600–940 (with pumping)	
Minimum total costs*–p m^{-3} at treatment rates of 400 m^3 m^{-2} day^{-1}	
0·04 (without pumping)	0·09
0·08 (with pumping)	

Costs of clarification

Humus tanks	Approximate total costs for population 50–100,000	0·12 p m^{-3}
Rapid sand filters	Typical total cost to treat 3 D.W.F. at 400 m^3 m^{-2} day^{-1} (without pumping)	0·11 p m^{-3}
	(with pumping)	0·23 p m^{-3}
Micro strainer	Typical total cost to treat 3 D.W.F. at 400 m^3 m^{-2} day^{-1}	0·26 p m^{-3}

* Assuming a constant flow and including cost of disposal of washwater.

It was agreed that micro-strainers cannot cope with a sudden load whether it be in volume or quantity (colloids). OAKLEY brought out that sand filters were of higher efficiency and that the advantage of sand lay in its flexibility that it can cope with almost anything—for higher loading coarse sand can be used in shallow beds.

WOOD: Pointed out that tertiary treatment may not be necessary at all times. Costs could be astronomical (£12,000 per ton) if actual necessary removal was costed. The purpose of this treatment is to provide a good effluent.

JENKINS: Queried Oakley's statements—laboratory work on a feed of 40–60 mg l^{-1} suspended solids on sand filters showed that their life is so short that it would be necessary to reduce loading to 25 mg l^{-1}.

OAKLEY: Said that it was wrong to instal micro-strainers where flow of bad quality; with micro-strainers a good effluent can be made better but the capacity of the micro-strainer falls off with heavy loading.

MIDDLETON: In his paper on the control of phosphorus spoke of the presence of phosphorus in detergents and that it was more advisable to remove and control the P content rather than eliminate it from the products. The chemicals which can be used in primary treatment are, Iron, Ferrous Chloride or Sulphate, Ferric Chloride or Sulphate. Lime or Sodium Hydroxide, Alumina or Lime and Alumina or Iron.

WILDI followed with his paper on operating experience on the precipitation of phosphates. The question of phosphates from agricultural sources entering watercourses was discussed and it was agreed that effluents from works should have as low a phosphorus content as possible. It was also agreed that the high cost of removal of phosphorus lay in the cost of the chemicals and that size of plant had no effect on this cost.

GREENLUND: Gave details of a requirement of 4 mg l^{-1} B.O.D. 0·2 mg as PO_4, 0·7 mg as P by precipitation with lime in primary tanks. With precipitation, if up to 95 per cent removal is required the costs will drop, but if over 95 per cent P removal is required costs will rise steeply as filtration required. JENKINS reported that in the United Kingdom phosphorus removal was to be used for effluent required for industrial purposes. He stated that since there was a 40–60 per cent removal during conventional normal treatment it was better to treat the effluent and the most economical flocculant was lime.

The use of lime and aluminium sulphate allowed a wide range of pH values to be obtained. Precipitation resulted in a sludge increase of 50–100 per cent. It was feasible to aim at design criteria of phosphorus content of 0·5–0·1 mg l^{-1} with a plant of the trickling filter type, the flocculent could be added before the humus tank, thus saving in sedimentation tank capacity. To obtain the full benefit of flocculation it seemed necessary to pass the effluent through sand filters.

ZEPER: Stated that on ferrous chloride being added in the oxidation ditch the resulting sludge when deposited in sludge bed to a depth of 1 ft thickness reduced to 3 in. in 24 h. LUDWIG introduced a paper on a pilot study of physical—chemical treatment of raw wastewater and the problems which arose the comparison of physical—chemical treatment with biological treatment. The question of cost of the plant on physical—chemical basis is shown as under:

Physical—chemical treatment costs [cents (1000 gal^{-1})]

Plant size (mg day^{-1})	5	10	100
Amortization, Operation, Maintenance	24–36	18–26	10–15

Cost distribution

Process	Percentage of total cost
Chemical clarification	35
Carbon adsorption	50
Filtration	15

Dr Imhoff in Chair

GUARINO: In discussing his paper on Control of the Anaerobic Digestion Process and Supporting Unit Processes stated that if the plant were to be re-built he would have circular tanks as against rectangular. There were doubts as to the accuracy of the measuring devices; it was suggested that there was no automated process for volatile acids and the Ranarex indicator was also criticized. It was suggested that colombmetry could be used as were infra-red detectors.

JENKINS: Brought practicality into the discussion when he advised that the experience of the operator should not be discounted—by looking at sludge a good operator could tell the degree of consistency. This can be done by an operator at a control panel by having a TV camera located at a bell-mouth from which sludge was discharging with a screen at a control panel—with the use of automatically controlled valves the TV operator can open a valve and decide the condition of the sludge from its appearance as seen on the TV screen. This system was in operation at the Upper Tame Main Drainage Minworth Works, Birmingham.

ZEPER: Paper discussed the use of centrifuges and belt-press (press)-filters for the dewatering of digested sludge from oxidation ditches. The cost of achieving dewatering is largely dependent upon the cost of the flocculation agent and, at present, these are expensive.

KALBSKOPF: Introduced the subject of dewatering of sludge by heating and pressure filter. It was shown that the filtrate of the heat-treated sludge is heavily polluted.

MEREDITH: In his paper looked at the economics of heat conditioning and pressing as compared with chemical conditioning and pressing. It was shown that initially the running costs of the heat conditioning and pressing plant were in the ratio of 2–3 for the chemical conditioning but actual operational costs showed that the heat conditioning and pressing was in the ratio of 9·8–9 to the chemical conditioning and pressing. While running costs of the two plants are comparable the chemical process does not have the aerial odour problem of the heating plant and there are no heat exchanger tubes to clog. In the plant being discussed the tubes were given an acid wash-out once each month and this required a close-down of the plant of 3 h.

WHEATLAND: Spoke of filtrate from heat treated sludge being evaporated and incinerated at a cost of £2–£3 per 1000 gal.

Dr. Niemitz in the Chair

ZINGLER: During the introduction to his paper on Thermic Sludge Drying stated that the vibrating channel had given trouble due to the stresses from heat and movement and that the plant could not operate at the design temperature of 800°C. This had had to be reduced to 600°C.

TENCH: Gave the back-ground to the decision for adopting filter-pressing and incineration of sludge at Sheffield—originally wet digested sludge was tipped on land, then pressing plant was introduced and the sludge tipped. The resulting odour could be smelt 2 miles away so incineration was the next step. It was interesting to note that on sludge being tipped on the land and left there was no growth as the iron content in the sludge immobilized phosphorus and so stifled plant life. It was admitted that lime for treatment was expensive but that there were sources of waste lime available in the Sheffield area. MALAN enquired as to fermentation—the answer was 11 days in winter, 7 days in summer and the thicker the sludge the better. IMHOFF suggested thickening the sludge with lime but it was pointed out that the tanks were too big—it took 3 h to feed into filter press. WHEATLAND suggested controlling the heavy metals by pre-treatment and this would be attainable by normal methods. JENKINS spoke of the problem of the variation in calorific value and the presence of oil in the sludge. He noted that under certain conditions of light, steam could appear to be black or grey in colour. TENCH stated that the plant was closed for 1 week in each year for maintenance—no complaints had been received from the general public but there were complaints about the adjacent Power Station. The reason for the steam appearing black was insufficient air (less than 50 per cent)—under certain conditions the steam plume comes down to earth 100 yds away. Variations in calorific value can be too quick for automatic control. The need for efficient management was stressed.

LYNAM: Paper on the use of "Liquid Fertilizer" to reclaim land and produce crops quoted four requirements on any system of solids disposal these are (1) the disposal method must not pollute the air or the water, (2) it must be economical, (3) it must conserve the organic matter in sludge for a beneficial purpose and (4) it must solve the disposal problem forever. The conclusions of the paper included these points (1) properly stabilized solids can be safely applied in the vicinity of residential areas without any nuisance, (2) stabilized sewage solids are an effectice source of nitrogen, phosphorus and micronutrients for crop growth, (3) Faecal coliforms in freshly digested sewage solids die rapidly when applied to soil surface, (4) swine enterovirus and probably other virus forms, do not survive for more than 4 days in an efficient anaerobic digester. WOOD in his paper on the application of liquid digested sludge to agricultural land gave a clear picture of the development of this form of disposal. He spoke of the difficulties which had to be overcome—the opposition initially of Central

Government who were concerned about the possible toxicity of metals, the difficulty of getting farmers to co-operate, the need for efficient organization of transport. One problem which arose was that of vehicle loading on land which caused anaerobic conditions 2–3 ft below ground.

KUGEL: Continued with a paper on a similar subject but in his case the vehicles used for transporting the sludge to the site for spraying do not pass over the fields and a sludge gun is used for spraying. From these papers followed details of transporting sludge by pumping through long pipe lines and the limits controlling this method. It was agreed that above 8 per cent of solid matter in the sludge pumping was difficult; below this characteristics were similar to the pumping of water.

The question of sludge spreading on land is controlled by seasonal considerations with storage of 6 months being not unusual. WOOD stated that he had areas of sludge drying beds for summer drying and used the digesters for storage until sludge required.

ANDREWS: Was concerned at the loss in head in pumping through long mains but was advised by LYNAM that multi-stage pumps with a velocity of 2–6 ft s^{-1} with large diameter pipe (20 in. in diameter) had proved satisfactory. GUARINO spoke on factors governing pumping—if the inorganic content is high as compared with the organic content friction losses are low and if the pumping velocity is less than 2 ft s^{-1} friction losses are high. Care had to be taken that pumping velocities are not excessive as scouring action can result especially at bends or elbows.

Various problems concerning spreading or spraying by pipe line were raised—the presence of solid matter such as plastic bags requires mechanical screening to be provided. The spray gun has to be manually operated because of clogging, feed pipes to tankers can be blocked by scum from digesters. JENKINS recalled that Birmingham (U.K.) had applied sludge to land in 1877. Sludge with 10 per cent dry solids content had been pumped over 4–5 miles for many years by reciprocating pumps. Field tests had been carried out to compare the results of using sludge in the U.K. It was recommended that 5 tons per acre of sludge was applied approximately once every 3 yr. At Birmingham the sludge has a high content of heavy metals which is precipitated before it enters the digesters. Digestion even with 0·5–1·0 per cent of these metals was satisfactory. He stressed the need for care to be exercised in applying sludge year after year—where soil received sludge for 5 yr desert conditions resulted but in other areas over a long term, controlled application of sludge can give soil with a good open texture, provided the capacity of the soil to absorb the sludge is not exceeded.

From KUGEL's paper on pasteurization it was brought out that tests were being carried out on the presence of salmonella—the effective pasteurization time was 10 min but for engineering purposes it required 20 min to effect complete elimination. Experience had shown the heat exchanger to be reliable and no duplication of plant had been required; layers of sludge build up on plates in heat exchangers but since the plates have a slope the condensed steam causes the sludge to flow.

GUARINO: Spoke of the opposition arising in U.S.A. to the disposal of sludge at sea and gave details of tests presently being carried out.

Professor Andrews in Chair

PROFESSOR ANDREWS: Introduced his paper on control systems for waste water treatment plants tracing the evolution of control, manpower to computer, and compared the investment, in the U.S.A., in automatic control between the Petro-chemical industry (17 per cent) and the Sewerage Industry (1–1·5 per cent). A significant statement from the paper should be quoted "Prospective Users should recognize that, although the potential use of computer control is great, the adaptation of a process to computer control is difficult and a time consuming task. A substantial amount of technical manpower is required to implement computer process control and it has been reported (9) that from 2 to 21 man-yr have been required to complete computer control projects in industrial plants". During the discussion WOOD was concerned at the cost involved in providing the required equipment and considered the costs to be unjustified—due to possible changes within, say 10 yr, control systems may have to be modified. He posed the question of the cost of employing electronic engineers in place of sewage works operators. OAKLEY also expressed concern and brought out two major difficulties—(1). Reliability of instruments—an intelligent operator can make judgements—an instrument cannot. (2). Cost benefit or effectiveness, quoting that at Milton Keynes automatic control will account for 2 per cent of cost.

GUIVER: Regretted that there was insufficient liaison between manufacturers and sewerage-authorities—because of the limited market, use is being made of instruments (sensors) presently in production. ANDREWS spoke of the lack of reliable sensors but questioned whether this was a serious draw-back. He stated that the position was not serious in the U.S.A.—plants are sluggish—analysis in 20–30 min is sufficient for the purpose. The difficulty is that analytical results are occasionally filed and rarely used. A computer asks for information, uses it and the operator obeys the instructions given.

GUARINO: Suggested instrumentation should be introduced gradually and referred to a decision taken in 1962 which justified the purchase of a computer for the water industry alone—since then more and more use has been made of that computer for other purposes. "An idea (a dream) must be followed up".

An idea must be followed up—the result of the idea must be considered carefully—Is it justified? Will it do the job it is intended to do?—Has it limitations?—Can the cost be justified throughout all the ranges of the process? Is man necessary to make a balanced judgement using the system which will provide the information on which to make the decision?

All these questions have still to be answered.

AUTHOR INDEX

Numbers in italic indicate the page on which references are listed

SUBJECT INDEX

Printed in Dunstable, United Kingdom

91898128R10190